2195

D0576106

2195

ESTIMATING RESIDENTIAL CONSTRUCTION

ESTIMATING RESIDENTIAL CONSTRUCTION

Alonzo Wass

Prentice Hall, Inc.
Englewood Cliffs, NJ 07632

Library of Congress Cataloging in Publication Data

Wass, Alonzo.
  Estimating residential construction.

  Includes index.
  1. Building—Estimates. 2. House construction—
Estimates. I. Title.
TH435.W383   692'.5   79-15099
ISBN 0-13-289942-6

© 1980 by Prentice-Hall, Inc., Englewood Cliffs, N.J. 07632

All rights reserved. No part of this book
may be reporudced in any form or
by any means without permission in writing
from the publisher.

Printed in the United States of America

10 9 8 7 6 5 4 3 2 1

Prentice-Hall International, Inc., *London*
Prentice-Hall of Australia Pty. Limited, *Sydney*
Prentice-Hall of Canada, Ltd., *Toronto*
Prentice-Hall of India Private Limited, *New Delhi*
Prentice-Hall of Japan, Inc., *Tokyo*
Prentice-Hall of Southeast Asia Pte. Ltd., *Singapore*
Whitehall Books Limited, Wellington, *New Zealand*

CONTENTS

Preface xiii

1 A GUIDE LIST FOR CONTRACTORS AND ESTIMATORS 1

1.1 *American Society of Professional Estimators, 1*
1.2 *Financial, 2*
1.3 *Sources of Credit, 2*
1.4 *Financial Definitions, 2*
1.5 *Credit Bureaus, 3*
1.6 *Builder's License, 3*
1.7 *Local and National Construction Associations, 3*
1.8 *Land Purchase, 4*
1.9 *Land Registry Office, 4*
1.10 *City Offices, 5*
1.11 *Survey, 6*
1.12 *Batter Boards, 6*
1.13 *Datum and Bench Marks, 6*
1.14 *Building Inspectors, 7*
1.15 *Apprenticeship and Tradesman Qualification Board, 7*
1.16 *Technicians, 7*
1.17 *Local Labor Costs and Availability, 8*
1.18 *Fringe Benefits, 8*
1.19 *Local Material Costs, 8*
1.20 *Railways and Freightage, 8*
1.21 *Demurrage, 8*
1.22 *Subtrades, 8*
1.23 *Workmen's Compensation Board, 8*

1.24 *Field Trips, 9*
1.25 *Standard Contract Documents, 9*
1.26 *New Materials, 9*
1.27 *The American Institute of Architects Publications, 9*
1.28 *Labor Legislation, 9*
1.29 *Government Services Available to Business, 10*
1.30 *Contractor's Fees and Progress Payments for Small Jobs, 11*
1.31 *Bank, 11*
1.32 *Depreciation, 12*
1.33 *Individual Ownership, Partnerships, Limited Companies, and Corporations, 15*
1.34 *Obtaining Drawings, Specifications, and Contract Documents, 17*
1.35 *Notices to Bidders: Specifications, 38*
1.36 *Examples of Tenders, 38*
1.37 *Drawings and Specifications and General Conditions, 38*
1.39 *Examination of the Site, 38*
1.39 *Excavation and Backfill, 38*
1.40 *Precedence of Documents, 38*
1.41 *Disputes, 38*
Questions, 39

2 THE CONSTRUCTION COMPANY CHAIN OF COMMAND AND PROGRESS PAYMENTS 40

2.1 *Bids and Estimates, 41*
2.2 *The Owner, 42*
2.3 *General Manager, 42*
2.4 *Manager, 42*
2.5 *The Architect, 42*
2.6 *The Architect's Progress Certificates, 42*
2.7 *The Engineers, 43*
2.8 *The Technician, 43*
2.9 *Attorneys and Lawyers, 43*

2.10 *The Estimator, 43*
2.11 *The Junior Estimator, 44*
2.12 *The Building Superintendent, 44*
2.13 *The Timekeeper, 44*
2.14 *The General Contractor's Foreman, 46*
2.15 *The Subcontractors, 47*
2.16 *Workmen, 47*
Questions, 47

3 BONDS, CONTRACTS, AND INSURANCE 48

3.1 *Bonds, 49*
3.2 *Tender or Bid Bond, 49*
3.3 *Performance Bond, 49*
3.4 *Labor and Material Payment Bond, 49*
3.5 *Liens, 49*
3.6 *Lien Bond, 49*
3.7 *Fire Insurance, 50*
3.8 *Bonds, 50*
3.9 *Stipulated-Sum Contract, 50*

3.10 *Cost Plus a Percentage, 50*
3.11 *Cost Plus a Fixed Fee, 79*
3.12 *Contract for a Cost Plus a Variable Sum, 79*
3.13 *Quantity Estimate, 79*
3.14 *Construction by Day Labor, 79*
3.15 *Checklist of Bonds and Insurances for Contractors, 79*
Questions, 81

4 NOTICE TO TENDERERS-GENERAL CONDITIONS-INDENTURE 83

4.1 *Notice to Tenderers, 83*
4.2 *Examples of Tenders, 84*
4.3 *Plan of Specifications—General Conditions, 84*
4.4 *Examination of the Site, 84*
4.5 *Precedence of Documents, 84*
4.6 *Disputes, 84*
4.7 *Arbitration, 85*
4.8 *Bonds, 85*
4.9 *Progress Schedule and Diary, 85*
4.10 *Progress Reports and Photographs, 85*
4.11 *Payments, 85*
4.12 *Temporry Office and Store Sheds, 85*
4.13 *Temporary Services, 85*

4.14 *Supervision, 85*
4.15 *Contingency Sum, 85*
4.16 *Prime Costs, 86*
4.17 *Subcontractor, 86*
4.18 *Fire Insurance, 86*
4.19 *Liens, 86*
4.20 *Cleanup, 86*
4.21 *Examining the Plans and Specifications, 86*
4.22 *Alterations and Additions, 86*
4.23 *Extras, 87*
4.24 *Delays, 88*
4.25 *Instructions to Bidders, 88*
Questions, 96

5 OVERHEAD EXPENSES AND PROFIT 97

5.1 *Individual Job Overhead Expenses, 97*
5.2 *General Overhead Expenses, 98*
5.3 *Profit, 98*
5.4 *Separate Job Expenses for Salaried Professionals, 98*
5.5 *Guide List of Overhead Expenses Chargeable to Each Individual Job, 98*
5.6 *Salaries, 99*
5.7 *Temporary Buildings, 100*
5.8 *Equipment Used on the Job, 100*

5.9 *Concrete Testing, 100*
5.10 *Progress Reports, 100*
5.11 *Attending on Other Trades, 100*
5.12 *Completion Date, 100*
5.13 *Guide List of Major Overhead Expenses not Chargeable to any one Job, 100*
5.14 *General Overhead Expenses, 101*
5.15 *Individual Job Overhead Expenses (Check), 101*
5.16 *Final Profit 101*
Questions, 102

6 ACCURACY TESTS AND QUICK CALCULATIONS 103

6.1 *Rapid Mental Calculations, 104*
6.2 *Percentages, 104*
6.3 *To Find the Area of Plane Figures, 105*
6.4 *To Find the Approximate Area of a Circle, 105*
6.5 *To Find the Circumference of a Circle, 106*
6.6 *To Find the Area of an Ellipse, 106*
6.7 *To Find the Circumference of an Ellipse, 106*
6.8 *To Find the Volume of a Sphere, 106*
6.9 *To Find the Surface Area of a Sphere, 106*
6.10 *To Check Multiplications, 107*

6.11 *To Check Additions, 107*
6.12 *To Add Columns of Dimensioned Figures, 107*
6.13 *The Twenty-Seven Times Table, 107*
6.14 *To Check Subtractions, 108*
6.15 *To Check Division, 108*
6.16 *Miscellaenous Capacities, Measures, and Weights, 108*
6.17 *Accuracy Tests, 108*
Questions, 129

7 BOARD AND LINEAL MEASURE, STOCK MOULDINGS, PLYWOOD, FIBERBOARDS, AND NAILS 132

7.1 *Dimension Lumber, 133*
7.2 *Transposition of Figures, 134*
7.3 *Work-Up Sheets, 135*
7.4 *How to Use Work-Up Sheets, 135*
7.5 *Checking Estimates, 135*
7.6 *New Lumber Standards, 135*
7.7 *Example of the Work-Up Sheet, 140*
7.8 *Pricing Lumber, 140*
7.9 *Transferring Numbers, 140*
7.10 *The General Estimate Sheet, 140*

7.11 *Allowance for Waste in Wood, 141*
7.12 *Table of Allowance for Waste in Wood, 141*
7.13 *Allowance for Waste in Wood for Hardwood Flooring ½" or ¹³⁄₁₆", 141*
7.14 *Stock Mouldings, 142*
7.15 *Lineal Feet, 158*
7.16 *Plywood and Fiberboards, 158*
7.17 *Nails, 158*
7.18 *Screws, 158*
Questions, 160

8   PERIMETERS, AREAS, AND EXCAVATIONS   162

8.1 *Perimeters, 162*
8.2 *Areas, 163*
8.3 *Excavations, 164*
8.4 *Excavations Mass, 165*
8.5 *Excavations: Service Trenches, 165*

8.6 *Drainage Trench, 167*
8.7 *Drainage Area, 167*
8.8 *Swimming Pool, 167*
     Questions, 168

9   TYPICAL WALL SECTION AND GUIDE LIST FOR RESIDENTIAL CONSTRUCTION   169

9.1 *Basic Data, 170*
9.2 *Perimeter Expressed in Lineal Feet—Exterior Walls, 171*
9.3 *Interior of Typical Wall Section, 172*
9.4 *Plan Areas, 172*
9.5 *Flat Roof Area Including the Overhang, 172*
9.6 *Pitched Roof Areas, 172*
9.7 *An Estimator's Guide List for Housing Projects, 172*

9.8 *Weeping Tile, 177*
9.9 *Damp-Proofing and Waterproofing Masonry Below Grade, 178*
9.10 *Installations with Sales Appeal for Speculating Builders, 179*
9.11 *The Appraiser, 180*
9.12 *A Guide List of Subtrades, 180*
     Questions, 181

10   EARTHWORK AND LAND LEVELLING   183

10.1 *A Glossary of Land-Grading Terms, 183*
10.2 *Cut and Fill, 184*
10.3 *Sanitary Fill, 184*
10.4 *Batter Boards, 185*
10.5 *Swell Percentage of Cut Earth, 187*
10.6 *Bench Mark or Data, 187*
10.7 *Decimal Equivalents of Inches in Feet, 188*
10.8 *Estimating Cut, 188*

10.9 *Interpolation of Datum Lines, 191*
10.10 *Grades: Cut and Fill, 195*
10.11 *Cut and Fill Estimating Sheet, 196*
10.12 *Compaction and Swell Factors in Cut and Fill, 198*
10.13 *Cut and Fill: Sand Model, 198*
10.14 *Estimating Cut and Fill for a Parking Lot, 200*
     Questions, 201

11   COMMON SOURCES OF ERROR IN ESTIMATING   203

11.1 *Main Sources of Error with Figures, 203*
11.2 *Main Sources of Error on Estimating, 204*
11.3 *To Avoid Errors on General Estimate Sheets, 205*

11.4 *Dimensions, 205*
     Questions, 206

12 PRELIMINARY, DETAIL AND UNIT ESTIMATES; PROGRESS SCHEDULES AND COST SUMMARY-SHEETS 207

12.1 *Preliminary Estimates, 208*
12.2 *The Square-Foot Method, 208*
12.3 *The Cubic-Foot Method, 208*
12.4 *Comparative-Appraisal Method, 208*
12.5 *The Detail Estimate, 209*
12.6 *Progress Schedules, 211*
12.7 *The Gantt Bar Progress Schedule, 212*

12.8 *The Critical-Path Method (CPM), 217*
12.9 *Quantity Surveying, 221*
12.10 *Unit Estimates, 223*
12.11 *Cost Summary Sheets, 223*
12.12 *Fire Damage Assessing, Renovating, and Alteration Work, 226*
Questions, 227

13 FLOOR ASSEMBLIES 229

13.1 *Supporting First-Floor Joists, 229*
13.2 *Floor Joists, 231*
13.3 *Bridging for Wood Joists, 232*
13.4 *Combination Tile and Concrete Floors on Grade, 233*

13.5 *Typical Floor-Bearing Details for Concrete Masonry Walls, 234*
13.6 *Structural I-Beams, 234*
Questions, 236

14 CONCRETE: PROPORTIONING, MIXING, AND PLACING 237

14.1 *Measuring Materials, 238*
14.2 *Mixing Concrete, 238*
14.3 *Ready Mixed Concrete, 238*
14.4 *Remixing Concrete, 239*
14.5 *Transporting and Handling Concrete, 239*
14.6 *Depositing the Concrete, 240*
14.7 *Finishing Concrete Slabs, 240*
14.8 *Striking Off, 240*
14.9 *Making Joints in Floors and Walls, 242*
14.10 *Water, 242*
14.11 *Water-Cement Ratio, 242*
14.12 *Storage of Cement, 243*
14.13 *Bank or Pit-Run Aggregates, 243*
14.14 *Surface Area of Aggregate 243*
14.15 *Design Mix by Volume, 243*

14.16 *Estimating Quantities of Dry Materials for a Concrete Floor and for Each Batch of the Mixer to be Used, 243*
14.17 *Measuring the Correct Volumes per Batch, 244*
14.18 *Measuring the Correct Volume of Dry Materials and Water per Batch, 244*
14.19 *Sizes of Concrete Mixers, 249*
14.20 *American and the old Canadian Gallon, 249*
14.21 *Runways and Ramps, 249*
14.22 *Machine Mixing Time, 249*
14.23 *Weight of Concrete, 249*
14.24 *Curing, 249*
14.25 *Demurrage, 250*
Questions, 250

15 CONCRETE FORMWORK AND REINFORCING STEEL 251

15.1 *Concrete Formwork: Footings, 251*
15.2 *Concrete Formwork: Basement Walls, 253*
15.3 *Semipermanent Form: Estimating, 253*
15.4 *Stripping and Reconditioning of Forms, 253*
15.5 *Concrete Walls: Estimating, 255*
15.6 *Reinforcing Steel for Concrete, 255*
15.7 *Reinforcing-Steel Contractors' Specifications, 255*
15.8 *Estimating Reinforcing-Steel Bars, 257*
15.9 *Concrete Column Footings Schedule, 257*
15.10 *Form Lumber Specifications, 258*

15.11 *Concrete Reinforcing-Steel Specifications, 260*
15.12 *Concrete Reinforced Columns, 260*
15.13 *Form Hardware, 262*
15.14 *Concrete-Form Tie Loading and Practical Form-Design Data, 274*
15.15 *Concrete Water Tank Estimate, 274*
15.16 *Gravel Base Under Concrete Floors, 276*
15.17 *Concrete Basement Floors, 276*
15.18 *Mesh: Steel-Welded Reinforcing, 277*
Questions, 277

16 CONCRETE BLOCK AND BRICK CONSTRUCTION 278

16.1 *Sizes and Shapes of Concrete Masonry, 279*
16.2 *Concrete-Block Constructed Warehouse, 279*
16.3 *Brickwork Definitions, 281*
16.4 *Brickwork Bonding, 282*
16.5 *Estimating Information for Canadian Clay Brick, 285*
16.6 *Estimating Tables, 286*

16.7 *Brickwork Estimating, 294*
16.8 *Brickwork Estimating: Cubic-Foot Method, 294*
16.9 *Reference Tables for Estimating Brick, Concrete Block, and Mortar, 296*
16.19 *Estimating the Number of Bricks in Chimneys, 298*
Questions, 299

17 FRAME CONSTRUCTION OF WALLS: EXTERIOR, INTERIOR, AND CLADDING 300

17.1 *Rough Carpenty Stud Wall Framing, 301*
17.2 *Staggered-Stud Wall, 301*
17.3 *Estimating Wood Wall Framing, 301*
17.4 *Girths, 301*
17.5 *Grounds, 301*
17.6 *Fiberboard, 305*
17.7 *Building Papers, 305*
17.8 *Sidings, 306*
17.9 *Labor for Applying Bevel Siding: First-Class Work, 311*

17.10 *Shingles, 311*
17.11 *Insulation, 311*
17.12 *Stucco Wire Mesh, 315*
17.13 *Brick Veneer for Frame Construction, 316*
17.14 *Glass Blocks, 316*
17.15 *Ornamental Masonry Units, 317*
17.16 *Carpentry, Joinery, and Millwork, 317*
Questions, 318

18 ROOFS: DEFINITIONS OF ROOF MEMBERS, BASIC ROOF DATA, AND ROOFING MATERIALS 320

18.1 *Types of Roofs, 321*
18.2 *Definitions and Roof Terms as Applied to Regular Roofs, 322*
18.3 *The Span, 322*
18.4 *The Run, 322*
18.5 *The Rise, 322*
18.6 *The Pitch, 323*
18.7 ***To Find the Area of a Gable Roof, 323***
18.8 *To Find the Area of a Regular Hip Roof, 324*
18.9 *To Find the Area of a Roof: Framing-Square Method, 326*
18.10 *Increased Percentage of Roof Area Over Ground Are, 327*
18.11 *Increased Percentage of Roof Area Over Ground Area for a Roof of Unequal Pitch, 328*

18.12 *To Find the Merchantable Length of a Gable Roof Rafter, 330*
18.13 *To Find the Length of Hip Rafters for Regular Hip Roofs, 330*
18.14 *To Find the Length of Jack Rafters Required, 331*
18.15 *Labor Costs for Roof Framing, 331*
18.16 *Flat Roofs, 332*
18.17 *One-Way Combination Tile and Concrete Roof Assembly, 333*
18.18 *Metal-Pan Formed Reinforced Concrete Floors and Roofs, 334*
18.19 *Coping and Parapet Walls, 334*
18.20 *Parapet Walls and Flat Deck Assemblies, 334 Questions, 336*

19 STAIRS: WOOD AND CONCRETE 337

19.1 *Wooden Stairs, 338*
19.2 *Stair Definitions, 338*
19.3 *Stairway Types, 339*
19.4 *Labor for Erecting a Closed-Stringer Staircase, 339*

19.5 *Small Apartment Block Stairs, 339*
19.6 *Concrete Steps, 344*
19.7 *Reinforced-Concrete Stair Problem, 345*
19.8 *Steel Stairs, 345 Questions, 347*

20 STRUCTURAL STEEL AND METAL UNITS 348

20.1 *Structural Steel, 348*
20.2 *Miscellaneous Metal Units, 349*
20.3 *Column Base Plates, 350*

20.4 *Wall Anchors for Beams, 350 Questions, 351*

21 FINAL INSPECTION AND HANDOVER OF THE PROJECT 352

21.1 *Extract of Specifications, 352*
21.2 *Sidewalk and Landscaping, 353*
21.3 *General Contractor's Inspection, 353*
21.4 *Window-Cleaning, 353*
21.5 *Janitor Services, 353*

21.6 *Architect's Final Inspection, 353*
21.7 *Maintenance, 353 Questions, 354 Glossary, 355*

Glossary, 355
Index, 365

PREFACE

Estimating Residential Construction is designed for building construction educator and students, general building contractors, builders of housing units, contractors for alteration and renovation work, fire damage assessors, building superintendents, technicians, foremen, builders' supply houses, and for all those ambitious men who are willing to continue their studies and accept the responsibilities in the fascinating and rewarding subject of estimating.

All material is presented in a step-by-step approach. After examples have been shown, similar problems which are designed to meet local conditions are given. An endeavor has been made to seize on those aspects of estimating which appear to offer the best guide to success in this field with the greatest economy of time. The approach used is as simple as the subject justifies, and in this respect I have been guided by Rutherford, who said: "I never consider a problem solved until it can be stated in simple terms."

The first five chapters are devoted to general information that an estimator or building contractor should know; these chapters are designed to help ambitious men to find their way quickly to the right people, places, and things for the information they require in estimating.

The hallmark of a technician is not only what he knows but also his knowledge of where to look for what he wants to know. There are a number of tables in the text (with explanations as to how they were composed); others outside the scope of this work may be found in the manufacturers' brochures and specifications often used by architects. Where labor time has been given, it is offered as a guide and should be amended to suit individual contractors' time-studied operations.

A number of guide lists and summary checklists have been included as a help for the estimator to marshal methodically all the component parts of the estimate in the order in which materials will be used on the job.

The main part of the book gives detailed examples and exercises from topsoil to rooftop to final handover of the property.

Where it has been proved with classes (and recommended by readers) that hand-written specimen sheets are helpful, they have been included. An endeavor has been made to simulate job experience with classroom instruction. *The theme of the book is organization, neatness, method, and morale.*

For those in the field of education, the book presents an organized approach to the subject. The work has been tried in class with satisfying results for both students and instructor. An excellent learning situation may be created where students work in small committees and give oral reports on assigned reading from the text one week ahead of supplemental lectures. In this respect I would particularly like to thank my former students, who by their searching questions have inspired me to write the original edition.

Topics covered include: the critical-path method; government services available to business; labor legislation; financial definitions and statements; bankruptcy; depreciation; partnerships; architectural drawing practices; contractors' associations and bid depository; sanitary fill; concrete form hardware; tie loading and practical form data; and estimating chimneys and fireplaces. Also included are: new lumber standards; demurrage; a glossary of land building terms; a guide list for subcontractors: the Gantt bar progress chart; contract change orders; a fuller treatment of insulation, and a glossary of building construction terms.

Fuller treatment, in response to specific requests, has been given in the following areas: bonds, contract documents, and types of insurance; rapid calculations of areas of plane figures and solids; cut and fill; interpolation of datum lines; and common sources of error in estimating. In addition, the whole text has been updated.

Further suggestions by readers will be very much appreciated.

I sincerely thank Mrs. Rosamond Gladwys Cox for her enthusiastic help in the preparation of the manuscript, to Mrs. M.A. Olivier for her cheerfulness and efficiency in the secretarial work in the production of the manuscript, and to my wife Joan Rose Wass for her continued patience and encouragement.

It would seem appropriate to conclude this preface with a quotation from antiquity:

He who is theoretic as well as practical is therefore doubly armed; able not only to prove the propriety of his design, but equally so to carry it into execution. *

ALONZO WASS

---

*Vitruvius (Marcus Vitruvius Pollio) first-century B.C. writer on architecture. His *De Architectura* was widely used by Renaissance architects in the classical revival.

1

A GUIDE LIST OF INFORMATION FOR CONTRACTORS AND ESTIMATORS

1.1 AMERICAN SOCIETY OF PROFESSIONAL ESTIMATORS

The following excerpt was taken from Engineering News Record.

The American Society of Professional Estimators (ASPE) has taken the first formal steps towards certifying construction estimators.

Meeting in San Francisco last month, the society's executive board approved testing procedures for certification examinations and directed its certification and registration committee to prepare tests for estimators based on the Construction Specifications Institute's (CSI) 16-division format for estimating various types of construction work.

Under the approved testing procedures, an ASPE member with five years of estimating experience would take written and oral exams on CSI section one, which

1

covers general estimating conditions, and on the section or sections covering his particular disciplines. He would also be required to prepare a take-off from plans and specifications. . . .

The board's actions will be presented to the ASPE's 10 chapters at its annual convention in Phoenix next January. If they are approved, the society expects to start administering tests on electrical, mechanical, carpentry, metal and concrete divisions by September, 1976, and on all CSI divisions by September, 1977.

ASPE hopes its proposed certification program will prompt state legislatures to pass laws requiring the formal registration and licensing of estimators.

In this chapter, you will be introduced to some of the many sources of information that you should be aware of as an estimator or as a person entering the contracting business. For the purpose of discussion, it is assumed that you are entering the construction field either as a beginner with a large company or as a building contractor on your own account. This, then, is a broad outline of the action you may take from the time you receive your builder's license until you have completed your first contract.

1.2 FINANCIAL

It is not the policy of lending institutions to finance just anyone into a new business venture; you must first prove to the lender's satisfaction that your business is already a success. Then you may obtain a loan for business expansion. In the early days of your business life, you may have to put up very sound collateral for every penny borrowed from a lending institution. Collateral is security—held against property, stocks, bonds, certificates, or notes having a monetary value—which is given by a borrower to a lender in return for a loan. Collateral may include physical objects such as real estate, machinery, furniture, and so on. A promissory note is a document comprising a promise to pay a specified sum of money to a named person on a given date, or it may promise payment "on demand." It is understandable that lenders require good security, since they are the custodians of other people's money. *Success in business breeds success.* As your business progresses, lenders will be only too happy to continue financing you.

1.3 SOURCES OF CREDIT

Following is a list of some of the sources of credit that may best suit your needs (remember that interest on borrowed money is tax deductible):

(a) commercial banks;

(b) finance and mortgage companies;

(c) insurance companies;

(d) real-estate companies;

(e) savings and loan associations;

(f) credit unions;

(g) friends and relations;

(h) government-sponsored loans;

(i) equity financing.

Government-Sponsored Loans

The Small Business Administration (SBA) is an independent agency of the Federal Government—established by Congress to help the nation's small businesses—with field offices in many cities. The Industrial Development Bank (IDB) was established by Parliament in Canada to help that nation's small businesses.

Equity Financing

A small businessman may be reluctant to accept new equity from outside the business because he wants to keep all the affairs in his own hands; but in some cases, the addition of equity capital and of the skills of another individual may be advantageous. Carefully study all methods of financing—this is the start of your estimating experiences.

Builders' Loans

Some companies specialize in making builders' loans. This is risk capital and you would have to be quite convincing to the loan company, not only on your ability as a builder, but also on your ability to administer your affairs with recorded clarity. Most contractors are excellent builders; however, many fail due to poor business knowledge. A short course in business management would be very advantageous for entrants into this field.

1.4 FINANCIAL DEFINITIONS

This section defines some of the most common financial terms used in the accounting of a business.

Balance sheet: All the assets of the company are listed on the left hand side; on the other side are listed all the liabilities. When the liabilities are subtracted from the assets, the remainder is a credit balance, and the business is solvent.
Current assets: These are items in various stages of being converted into cash, such as cash on hand, cash in the bank, and cash due from sales.

Inventory: Merchandise on hand such as hardware, bricks, lumber, partly completed speculative buildings, and so on.

Fixed assets: These are properties used in the business and not intended for resale, such as real estate used for the business office, store yard, and so on.

Machinery: This category includes equipment and motor vehicles used in the business. Such items are listed at the original cost and deductions are made for depreciation. (See the Inland or Internal Revenue Department for allowances for depreciation on builder equipment and vehicles.) *This is extremely important.* (See also Sec. 1.32.)

Net fixed assets: This is the cost of fixed assets less depreciation; the difference is called net fixed assets. This is the amount realizable if the items were sold.

Assets: Assets are what the business itself owns. The business must always be regarded as a separate entity; it is not you or your associates, it is a business!

Liabilities: These are the claims against the business (what it owes), including the owner's claim (equity) against the business.

Notes payable: The agreements signed in favor of claimants against the business, such as bank loans or loans from others.

Income tax: An amount set aside out of current earnings, which is to be paid to the government.

Long-term liabilities: Debts due to be paid after twelve months or more.

Net worth: The owner's claim against the business at any given time.

Surplus paid in: Earnings from the company by the owner(s) and left in the business as additional equity money. *This has been earned by way of wages, profit, or both and is subject to income tax.*

Accounts receivable: Accounts due to the business, such as proceeds from the sale of a newly built home and also progress payments from work accomplished on contracted building operations.

Working capital: Available cash for meeting current wages and material suppliers' accounts.

Reconciliation of net worth: This is a statement that shows how the net worth of the business has increased or decreased for the statement period covered. Gross worth is the value before expenses have been deducted; for example, a house is worth $55,000.00 before real estate expenses have been deducted. If the house is $52,750.00.

Profit and loss (or Income statement): This document shows the results of the business operations during a certain time.

Financial statement: This lists the balance sheet, the profit and loss statement, and the income statement.

The last three statements are required by lenders in the appraisal of a business that is applying for a loan.

Mortgage Companies

If you are building speculative residential properties, it is imperative that you get large conventional mortgages on every home that you build. Without this provision, you would soon have all your working capital tied up in mortgages. As with banks, mortgage companies are very circumspect with entrants into the construction field. The first three or four years are probably the most difficult for beginners in any business.

The percentage of construction industry failures between 1956 and 1965 as shown in the *Statistical Abstract of the United States* was as follows:

First year	1.7	Fourth year	12.8
Second year	11.1	Fifth year	10.4
Third year	14.8	Tenth year	3.6

1.5 CREDIT BUREAUS

Be aware that merchants with whom you may wish to do business will probably obtain a credit rating on you. Likewise, you should consider getting a credit rating on the newer merchants with whom you may intend doing business. If you intend to sell speculatively built homes and hold some part of the mortgages yourself, you most certainly should get a credit rating on the prospective purchasers.

1.6 BUILDER'S LICENSE

In many areas before a person starts to operate as a builder, it is necessary that he obtain a builder's license from city authorities. There is a different charge for builders living and contracting within the city limits and for builders living outside of city limits and working in the city. There is also a difference in charges for subcontractors. The purchase of a builder's license is not an indication that the purchaser is a good builder. His work has to stand up to progressive inspection by the city building inspectors during all phases of construction. Also some lending institutions make building inspections to see how their money is being used.

The builder's license will enable the holder to make building material purchases at normal trade discounts. In business there are several different types of discounts which are regulated against the number of days within which payment for delivered goods must be made. The differences are considerable and you should make very careful enquiries of the merchants with whom you may deal. A license holder may also, upon nomination, be accepted into the local and national construction associations.

1.7 LOCAL AND NATIONAL CONSTRUCTION ASSOCIATIONS

These associations operate at the city level for local convenience, at the state or provincial level for area affairs, and at the federal capitals for affairs of national importance to allied members.

Several such nonprofit associations are as follows:

Home Manufacturers' Association, Barr Building, 910 17th Street N.W., Washington, D.C. 20006.

National Association of Home Builders of the United States, 1625 L Street, Washington, D.C. 20036.

The Associated General Contractors of America, Inc., 1957 E Street N.W., Washington, D.C. 20006.

Canadian Construction Association, Ottawa, Ontario, Canada.

Amalgamated Construction Association of British Columbia, 2675 Oak Street, Vancouver 9, B.C., Canada.

The objectives of all these organizations are similar.

Membership. Upon the sponsorship of two (or three) members, membership in these nonprofit associations is open to small, medium, or large general contractors; individuals; partnerships; firms; manufacturers and suppliers to the construction industry; and corporations and associations actively engaged in industrial and institutional construction.

Objectives. To maintain a strict code of ethics; to lobby for beneficial industrial construction legislation; to promote better relations between its members and owners, architects, and engineers; to establish high professional standards among contractors and subcontractors in the construction industry; to encourage methods of contracting which relieve the contractor of improper risks; to promote educational programs with members of the association and labor organizations; and to ensure to the public the benefit of competitive contracting.

Services and Facilities. Each of these organizations provides plan room service; bid depository; members' roster; administration of oaths; contract and tender documents; integrated membership in local, area, and national organizations; bulletins about construction projects; library; a list of latest tenders called and times of closing (be aware of time zones when bidding interstate); construction-news magazines and releases of latest government regulations and local ordinances pertinent to the construction industry; and committees, workshops, and social activities.

Plan Room Service. Large plan rooms display drawings and specifications calling for tenders for projects in the vicinity and neighboring areas. Smaller rooms are provided for subtrade estimators to take-off quantities on the spot. They then place their bids to as many general contractors in the bid depository as they wish.

Subcontractors may place their bids to any or all of the general contractors, but it should be noted that they are not bound to give the same bid to all general con-

tractors. If they have had difficulty in finalizing their accounts with a particular general contractor, they tend to present him with increased bid pricings.

Bid Depository. A bid depository is maintained in each branch office. The bid depository system for the reception of sealed tenders by subcontractors to general contractors protects the integrity of bidding and ensures that those receiving tenders obtain firm quotations in writing with adequate time for them (the general contractors) to complete their tenders on new projects.

1.8 LAND PURCHASE

The land in North America is owned either by government or by private parties. The title to all parcels of land is recorded in the land registry office of the area in which the land is located.

Assume that you wanted to purchase a small parcel of vacant land within the city limits and that you did not know the owner. Take the following steps:

(a) Obtain a map of the area from the local authority and identify the land and the address of the nearest adjoining property.

(b) Inquire at the city taxation department for the name of the person paying the taxes—the owner.

(c) Obtain from the taxation department the official description of the property. This bears no relation to the postal address. It may be something like: Lot 7 of Block 14 City of_____AP 2816.

(d) Proceed to the city planning department and check for zoning restrictions on the property.

(e) Using the official description of the property, go to the land registry office (or have your agent go for you) and search the title.

(f) Write to the owners of the property asking if they wish to sell the land. They may live miles away, possibly even abroad.

1.9 LAND REGISTRY OFFICE

In every state and province there is a land registry office where all the land in that particular registration district is registered according to ownership on a document called a *title*. All the land in North America is registered by individual parties or by the government.

Searching the Title

For a very nominal fee (about one dollar) any citizen, or his lawyer or agent, may scrutinize any title to any land registered in any particular office. This is called "searching the title." The prospective purchaser of a parcel of land may wish to search the title with a view to having any encumbrance removed from it. If there are no encumbrances registered on the title, the property is said to have a *clear title.*

Some of the encumbrances registered on the title may be as follows:

(a) A reservation by any party or the government for the mineral rights on the property.

(b) A first, second, third, or more mortgages registered against the property.

(c) *Mechanic's Lien.* This is a statement of claim registered against the property for work alleged to have been done on the property for which the mechanic has not been paid. *Such liens must be registered by the individual who did the work within a certain number of days after completion of the work.* You should check in your area. Why not arrange a field trip by your local association to visit one of these registry offices and get all the information firsthand?

The mechanic's lien may have to be proved in a court of law. It is of prime importance that a very careful record be kept of the date and time that the last work was done; this record should be witnessed. Some contractors will leave a small portion of work uncompleted until they feel sure that they will be paid for all the work. If they have real doubts about being paid, they then go back and complete the job and immediately file a mechanic's lien in the land registry office.

(d) *Easements.* An easement is the right of one party to enjoy some privilege on the land of another. Easements are often registered for such purposes as sewer, gas, power, and telephone lines, and the owners of these have the right to enter the property for the maintenance and servicing of such lines. Sometimes temporary easements may have to be taken out by a contractor building in a metropolitan area where he wants the use of the adjoining property (if vacant) for temporary storage of his materials and machines. There are, of course, many other types of easements.

(e) *Power of Attorney.* This entry on a title shows that the title owner has given another party the authority to act for him in any dealings in connection with the land specified. These powers are very sweeping and are usually given to highly reputable law firms, but the power of attorney may be given to almost anyone over the age of twenty-one. This instrument is useful where the owner is living at a great distance from his holding.

(f) *Writ or Judgment.* This is a recorded statement on the title deed showing that at a certain time and place a judgment was handed down by a court of law to the effect that the party in whose name the title stands was indebted at that time to another party for some services or goods that were not paid for. These services or goods could be indebtedness to several stores, garages, and so on.

(g) *Caveat.* This instrument is a "notice to beware" that the party who has filed the caveat has an interest in this property; this interest must be taken into account before any change in ownership is made.

Abstract of Title

This is a written official statement of any/all the encumbrances appearing on a title at the stated date and hour of the day shown. It may be used in a court of law. You may get an abstract of title from the land registry office in regard to any land; in fact, you may get a photostat copy of the title. All these services are performed for a very nominal charge, since government offices do not work on a profit motive. An abstract may be very useful for a person who lives some distance from the land registry office.

1.10 CITY OFFICES

The estimator or contractor should be very familiar with all the city departments which may be helpful to him in his business.

Trouble Calls

Some city departments provide twenty-four-hour service. These include: electrical light and power, gas, sewer, garbage, streets, fire, police, and ambulance.

You as a builder may want at any time one or more of these services. Their telephone numbers should be kept in a ready file. Assume that your street has caved in because of a flash flood. What are you going to do about it?

City Planning Department

Before purchasing building land, the contractor should visit the planning department. Here may be seen the proposed plans for the future development of the city showing future freeways, expressways, airport developments, bridges, recreation grounds, suburban development for residential construction, schools, shopping centers, churches, libraries, hotels, and motels. Also shown will be the very important zoning regulations such as districts for single-dwelling units, two units, three units, and multiple-dwelling units such as apartment blocks. In some areas the heights and floor areas of buildings will be restricted. In the better-class districts, residential areas may have all the power lines buried and have very decorative standard street lighting. These services will be reflected in the city taxes.

Future industrial developments will also be shown with road and railway trackage and available services.

Some housing project promoters purchase and develop, with roads and city utilities, large areas of land, which they then divide into separate building lots, and offer for sale to the public, complete with a modern house on each lot. All this development must conform and be approved by the city engineers department and the city planning department. An advantage to both the vendor and the purchaser of such a unit is that the utilities—that is, roads, sidewalks, boulevards, gas, power, water, telephone, and cable television—are an inclusive cost in the original purchase of the property. The purchaser is only responsible for making payments to one party each month.

When a smaller housebuilder offers a new home for sale, it is quite usual that the purchaser pays to the vendor each month a certain sum to cover the mortgage interest and repayment plus a separate annual tax bill to the city. In many instances, however, the purchaser may have to wait a long period of time before all the utility amenities are installed.

It is from the city planning department that a contractor might want to seek to have a parcel of land rezoned from a single-dwelling-unit area to a multiple-dwelling area for the erection of an apartment block. *You must visit your city planning department. The officers there will be very happy to show you through and answer your questions.* From the information given and a further talk with the taxation department, you may get a fair assessment of future property taxes in the area in which you wish to build.

City Engineers Department

The city electrical engineers department is usually separate from the remainder of the engineering services. The city engineers are responsible for streets, sewers, garbage, building inspections, waterworks, testing of building materials such as concrete, and so on. They will have their equipment yard and also an industrial development coordinator, license and building permit bureau, construction and maintenance division, and a traffic engineers department.

Building Permits

A proposal for a new building permit must first be submitted to the city planning department, which examines the plans and specifications and decides whether or not they will conform to the planned pattern of the area in which the proposed building is to be erected.

When approved by the planning department, the proposal is passed to the city engineers for structural examinations and, when approved, a building permit is issued upon payment of a scaled fee depending on the type of building. This building permit must be prominently displayed at the new building site.

Other permits may include water, street closing, permission to use city water hydrants, and connections to city sewer line.

1.11 SURVEY

Some cities may, for a charge, officially survey your land. *Where this is not done and there is the slightest doubt as to the location of the original survey stakes, employ the services of a registered surveyor.* While you are reading this chapter, someone is having a partly (or completely) finished house removed from one building lot to a correctly located one. This situation is avoidable, and building inspectors may tell you of many instances where people have built on the incorrect lot.

Occasionally even the surveyor can make a mistake, but he is bonded against such an eventuality, and thus you are protected. The best advice, therefore, is to employ a registered surveyor who will correctly locate your lot, put up the batter boards, and establish your levels. When this is done professionally, you can commence your building operations with every confidence as to location and levels.

1.12 BATTER BOARDS

Batter boards are frames placed adjacent to (but on the outside corners of) proposed excavations over which a taut mason's line (or wire for larger jobs) is strung for delineating building lines when excavations are completed. See Fig. 1.1.

Let us assume that we have established the front building line stakes for a residence having a rectangular plan of $30'-0'' \times 48'-0''$. Note carefully that for small offsets from an otherwise rectangular plan, the wooden square may be used.

1.13 DATUM AND BENCH MARKS

A very important post held by an official in the city engineer's department is that of *official grade striker*. There is probably more grief over incorrect grades in building construction than in any other phase of operations. After the survey of the land, and before the contractor builds, the bench mark or datum must be established. (See Sec. 10.4.)

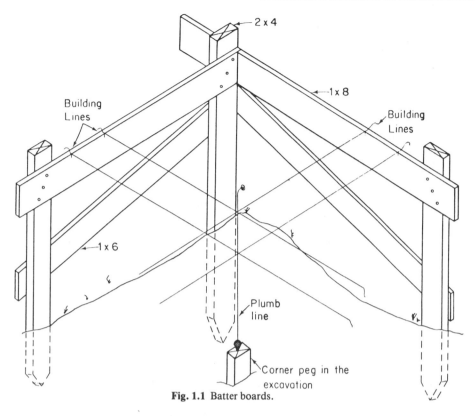

Fig. 1.1 Batter boards.

The official grade striker will come to the land and set up a peg showing the depth of the sewer line and another peg showing the height of the city sidewalk curb. This is assuming that you are building on virgin ground. If you set your house too low in the ground, you may have difficulty in getting a fall to the city sewer line. If your house is built too high, you may find later that the city will cut a road and leave your house far too high, making it difficult to drive onto your lot.

Assume that you intend to erect a building close to a proposed expressway or freeway. Be aware that the grade on the outside curve of the road will be higher than that on the inside of the road.

Note: The local authorities make certain (and we must be aware) that the grade will be higher on the outside. The road may be built after we have built the project.

1.14 BUILDING INSPECTORS

It is in the public interest that progressive official inspections be made of new structures; for example, faulty plumbing could be a severe public hazard. It is obligatory for contractors to request that inspections be made at prescribed times. All work—such as on the sewer line, electric circuits, and so on—must be left open until an official certificate of approval has been given by the building inspector.

Building inspectors specialize in their own fields and you must be aware that inspections have to be arranged ahead of time so that the work schedule will not be delayed.

1.15 APPRENTICESHIP AND TRADESMAN QUALIFICATION BOARD

None but the best tradesmen should be employed. As in all fields or purchase, you will generally get what you pay for. An incompetent tradesman can be a very expensive proposition. It is recommended that only qualified men be employed at the accepted contracted rates of pay. The state or province has a department which deals with apprenticeship training. You should visit the school and see for yourself what is being done. Every endeavor is being made by all levels of government to educate people to adjust to this new and highly technical age. You should encourage good workmanship and be aware that an architect or his inspector may call for the removal of incompetent workmen.

1.16 TECHNICIANS

It is estimated that six technicians are needed for every engineer or other professional man in the construction industry. These men are specially educated to stand between tradesmen and the original designers of the structures to be built. (See Sec. 2.8.)

1.17 LOCAL LABOR COSTS AND AVAILABILITY

Before tendering for work in outlying rural areas, be sure to contact the local unemployment office and the trade union office regarding the availability of local labor. You may have to take key men from your home base who may demand that they work overtime with all wages at premium rates.

1.18 FRINGE BENEFITS

Contact the local Social Security office or the trade unions for the fringe benefits applicable in your building area. These items will appear on the workmen's pay statements.

1.19 LOCAL MATERIAL COSTS

Material costs will vary from place to place according to the availability of and haulage necessary for raw materials. Costs may be obtained from:

(a) builders' supplies merchants;

(b) hardware merchants;

(c) concrete manufacturers;

(d) brick manufacturers;

(e) sash and door manufacturers, and so on.

Secure a list from the local telephone directory and enquire at the local builder's exchange.

You should be aware of the abbreviations on the merchants' quotations that you receive. Where it states "F.O.B.," this means "free on board," but you must know *where* it is free on board. If you are operating at a distance from Pittsburgh, and the quotation is F.O.B. Pittsburgh, this means that you pay the freight costs from Pittsburgh. On the other hand, if it is quoted F.O.B. at the place where you are building, it means the freight is paid by the supplier direct to the nearest railhead.

1.20 RAILWAYS AND FREIGHTAGE

Shippers of goods (in quantity) will send you a loading list giving the boxcar numbers. With this information, you may check through the local railway as to where these boxcars are at any time. The boxcars containing your goods may be derailed, and this could upset your progress schedule.

1.21 DEMURRAGE

This term refers to the detention of materials by unloading vessels, boxcars, transit concrete-mix vehicles, or any other carriers beyond a stipulated time. This incurs a charge against the violator.

1.22 SUBTRADES

The general contractor is financially responsible only for about 15 to 25 percent of the cost of a new midtown building. The remainder is taken up by the subcontractors.

The address of every subcontractor should be kept on file. For a list of subcontractors, see the local housebuilders' association or builders' exchange, and read the classified advertisements in the telephone directory. Some governments issue a gazette showing all trades in the state or province.

1.23 WORKMEN'S COMPENSATION BOARD

The supervisor or a senior assistant of the local workmen's compensation board would make an ideal guest speaker at one of your association meetings. He will also show you some very impressive films. The board issues some very striking posters which should be prominently displayed around your jobs. A representative of the board has the right to come onto your project at any time to check that all safety precautions are being observed and that your medical kit or first-aid center is adequate for the job. Minimum standard kits are required according to the number of employees.

The question often arises with both workmen and employers as to how serious an injury must be before a report is made and first-aid care is sought. *All accidents should be reported.* Many workmen's compensation boards will honor any claim up to three years providing that an entry has been made at the time of accident in the accident report book or the builder's diary. Assume that a person slips and hurts the lower part of his back. No one knows whether or not this will develop into a serious problem. It would be very difficult for the individual to register a claim unless a record was made at the time of the mishap. In the case of a married man with children (unless he could register a successful claim), it would be a tragedy. *All accidents, however minor, should be reported at the time of accident; check with the local workmen's compensation board.*

On some jobs, it is a regular Monday morning feature for the superintendent to call all the workmen together for a five-to-ten-minute talk on safety. One of the prime causes of accidents in the building industry is untidy surroundings. Usually, the workmen will be just as untidy as the superintendent will permit them to be. Keeping a job clean is almost a trade in itself. *Don't talk over the phone to the workmen's compensation board or to any other board. Go in and see them.* The charges of the board for workmen's compensation are about 2 percent of the wage bill for ordinary construction. Do not start to work on a building site until you have covered every workman by this insurance. If the subcontractor has not legally covered his own men, he and his men are in your employ and you will be held responsible. Check that each subcontractor has insurance for his own men.

1.24 FIELD TRIPS

You should go on your own, or in company with other interested persons in your organization, to visit one by one all the offices mentioned in this chapter. Find out what happens on the other side of the counter. A visit through a friendly architect's office will help you understand the correlation of sketch plans, rough specifications, and tentative costs. Nearly all people are pleased to tell what they do for a living. Find out!

1.25 STANDARD CONTRACT DOCUMENTS

It is very important to you as a beginning estimator that you make a close study of all contractual documents that are used in the building industry.

Most contracts in the United States and Canada call for the use of standard forms. Such forms are published by the American Institute of Architects and by the Royal Architectural Institute of Canada and may be obtained (very cheaply) from the head office of either of these professional organizations. Respectively, the documents are called:

The Standard Documents of the American Institute of Architects

The Standard Documents of the Royal Canadian Institute of Architects

Included among these forms is "A Guide to Bidding Procedure," which is recommended when competitive bids are requested. *You should get these standard forms and study them.*

1.26 NEW MATERIALS

Once a contractor meets with success in his business, he becomes known in the construction industry. He then begins to receive a great deal of manufacturers' advertising material. Most of this information is vital. Architects frequently specify a certain manufacturer's product,

and further specify that it must be applied to manufacturer's specifications. This in turn means that the product has met all the requirements of the architects and testing laboratories such as ASTM (American Society for Testing Materials) and the CSA (Canadian Standards Association). Unless a contractor keeps abreast of this type of information, he will soon lose touch with current trends, materials, and practices. It is suggested that you write for all the manufacturers' publications that you think you may need. You will find them advertised in most trade magazines.

1.27 THE AMERICAN INSTITUTE OF ARCHITECTS PUBLICATIONS

Write to: The American Institute of Architects, 1735 New York Avenue, Washington, D.C. 20006, and ask for a copy of their Catalog and Order Form of Publications and Documents. It is specifically suggested that you purchase all the A Series: Owner-Contractor Documents; also a publication entitled "Uniform System for Construction Specifications, Data Filing and Cost Accounting."

The Royal Architectural Institute of Canada uses the same filing system as the American Institute of Architects so that all manufacturers' brochures, coming from either country, are filed under the same code numbers. You will need several metal filing cabinets. Clearly title and index all file covers from the Standard Filing System. File new material and discard superseded matter. It is just as important to discard dated technical brochures as it is to file new ones.

1.28 LABOR LEGISLATION

It is important that you be alert to the present and projected labor legislation for the industry in your area, and you must know the minimum requirements regarding the employment of labor.

Labor legislation is always in a state of progressive flux. The law is enacted through briefs presented by unions, management, national builders' organizations, interested citizens, and sometimes unfortunately by startling disasters that no one could foresee. You must be able to quickly assess what additional cost will be reflected on your business by any new legislation and be ready to adjust the price of your product to meet it. You must be prompt and decisive in all your business activities.

Here are some of the things of which you must be aware:

(a) fair minimum wages;

(b) hourly rates;

(c) overtime rates of pay and how much overtime is permitted (see the local union business agent);

(d) present and projected trade union negotiations for your own and associated industries;

(e) adjustments to existing fringe benefits;

(f) public and annual holidays;

(g) changes in the Workmen's Compensation Law;

(h) washroom facilities for the staff.

For the most part, labor legislation is a matter for state or provincial governments. Possibly the most important piece of legislation is that dealing with unionization. Freedom of organization is guaranteed; employers are required by law to recognize and bargain with the trade union representing the majority of their employees.

1.29 GOVERNMENT SERVICES AVAILABLE TO BUSINESS

The North American federal governments, state and provincial governments, and individual city councils make available to citizens and foreigners comprehensive statistical publications at nominal costs. All levels of government are interested in attracting orderly, developed industries and highly sophisticated residential construction into their areas.

Government statistics enable people to plan intelligently where they want to work, buy, sell, live, play, and have foreknowledge of such things as weather, labor conditions, population trends, and earnings by industries. These statistics also indicate the number of persons owning, renting or purchasing their living quarters; social amenities such as number of housing starts over a given period; business failures by different industries; and other vital statistics useful to you. *It must be remembered that statistics are never quite up to date. How can they be?*

Here is a list of publications of which you should avail yourself:

Statistical Abstract of the United States

This book covers a wide variety of subjects. The *Abstract* is published yearly in hard cover and is sold for $4.00 by the Superintendent of Documents, Washington, D.C. 20402.

In addition to selected statistics, the *Abstract* has a reference section for other sources of information. Because of its wide coverage, low price, and convenient size, the *Abstract* is probably the most useful single publication of government statistics. The degree of detail shown in the *Abstract* is naturally restricted. Most of the statistics shown are national, although some data are given on a regional or state basis, and a few for large cities and metropolitan areas.

County and City Data Book

For more local, geographic detail, the Department of Commerce publishes this volume. It presents a selection of available statistics for all counties and for cities over 25,000 population. It is a hard-cover book, printed at $4.50, published about once every three years and sold through the Superintendent of Documents, Washington, D.C. 20402.

These publications may be seen at your public library.

Small Business Management Series of Publications

Examine the following Small Business Management Series of Publications, which are obtainable from the U.S. Government Printing Office. Many of these publications would be invaluable for your library. The government is continually updating them.

1. *An Employee Suggestion System for the Small Plant.*
2. *One Hundred and Fifty Questions for a Prospective Manufacturer.*
3. *Human Relations in Small Industry.*
4. *Improving Materials Handling in Small Plants.*
5. *Public Accounting Services for Small Manufacturers.*
6. *Cutting Office Costs in Small Plants.*
7. *Better Communications in Small Business.*
8. *Making Your Sales Figures Talk.*
9. *Cost Accounting for Small Manufacturers.*
10. *Design Is Your Business.*
11. *Sales Training for the Small Manufacturer.*
12. *Executive Development in Small Business.*
13. *The Small Manufacturer and His Specialized Staff.*
14. *The Foreman in Small Industry.*
15. *A Handbook of Small Business Finance.*
16. *Health Maintenance Programs for Small Business.*
17. *New Product Introduction for Small Business Owners.*
18. *Profitable Advertising for Small Industrial Goods Producers.*
19. *Technology and Your New Products.*
20. *Ratio Analysis for Small Business.*

21. *Profitable Small Plant Layout.*
22. *Practical Business Use of Government Statistics.*
23. *Research Relations Between Engineering Educational Institutions and Industrial Organization.*
24. *Equity Capital and Small Business.*
25. *Guides for Profit Planning.*
26. *Personnel Management Guides for Small Business.*
27. *Profitable Community Relations for Small Business.*
28. *Small Business and Government Research and Development.*
29. *Management Audit for Small Manufacturers.*
30. *Insurance and Risk Management for Small Business.*

Orders for these publications should be placed with the Superintendent of Documents, U.S. Government Printing Office, Washington, D.C. 20402. In lots of 100 or more copies mailed to a single address, a discount of 25 percent is offered. Check the latest catalog for availabilities and prices.

1.30 CONTRACTOR'S FEES AND PROGRESS PAYMENTS FOR SMALL JOBS

Where a contractor undertakes small jobs such as repairs, alterations, or additions to existing structures (especially residential units), it is recommended that such work be done on a cost-plus-percentage of say 12½ to 15 percent basis.

In the first instance a very careful assessment of the work involved must be made. An *estimate* is then given—this is not a firm bid or actual cost—it is merely an estimate. Before an agreement is made, a credit rating of the party requiring the work to be done should be obtained, and also a search of the title should be made to see if there are any encumbrances registered against the property. Assuming all reports are good and if the client is agreeable to the estimated cost-plus-percentage, an agreement may be drawn up and notarized and work commenced. It should be noted that clients are always very happy if they find the final cost to be less than the estimate.

The method of progress payments will be shown on the agreement. It is suggested that the client pay for all materials upon delivery and a biweekly progress payment for work performed. In this manner, the contractor can operate with a smaller amount of working capital.

One of the most difficult estimates to be made is in the tearing down of old work. A contractor can never be quite sure as to how much old work will have to be torn down in order to attach new work. All debris will have to be hauled away and *legally* dumped. Careful thought must be given to the distance from job to dump, what the traffic conditions may be at certain times of the day, and how much interference will be caused to workmen on the job for allowing the passage of persons or things through the area of operations. It must also be determined whether or not the job will require late night, overnight, or weekend work to be done at overtime rates. On some large alteration jobs there may be a clause concerning the finding of treasure trove (valuable articles such as coins built into the original structure); such finds usually become the property of the owner. All these points should be discussed thoroughly with the prospective client.

1.31 BANKRUPTCY

It is important that a contractor and estimator be aware of the implications of a company becoming insolvent.

Bankruptcy may be defined as: the adjudication of a debtor's inability to pay his debts. Bankruptcy proceedings are of two kinds: voluntary and involuntary, i.e., instituted by the insolvent debtor or by his creditors. Two of the main purposes of these proceedings are to secure equitable division among creditors of the bankrupt's available assets, and to release or discharge the bankrupt from his obligations if he has complied with the law.

Bankruptcy Cases Filed and Pending. A bankruptcy case is a proceeding in a U.S. District Court under the National Bankruptcy Act. *Filed* means the commencement of a proceeding through the presentation of a petition to the clerk of the court. *Pending* is a proceeding in which the administration has not been complete.

It is an offense for a person to continue to trade after he knows that he is insolvent: *as soon as he believes himself to be insolvent he should see an attorney for advice.* Sometimes the official receiver (person appointed by a court to manage property in controversy, especially in bankruptcy) may retrieve a situation in small building constructions by managing the business until all properties are completed and sold.

The duties of a bankrupt include: submitting to an examination, attending on creditors at the first meeting, fully disclosing all assets, and assisting in every way in the official administration of the estate. Punishable offences include: failing to comply with a bankrupt's duties, fraudulently disposing of property, concealing or falsifying books or documents relating to the business, refusing to answer fully and truthfully any question asked at an examination, and obtaining credit or property by false representation after or within twelve months preceding bankruptcy.

1.32 DEPRECIATION

The subject of depreciation should be studied well by all building construction estimators. It is important to know not only the several methods of depreciation acceptable to governments, but also which method will be most advantangeous to use at any given time. In this section we shall discuss how depreciation is determined and those methods that are acceptable to the Internal Revenue Service of the country in which it is used.

In accounting for contractors' equipment, the depreciation of an asset represents a loss in its worth from its last *assessed book value*. The book value of an asset is not necessarily the actual worth of said asset on the market. Equipment such as woodworking machines, tubular scaffolding, trucks, and so on, are all subject to allowable depreciation for income tax purposes. It should be noted that there is a different between "depreciation" and "depletion"; the latter refers to a wasting asset through exhaustion, such as a mine or an oil well that are worked to complete exhaustion.

There are several acceptable methods used to evaluate depreciation, and each year governments make tax guides available, outlining such methods. However, it must be remembered that *tax laws may be changed at any time by an Act of Congress in the United States, or an Act of Parliament in Canada.* You may obtain a copy of *Depreciation Guidelines and Rules* (Revenue Procedure 62–21) from the Superintendent of Documents, U.S. Government Printing Office, Washington, D.C. 20402. Price 35 cents, U.S. currency. Since Section 521 of the Tax Reform Act of 1969 made changes in U.S. law relating to depreciation with respect to taxable years ending after 24 July 1969, a copy of Public Law 91–172 may be obtained from the same source for one dollar, U.S. currency.

The Canadian Government issues a General Tax Guide (free) that may be obtained from any federal taxation office. For more details see one of the Revenue Department officials of the country concerned.

There are different taxation schedules showing the lifetime expectancy for different kinds of assets. Certain equipment may have a lifetime expectancy of five years *or less*. On the other hand, a building may have an allowable lifetime of fifty years. Using the "straight-line method of depreciation," the former allows for a 20 percent per annum depreciation (over a period of five years) from the original cost less the salvage value. The latter allows for a 2 percent depreciation (over a period of fifty years) from the original cost less the salvage value.

In all cases the salvage value of an asset is not allowed "by law" to fall below a reasonable value. In case of doubts as to what is meant by a "reasonable value," contact an Internal Revenue Service official.

In cases where major repairs and/or modifications have been made to an asset, which have increased its value and utility, the remaining book value at the end of the year may be more than its value at the previous year's end. The improvement then becomes a further capital cost, and the asset *appreciates* in the year-end value.

In unusual cases, depreciation can result from an asset becoming obsolete. Assume that a new piece of machinery could outperform older, existing equipment in quality, quantity of work performed, relative prime cost, and maintenance expenses. The older machine may be declared obsolete because the owner could not possibly compete on equal terms with the owners of new equipment. However, a very convincing argument would have to be presented before the officials of the Internal Revenue Service would allow it.

The basic premise for depreciation allowances arises from the fact that in using a piece of equipment to achieve a desired end, such equipment is subject to wear, tear, and cost of maintenance, and that such costs are recoverable by the owner from the revenue of the work performed. Thus the owner recovers the depreciated value of the equipment for the period of time during which it is operated.

Depreciation may be stipulated in terms of a year, a month, a day, or even an hour. Assume that the installed cost of a piece of equipment is $9,250.00, with a salvage value of $750.00 over a five-year life. Using the "straight-line method of depreciation" (which follows), the asset would depreciate at the rate of 20 percent per annum from its original cost less salvage value. Thus in five years $8,500.00 ÷ 5 = $1,700.00 per annum depreciation. Further, suppose the working time for the equipment to be 2,000 working hours per annum. Then the actual depreciation of the equipment would be $1,700.00 ÷ 2,000 = $0.85 per hour. Estimators use this method when pricing certain jobs.

Straight-Line Method of Depreciation

Assume the following charges for the purchase and installation of a new piece of equipment: original cost $8,450.00; taxes $173.00; transportation charges from point of ownership between vendor and purchaser $272.00; unloading $150.00; and cost of assembly and installation $205.00. The total cost would be $9,250.00. Note that all the above costs would be incurred legitimately by the purchaser before he could start to earn money from his investment. Assuming the life expectancy of the asset to be five years and allowing for a final salvage value of say $750.00, the straight-line method of depreciation would be as follows.

Step 1: List and find the total cost of the equipment to be depreciated.

Purchase price	$8,450.00
Taxes	173.00
Transportation	272.00
Unloading	150.00
Installation	205.00
	$9,250.00

Step 2: Deduct from the total cost the final salvage value of the asset, say $750.00; then $9,250 − 750 = $8,500.00.

Step 3: Divide the cost of the asset less its salvage value by the number of years over which the equipment is to be depreciated. Thus $8,500.00 over a life expectancy of five years would be $8,500 ÷ 5 = $1,700.00 depreciation per annum. This is the straight-line method of depreciation.

Step 4: Lay out a table as follows:

Year	Depreciation per annum	Cumulative depreciation	Year end value
0	$1,700	$ 0	$8,500
1	1,700	1,700	6,800
2	1,700	3,400	5,100
3	1,700	5,100	3,400
4	1,700	6,800	1,700
5	1,700	8,500	0

Original total gross cost of the asset, $9,250.00; salvage value, $750.00; life expectancy, five years.

Cumulative depreciation	$8,500.00
Salvage value	750.00
Original cost	$9,250.00

Summary: The straight-line method of depreciation may be adopted where a regular flow of business is expected and where recovery of the cost of the asset is made in annual equal amounts during the life of the equipment.

Declining-Balance Method of Depreciation

This method differs from the straight-line method of depreciation in that, instead of depreciating the value of the asset by a fixed dollar value each year, the depre-ciation is made by a fixed annual percentage reduction of the year-end value of the asset. Using the same cost and condition of installing a piece of equipment as before, the declining-balance method of depreciation is as follows.

Step 1: Find the rate percentage that you would use on the same asset by the straight-line method of depreciation as before. This was 20 percent per annum of the original installed cost (less salvage value) spread over a period of five years.

Step 2: Double that rate, making 20 percent into 40 percent.

Step 3: Depreciate each *year-end book value* of the asset by 40 percent.

Step 4: Using this method, *the salvage value of the asset is not immediately taken into account,* but the balance remaining at the end of the fifth year will represent the salvage value with this qualification: the Internal Revenue Service says that you are never allowed to depreciate the value of an asset below a reasonable salvage value. The term "reasonable value" may have to be determined through an interview with an official from the taxation department.

Step 5: Using the same total cost and life expectancy of the asset as that shown in the straight-line method, the following table would result.

Step 6: Lay out the table as follows:

Year	Depreciation per annum from year end book value (40%)	Cumulative depreciation to salvage value	Year end value of asset
0			$9,250
1	$3,700	$3,700	5,550
2	2,220	5,920	3,330
3	1,332	7,252	1,998
4	799	8,052	1,198
5	479	8,531	719

Original total gross cost of the asset, $9,250.00. Salvage value equals the remaining balance at the end of five years of depreciation.

Cumulative depreciation from column 3	$8,531.00
Salvage value remaining in column 4	719.00
Original cost	$9,250.00

Analysis: Depreciation for the first year, 40 percent of $9,250.00 taken without any deduction at this time for salvage value of the asset, is $3,700.00. Deduct this amount from the original cost of the asset as shown at column 4, leaving $5,550.00 as the book value of the asset at the end of the first year.

The second year depreciation, 40 percent of $5,550.00, is $2,220.00. Deduct this amount from the year-end book value of the asset as shown at the end of the second year, column 4, leaving $3,330.00.

The cumulative depreciation for the first two years is $3,700.00 plus $2,220.00, which equals $5,920.00 as shown for the second year in column 3, and so on until there is a final year-end value of the asset (salvage) remaining of $719.00 as shown at the fifth year, column 4.

Summary: The declining-balance method of depreciation may be adopted where a contractor wishes to take the greatest amount of depreciation during his early years of ownership of the asset, to compensate for high annual profits during that projected period.

Sum of the Years-Digits Method of Depreciation

This method yields high rates of depreciation during the early years of the asset's life. Using the same equipment and conditions as before, the method of calculating the sum of the years-digits is as follows:

Step 1: Find the sum of the digits of the years of life expectancy of the asset.

Step 2: the sum of the years digits = 1 + 2 + 3 + 4 + 5 = 15 digits.

Step 3: *After deducting the salvage value of the asset,* the depreciation is calculated by deducting the following fractional amounts each year.

First year: 5/15 of the original cost of the asset
Second year: 4/15 of the original cost of the asset
Third year: 3/15 of the original cost of the asset
Fourth year: 2/15 of the original cost of the asset
Fifth year: 1/15 of the original cost of the asset

Total digits: 15/15

Step 4: Lay out the table as follows:

Year	Depreciation per annum	Cumulative depreciation	Year end book value
0			$8,500
1 5/15	$2,833	$2,833	5,667
2 4/15	2,267	5,100	3,400
3 3/15	1,700	6,800	1,700
4 2/15	1,133	7,933	565
5 1/15	567	8,500	0

Original total gross cost of the asset, $9,250.00; salvage value, $750.00; life expectancy of five years.

Cumulative depreciation	$8,500.00
Salvage value	750.00
Original cost of asset	$9,250.00

Summary: Using the sum of the years-digits method of depreciation, the greatest recovery is made during the early years of the life of the asset. This method may be adopted where the contractor wishes to take advantage of prospective high earnings during the first few years of the life of the asset (and thus save in income tax charges).

Comparative Depreciation Table

The following comparative depreciation table shows the relative amounts that may be depreciated for the same equipment for the same period of time as outlined in all the previous examples in this chapter.

COMPARATIVE DEPRECIATION TABLE

Year	Straight-line method	Delining balance method	Sum of years-digits method
	20%	40%	
1	$1,700	$3,700	$2,833
2	1,700	2,220	2,267
3	1,700	1,332	1,700
4	1,700	799	1,133
5	1,700	479	567
Salvage	750	719	750
	$9,250	$9,250	$9,250

Analysis: The depreciation for the first two and the last two years of the asset are as follows.

Straight-line 20%	$3,400	$3,400
Declining balance	$5,920	$1,278
Sum of the years-digits	$5,100	$1,700

Summary: The highest depreciation occurs during the first two years (of a five year period) on the declining-balance method, and the same method also shows the lowest depreciation for the last two years. Where earnings are expected to be high during the early life of an asset, either the declining-balance method or the sum of the years-digits method would be advantageous from a tax point of view. Where a steady turnover is expected, a regular straight-line method may be adopted.

1.33 INDIVIDUAL OWNERSHIP, PARTNERSHIPS, LIMITED COMPANIES, AND CORPORATIONS

Since many estimators ultimately join forces with others such as engineers/architects, or building superintendents to form partnerships and so on, it is important that they be made aware of some of the legal aspects, rewards, or possible penalties of such an undertaking. The following section outlines some aspects of individual partnerships and some joint ventures.

Individual Proprietorship

With individual ownership the total operation of the business is conducted by one person (frequently with the help of the family or employees). Legally, only one person is responsible for the indebtedness or profits of the business.

Some of the advantages and disadvantages of owning your own business are:

Advantages
(a) The profits and business expansions are yours.
(b) As your own boss you direct your own policy.
(c) There are no complications with others about responsibilities and rewards.
(d) You do not have to meet in committee to formulate policy.
(e) You set your own working days and hours.
(f) You may cut overhead expenses by operating the business from your own home.

Disadvantages
(a) Your liabilities include your business and *personal assets.*
(b) In case of business failure your personal and business assets are subject to seizure and sale.
(c) Assets that you may acquire from the time of your declaration of bankruptcy to the time of discharge are also subject to seizure.
(d) The liability of married men is restricted to their own personal property; that which is listed in the wife's name cannot be touched. *Married persons entering business in the Province of Quebec must go through legal proceedings to determine to what extent the property of the one is separate and distinct from the other.*
(e) The business is usually small in size and will have to compete with large organizations with specialists in every field.

(f) You cannot purchase materials in the same bulk with the same discounts available to larger companies.
(g) After your death, the administration of your estate may carry on the business under letters of administration until the estate is settled and the business is handed over to your heirs. The business may suffer as a consequence.

If you wish to add "and Company" to your name or to use an entirely different name from your own, such as a business name, it must be registered in the state or province in which you are conducting business. This is understandable since the public must always be protected by having access to the correct name of any organization with which any person wishes to do business or present claims.

Partnerships

In a partnership, at least two people pool their resources and abilities to conduct a particular business enterprise. A partnership is created by entering into a contract that has two main points: (a) it states the contribution to be made to the business by each partner; and (b) it specifies the manner in which earnings of the enterprise are to be shared by the partners.

General Partnerships

In general partnerships, the members are jointly and severally (individually) liable for the debts of the entire business. That is to say, if the creditors cannot get satisfaction from the business, they can enforce their claims against any partner or partners.

It has been said that partnerships are the easiest things to get into and the most difficult from which to withdraw. Any partnership agreement should be given most serious thought. It should be drawn up and registered by an attorney who is acquainted with your type of business. *Consult an attorney before (not after) you enter into partnership.*

Limited Partnerships

In a limited partnership, one or more partners operate the business and one or more invest cash only. The former are *general partners* and are liable to creditors for both their business and private assets. The latter are *limited* to creditors' claims only to the amount of cash they have invested in the business; no claim may be made against their personal assets.

Individuals considering entering into partnership should itemize those points they wish to have included in the agreement. They should meet in committee once or twice and present a rough draft in legal terminology. The participants should again meet in committee and then present the final draft to the attorney for final drafting and notarization.

Partnership Agreement

A partnership agreement should contain a number of details. Among them are:

(a) The purpose of the agreement.

(b) The name and permanent address of the firm and of each partner.

(c) The commencement date of the partnership.

(d) The duration of the partnership.

(e) The manner in which the partnership is to be terminated.

(f) The name and address of the bankers.

(g) The name and address of the attorney.

(h) The procedure for continuing the agreement;

(i) The amount of capital to be invested by each partner, with a statement as to whether or not the amount or part of it is to be paid in property, and if in property, a full description thereof.

(j) The starting date of the fiscal year.

(k) The amount of interest, if any, that is to be paid on capital.

(l) The powers and duties of the partners.

(m) Whether or not the partners are to devote full, normal business hours to the partnership.

(n) Whether or not partners are allowed to enter into any other business that is in any way connected with the building industry.

(o) The manner in which profits or losses of the business are to be divided.

(p) The name of the person who will keep the books of the business to which all members will have access at any reasonable time.

(q) The percentage division of partnership assets at dissolution.

(r) The method of continuing the partnership in the event of the death or incompetence of any partner.

(s) In case of dispute, all differences in regard to the partnership affairs shall be referred to the arbitration of a single arbiter, if the parties agree upon one; otherwise to five arbitrators, one to be appointed by each party, and the fifth to be chosen by the four first named before they enter upon the business or arbitration. The award and determination of such arbitrators, or any three of such arbitrators, shall be binding upon the parties hereto and their respective executors, administrators and assigns, always providing that the recommendation of the board shall not preclude any part the right of access to a court for a legal ruling;

It is very important that you know that in the absence of express agreement the law provides that the profits shall be equally divided, regardless of the ratio of the partners' respective investments.

The foregoing list is not exhaustive.

Advantages of Partnerships

(a) Partners can specialize in their own fields.

(b) Each can direct workmen in his own area.

(c) There may be more collective working capital.

(d) The company may obtain better trade discounts.

(e) The company may operate on a larger scale than an individual enterprise.

(f) The company can maintain and keep operating more equipment.

(g) The company may more easily undertake projects in different places at the same time.

(h) Partners pay income tax on net profits. Net profits are those remaining after all operating expenses have been deducted, and are subject to personal income tax whether or not the proceeds are taken out of, or remain in the business.

Disadvantages of Partnerships

(a) The partnership is not a tangible entity for the person who enters it.

(b) Each *general partner* is individually responsible for the whole business.

(c) If creditors cannot get satisfaction from the business partnership, they may press the total claim against any partner having private assets.

(d) The dishonesty of a partner.

(e) The incompetence of a partner.

(f) The moral lapse of a partner.

(g) The neglect of the business by a partner.

(h) The repeated bad judgment of a partner.

(i) The partnership ceases to exist if a partner severs his connection with the business or if a partner dies.

Corporations

A corporation is an association of three or more persons that are able, legally, to act as one entity under a common name. A corporation continues to exist even though its members may change. It may engage in business, or it may be a charitable enterprise or social religious association. It can own property, can be sued, and can sue. The extent of its activities are shown in a charter that is given by federal, state, or provincial governments. (A charter is a document granting certain rights to a person, group of persons, or a corporation.)

1.34 OBTAINING DRAWINGS, SPECIFICATIONS, AND CONTRACT DOCUMENTS

Information for obtaining building contract documents may be obtained from architects or other authorities through any of the following mediums:

(a) Newspaper advertisements (it is mandatory for governments to solicit bids through open advertisements).

(b) Trade organizations and magazines.

(c) Builders' and contractors' associations and exchanges.

(d) Invitations to bid given by architects, restricted to five or six general contractors.

In all instances tenderers will be informed of where the drawings and specifications may be obtained and of what amount of refundable deposit must be made for the use of them. Copies may also be displayed at the local Builders' Exchange for the use of subcontractors.

Drawings and specification sheets are numbered and they should be checked at the time of receipt to be sure that none is missing. It is also necessary to check that all drawing revisions and addenda to the specifications are included. *This is very important.*

Estimating for a one-family residence may be a one-person job, but estimating for a large tenement block may require the united efforts of several people to get the estimate completed in time to make a proposal bid. As an example, the work load may be divided as follows, where different people take care of:

(a) obtaining the coverage and cost of all insurances bonds and overhead expenses;

(b) estimating all cut and fill and excavations and trenching;

(c) estimating all concrete for underpinning and framing of the proposed building;

(d) all steel;

(e) contacting all subcontractors to submit a bid to the G.C. (general contractor).

If the subcontractors cannot see the drawings and specifications at the Builders' Exchange, the G.C. must allow them time to see his own copies. There are many other ways of distributing the estimating work load, but it is important that all participants in this endeavor meet their target date for the time allotted so that results may be collated and the proposal bid be deposited by the designated time at the designated place.

FOREWORD

ARCHITECTURAL WORKING DRAWINGS HAVE THE IMPORTANT FUNCTIONS OF
(A) RECORDING CLEARLY THE CLIENTS' REQUIREMENTS SO THAT COST
ESTIMATING AND BIDDING ARE FACILITATED, (B) FORMING A PART OF
THE CONTRACT BETWEEN THE CLIENT AND BUILDER AND (C) PROVIDING
INSTRUCTION TO THE BUILDER FOR THE PURPOSE OF CONSTRUCTION. TO
FULFIL THESE FUNCTIONS MOST EFFICIENTLY THE DRAWINGS MUST BE
COMPLETE, ACCURATE AND CONCISE. THE USE OF UNIFORM PRACTICES IN
PRODUCING THE DRAWINGS IS A VALUABLE AID TOWARD ACHIEVING THIS GOAL

THIS STANDARD IS PRESENTED AS A GUIDE TOWARD SUCH UNIFORMITY
AND TOWARD CLARITY AND SIMPLICITY RATHER THAN ARTISTRY IN
ARCHITECTURAL DRAWING. IT IS INTENDED PRIMARILY FOR USE WITHIN CANADIAN
GOVERNMENT AGENCIES, BUT IS AVAILABLE FOR USE GENERALLY AND,
INDEED, ITS GENERAL AND WIDESPREAD USE IS ENCOURAGED.

THE STANDARD IS NOT EXHAUSTIVE IN ITS TREATMENT OF ARCHITECTURAL
DRAWING PRACTICES, PARTICULARLY AS REGARDS SYMBOLS FOR ELECTRICAL,
PLUMBING AND HEATING APPLICATIONS, BUT DEALS WITH WHAT ARE CONSIDERED
TO BE THE MOST BASIC AND COMMONLY ENCOUNTERED ELEMENTS OF
ARCHITECTURAL DRAWING.

TITLE AND REVISION BLOCKS

THE TITLE BLOCK SHOULD BE POSITIONED IN THE LOWER RIGHT-HAND CORNER OF THE DRAWING SHEET AND SHOULD NORMALLY PROVIDE THE FOLLOWING INFORMATION:

- NAME AND LOCATION OF THE PROJECT
- NAME OF THE AGENCY RESPONSIBLE FOR THE DRAWING
- TITLE OF THE DRAWING
- NUMBER OF THE DRAWING
- DRAWING SCALES
- INITIALS OF THE DESIGNER, CHECKER, DRAFTSMAN AND SUPERVISOR, TOGETHER WITH APPROPRIATE DATES

THE REVISION BLOCK SHOULD BE ADJACENT TO THE TITLE BLOCK, EITHER DIRECTLY ABOVE IT OR ON ITS LEFT, AND SHOULD READ FROM THE BOTTOM UPWARD.

LETTERING

IT IS NOT CONSIDERED PRACTICAL NOR DESIRABLE TO SUGGEST A SPECIFIC STANDARD FORM OF LETTERING FOR ARCHITECTURAL DRAWINGS. THE LOGICAL GOAL IS TO PROVIDE DISTINCT, UNIFORM LETTERS AND FIGURES THAT WILL ENSURE THE PRODUCTION OF CLEAR, LEGIBLE PRINTS.

IT IS RECOMMENDED HOWEVER THAT VERTICAL UPPER CASE LETTERING BE USED AND THAT PREFERENCE BE GIVEN TO A MINIMUM NUMBER OF BASIC LETTER SIZES. IT IS CONSIDERED THAT THREE BASIC SIZES ARE SUFFICIENT FOR MOST ARCHITECTURAL DRAWING APPLICATIONS, THE LARGER SIZE BEING RESERVED FOR MAJOR HEADINGS, AN INTERMEDIATE SIZE FOR SUBHEADINGS AND THE SMALLER SIZE FOR NOTES. IT IS RECOMMENDED THAT THE LETTER SIZE FOR NOTES AND FOR DIMENSIONS SHOULD BE NOT LESS THAN $\frac{3}{32}$ INCH.

MICROFILMING

DRAWINGS THAT ARE TO BE MICROFILMED WILL REQUIRE SPECIAL CARE IN PREPARATION AS OUTLINED IN CANADIAN GOVERNMENT SPECIFICATIONS BOARD SPECIFICATION 72-GP-1: 35 MM MICROFILMING OF ENGINEERING AND ARCHITECTURAL DRAWINGS.

DOOR AND ROOM FINISH SCHEDULES

IT IS RECOGNIZED THAT NO STRICT RULES CAN OR SHOULD BE PROPOSED AS TO THE EXACT FORM AND ARRANGEMENT OF DOOR AND ROOM FINISH SCHEDULES. THE EXAMPLES SHOWN ON THE FOLLOWING PAGES ILLUSTRATE DOOR AND ROOM FINISH SCHEDULES THAT ARE IN COMMON USE IN SOME AREAS OF ARCHITECTURAL DRAWING, AND THEY ILLUSTRATE THE KIND OF INFORMATION THAT SHOULD BE PROVIDED BY SUCH SCHEDULES.

DOORS

WHEN USED IN CONJUNCTION WITH A DOOR SCHEDULE, DOOR
SYMBOLS INCLUDE THE APPROPRIATE DOOR SCHEDULE REFERENCE.
SEE EXAMPLES 'A' & 'B' BELOW.   IT IS OPTIONAL WHEN USING THIS
METHOD TO GIVE THE DOOR SIZE AS SHOWN IN EXAMPLE 'B'

SINGLE SWING DOORS

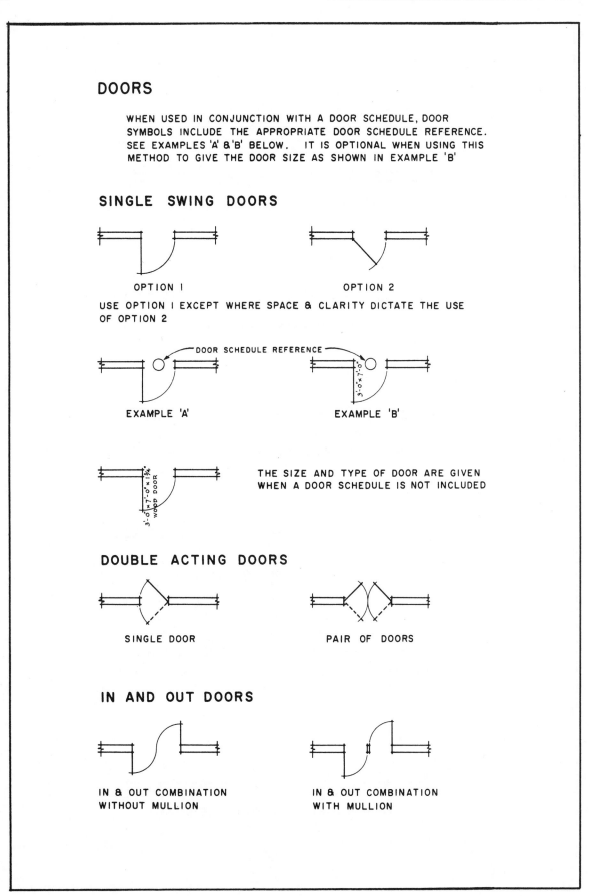

OPTION 1   OPTION 2

USE OPTION 1 EXCEPT WHERE SPACE & CLARITY DICTATE THE USE
OF OPTION 2

EXAMPLE 'A'   EXAMPLE 'B'

THE SIZE AND TYPE OF DOOR ARE GIVEN
WHEN A DOOR SCHEDULE IS NOT INCLUDED

DOUBLE ACTING DOORS

SINGLE DOOR   PAIR OF DOORS

IN AND OUT DOORS

IN & OUT COMBINATION   IN & OUT COMBINATION
WITHOUT MULLION   WITH MULLION

FOLDING DOORS AND PARTITIONS

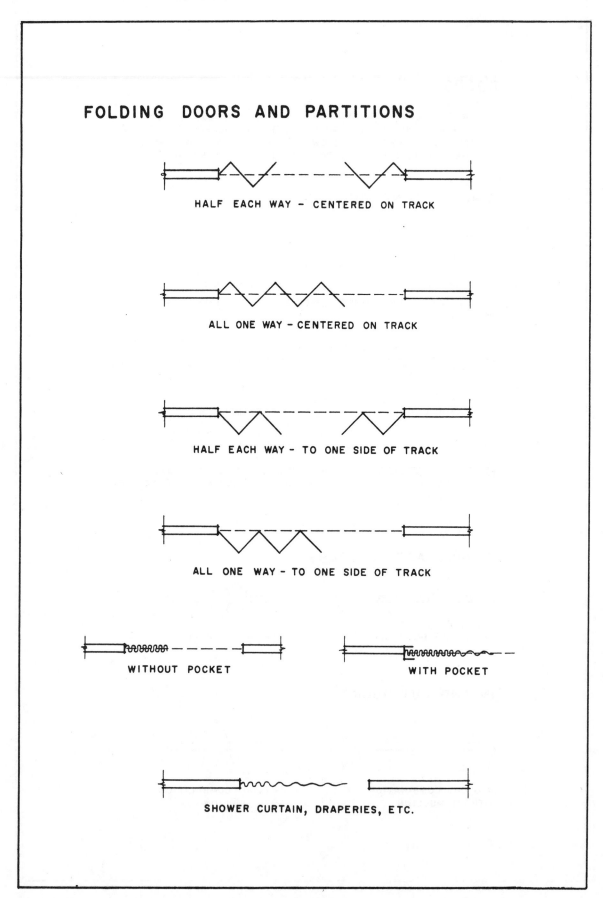

HALF EACH WAY - CENTERED ON TRACK

ALL ONE WAY - CENTERED ON TRACK

HALF EACH WAY - TO ONE SIDE OF TRACK

ALL ONE WAY - TO ONE SIDE OF TRACK

WITHOUT POCKET

WITH POCKET

SHOWER CURTAIN, DRAPERIES, ETC.

DOOR & HARDWARE SCHEDULE
EXAMPLE I

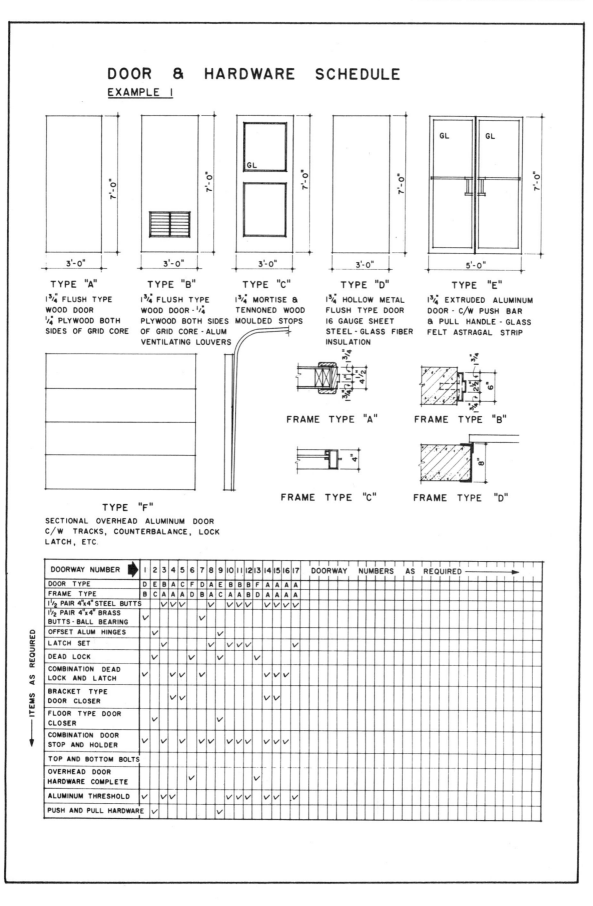

TYPE "A"
1¾ FLUSH TYPE WOOD DOOR ¼ PLYWOOD BOTH SIDES OF GRID CORE

TYPE "B"
1¾ FLUSH TYPE WOOD DOOR - ¼ PLYWOOD BOTH SIDES OF GRID CORE - ALUM VENTILATING LOUVERS

TYPE "C"
1¾ MORTISE & TENNONED WOOD MOULDED STOPS

TYPE "D"
1¾ HOLLOW METAL FLUSH TYPE DOOR 16 GAUGE SHEET STEEL - GLASS FIBER INSULATION

TYPE "E"
1¾ EXTRUDED ALUMINUM DOOR - C/W PUSH BAR & PULL HANDLE - GLASS FELT ASTRAGAL STRIP

TYPE "F"
SECTIONAL OVERHEAD ALUMINUM DOOR C/W TRACKS, COUNTERBALANCE, LOCK LATCH, ETC.

FRAME TYPE "A"   FRAME TYPE "B"   FRAME TYPE "C"   FRAME TYPE "D"

DOORWAY NUMBER	1	2	3	4	5	6	7	8	9	10	11	12	13	14	15	16	17	DOORWAY NUMBERS AS REQUIRED
DOOR TYPE	D	E	B	A	C	F	D	A	E	B	B	B	F	A	A	A	A	
FRAME TYPE	B	C	A	A	A	D	B	A	C	A	A	B	D	A	A	A	A	
1½ PAIR 4"x4" STEEL BUTTS			✓	✓	✓		✓		✓	✓	✓			✓	✓	✓	✓	
1½ PAIR 4"x4" BRASS BUTTS - BALL BEARING	✓					✓												
OFFSET ALUM HINGES		✓						✓										
LATCH SET			✓					✓	✓	✓	✓					✓		
DEAD LOCK	✓				✓			✓				✓						
COMBINATION DEAD LOCK AND LATCH	✓			✓	✓		✓							✓	✓	✓		
BRACKET TYPE DOOR CLOSER				✓	✓									✓	✓			
FLOOR TYPE DOOR CLOSER		✓						✓										
COMBINATION DOOR STOP AND HOLDER	✓		✓		✓	✓		✓	✓	✓				✓	✓	✓		
TOP AND BOTTOM BOLTS																		
OVERHEAD DOOR HARDWARE COMPLETE						✓							✓					
ALUMINUM THRESHOLD	✓		✓	✓					✓	✓	✓			✓	✓		✓	
PUSH AND PULL HARDWARE	✓							✓										

ITEMS AS REQUIRED

HARDWARE SCHEDULE

EXAMPLE 2

DOOR SCHEDULE

FLOOR	NUMBER OF DOORS	REVISIONS	LOCATION OF DOOR		DOOR							FRAME			HARDWARE				REMARKS
			OUTSIDE ROOM	INSIDE ROOM	SWING	TYPE	WIDTH	HEIGHT	THICKNESS	MATERIAL	FINISH	TYPE	THICKNESS	MATERIAL					
FIRST FLOOR	2		EXTERIOR	VESTIBULE 1-01	PD	D-1	3'-0"	7'-0"	1½"	AL		F-1		AL					SEE DWG #14
	2	*A	VESTIBULE 1-01	FOYER 1-02	PD	D-2	3'-0"	7'-0"	1¾"	WOOD	LAQ	F-2	14GA	MET					
			FOYER 1-02	GEN OFFICE 1-04	RH	D-4	2'-8"	6'-8"	1¾"	WOOD	LAQ	F-3	=	=					
			FOYER 1-02	CORRIDOR 1-03	RHR	D-3	3'-6"	6'-8"	1½"	MET		F-5	=	=					
			CORRIDOR 1-03	PUBLIC LAV 1-07	LH	D-6	2'-6"	6'-8"	1¾"	WOOD	LAQ	F-3	=	=					
			COFFEE SHOP 1-10	KITCHEN 1-12	DA	D-5	3'-0"	6'-8"	1½"	AL		F-7		AL					

SWINGS PD – PAIR OF DOORS
DA – DOUBLE-ACTING
RH – RIGHT HAND
LH – LEFT HAND
RHR – RIGHT HAND REVERSE
LHR – LEFT HAND REVERSE

✱ REVISION TABLE ADDED ONLY WHEN NECESSARY

	DATE	REVISIONS
A	15·2·61	SWING CHANGED TO LH FROM RH

ROOM FINISH SCHEDULE

EXAMPLE I

| ROOM | | FLOOR | | | BASE | | WALLS | | | | CEILING | | NOTES |
| | | | | | | | DADO | | FIELD | | | | |
NAME	NO.	VINYL TILE	TERRA-ZZO	LINO TILE	RUBBER	TERRAZO	MATL	HT	PAINT	BLOCK	ACOUSTIC	FIELD PAINTED	
DINING RM	1	•			•				•		•		
KITCHEN VEST.	2		•		•				•			•	
KITCHEN	3		•		•				•			•	
SNACK BAR	4	•			•				•		•		
OFFICE	5			•	•				•		•		
BAGGAGE RM	6	•			•					•		•	
CONCESSIONS	7		•		•				•		•		
"	8		•		•				•		•		
"	9		•		•				•		•		
WAITING RM	10		•		•				•		•		
AIRPORT ATTEND.	11			•	•				•			•	

EXTEND DOWNWARD AS REQUIRED BY NO. OF ROOMS

THE SCHEDULE IS USUALLY EXTENDED IN WIDTH TO INCLUDE MORE DETAILED SUBDIVISIONS UNDER FLOORS, WALLS, ETC.

ROOM NUMBER SHOULD BE INDICATED ON PLAN THUS ☐

ROOM FINISH SCHEDULE
EXAMPLE 2

LEAVE EXTRA SPACES IN EACH COLUMN FOR ADDITIONS

NO	NAME	FLOOR					BASE			DADO		WALL			CEILING			ACCESSORIES		REMARKS
		CONCRETE	TERRAZZO	QUARRY TILE	VINYL TILE .080"	BROADLOOM BY OWNER	TERRAZZO	QUARRY TILE	VINYL	VINYL FABRIC 4'-6"	CERAMIC TILE 7'-0"	UNFINISHED	PLASTER	WALNUT	PLASTER	GYPSUM BOARD	ACOUSTIC TILE	COUNTER CUPBOARDS	LOCKERS	
1	VESTIBULE			•				•		•			•		•					INSET DOOR MAT
2	DISPLAY AREA		•				•						•				•			SEE DWG #7 FOR TERRAZZO
3	INFORMATION OFFICE				•				•				•				•	•		
4	CONFERENCE ROOM	•				•		•	•					•			•			
5	CLOAK ROOM				•								•			•			•	
6	PROJECTS OFFICE				•				•				•				•			
7	WOMEN'S LAV.		•								•		•		•					

ROOM FINISH SCHEDULE

EXAMPLE 3

SCHEDULE OF FINISHES

ITEM NUMBER	ROOMS AND AREAS	FLOOR BORDER MATERIAL	FLOOR BORDER FINISH	FLOOR FIELD MATERIAL	FLOOR FIELD FINISH	BASE MATERIAL	BASE FINISH	WALLS DADO MATERIAL	WALLS DADO FINISH	WALLS FIELD MATERIAL	WALLS FIELD FINISH	CORNICE MATERIAL	CORNICE FINISH	CEILING BORDER MATERIAL	CEILING BORDER FINISH	CEILING FIELD MATERIAL	CEILING FIELD FINISH	TRIM DOOR MATERIAL	TRIM DOOR FINISH	TRIM WINDOW MATERIAL	TRIM WINDOW FINISH
COLUMN SYMBOL		a	b	c	d	e	f	g	h	i	j	k	l	m	n	o	p	q	r	s	t
1.	ENTRANCE VESTIBULE, LOBBY, WAITING ROOMS AND CORRIDORS	To	Pd	To	Pd	To	Pd	Pl	VP1	Pl	PT1	-	-	-	-	At	PT1	St	PT2	Al	
2.	BELOW GRADE, OFFICES, LABORATORY, EXAM ROOM, CANTEEN, LOCKER ROOM, PHARMACY, PACK STORES, C.R., REST ROOMS	To	Pd	Va	Wx	To	Pd	Pl	VP1	Pl	PT1	-	-	-	-	At	PT1	St	PT2	Al	
3.	ABOVE GRADE, WARDS	To	Pd	Hv	Wx	To	Pd	Pl	VP2	Pl	PT1	-	-	Pl	PT1	Pl	PT1	St	PT2	Al	
4.	UTILITY ROOMS	To	Pd	To	Pd	To	Pd	Pl	VP1	Pl	PT1	-	-	-	-	At	PT1	St	PT2	Al	
5.	TOILETS & BATH ROOMS	Ct	-	Ct	-	Ct	-	Gt	-	Kp	PT1	-	-	-	-	At	PT1	St	PT2	Al	
6.	AUDITORIUM	Wd	Vd	Wd	Vd	Wd	Vd	Pl	VP1	Pl	PT1	-	-	-	-	At	PT1	St	PT2	Al	
7.	DINING ROOM	To	Pd	Hv	Wx	To	Pd	Pl	VP1	Pl	PT1	-	-	-	-	At	PT1	St	PT2	Al	
8.	ARTS & CRAFTS	To	Pd	Hv	Wx	To	Pd	Pl	VP1	Pl	PT1	-	-	-	-	At	PT1	St	PT2	Al	
9.	STAIRWAYS	To	Pd	To	Pd	To	Pd	-	-	Tg	-	-	-	-	-	-	PT1	St	PT2	Al	
10.	SERVERIES	Qt	-	Qt	-	Qt	-	-	-	Gt	-	-	-	-	-	At	PT1	St	PT2	Al	
11.	OVERHEAD PASSAGE	To	Pd	Hv	Wx	To	Pd	Pl	VP1	Pl	PT1	-	-	-	-	At	PT1	St	PT2	Al	

KEY

SYMBOL	MATERIALS & FINISHES
At	ACOUSTIC TILE
Cp	CEMENT PLASTER
Ct	CERAMIC FLOOR TILE
Gt	GLAZED WALL TILE
Hv	HOMOGENEOUS VINYL
Kp	KEENES CEMENT PLASTER
Pl	GYPSUM PLASTER
PT1	PAINT - GLOSS
PT2	PAINT - ENAMEL
Pd	POLISHED
Qt	QUARRY TILE
St	STEEL
Tg	GLAZED STRUCTURAL TERRA COTTA
To	TERRAZZO
Vd	VARNISHED
Va	VINYL ASBESTOS
VP1	VINYL PLASTIC ·020"
	CLEAR VINYL
VP2	VINYL PLASTIC ·012"
Wd	WOOD HARDWOOD
Wx	WAX

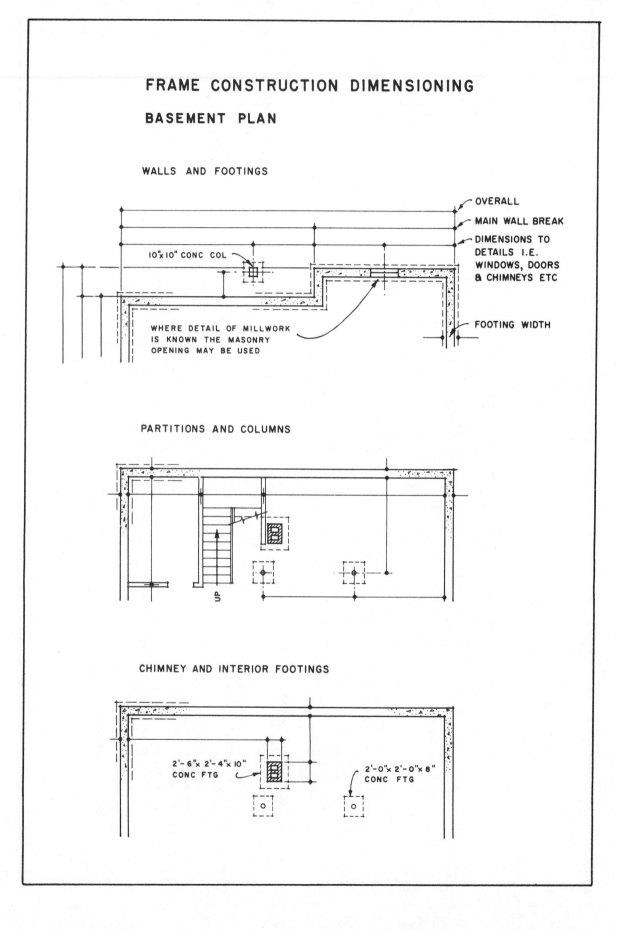

FRAME CONSTRUCTION DIMENSIONING

BASEMENT PLAN

WALLS AND FOOTINGS

10"x10" CONC COL

OVERALL

MAIN WALL BREAK

DIMENSIONS TO DETAILS I.E. WINDOWS, DOORS & CHIMNEYS ETC

FOOTING WIDTH

WHERE DETAIL OF MILLWORK IS KNOWN THE MASONRY OPENING MAY BE USED

PARTITIONS AND COLUMNS

UP

CHIMNEY AND INTERIOR FOOTINGS

2'- 6"x 2'- 4"x 10" CONC FTG

2'-0"x 2'-0"x 8" CONC FTG

FLOOR PLAN

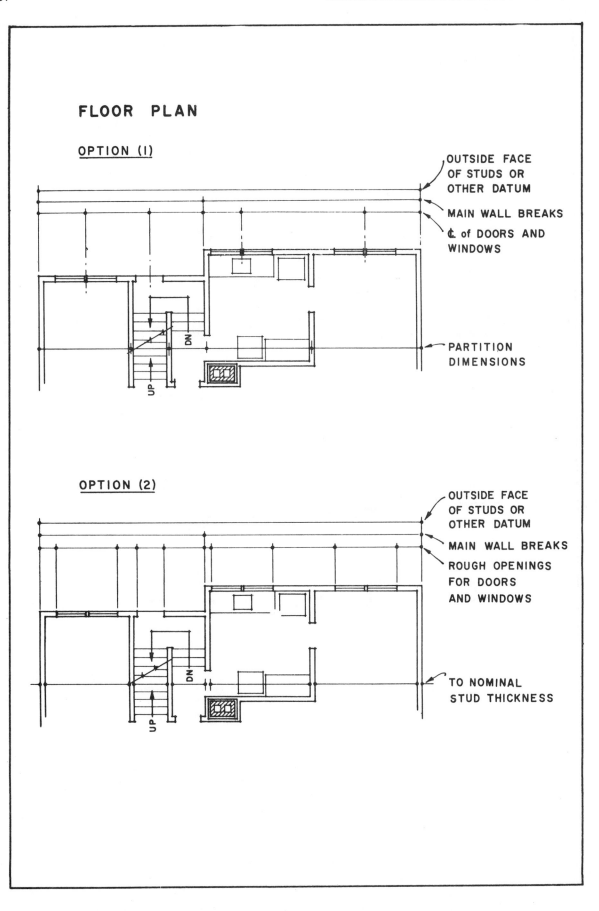

OPTION (1)

OUTSIDE FACE
OF STUDS OR
OTHER DATUM

MAIN WALL BREAKS

℄ of DOORS AND
WINDOWS

PARTITION
DIMENSIONS

DN

UP

OPTION (2)

OUTSIDE FACE
OF STUDS OR
OTHER DATUM

MAIN WALL BREAKS

ROUGH OPENINGS
FOR DOORS
AND WINDOWS

TO NOMINAL
STUD THICKNESS

DN

UP

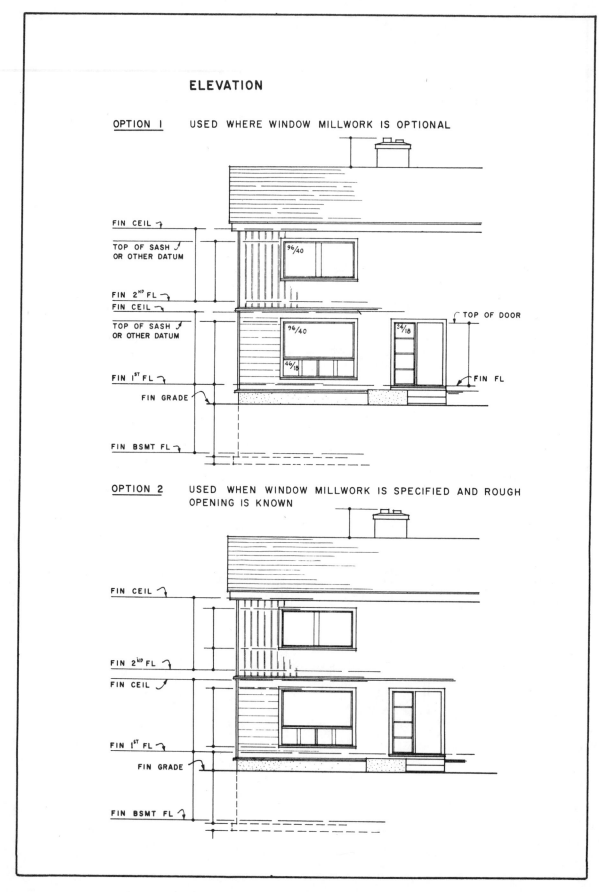

ELEVATION

OPTION 1   USED WHERE WINDOW MILLWORK IS OPTIONAL

OPTION 2   USED WHEN WINDOW MILLWORK IS SPECIFIED AND ROUGH
OPENING IS KNOWN

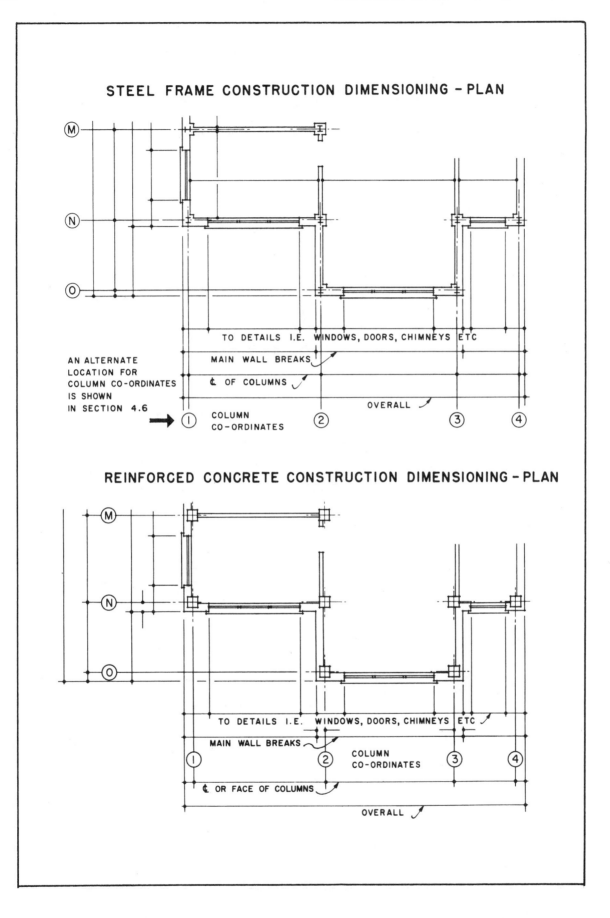

STEEL FRAME CONSTRUCTION DIMENSIONING – PLAN

TO DETAILS I.E. WINDOWS, DOORS, CHIMNEYS ETC

MAIN WALL BREAKS

AN ALTERNATE
LOCATION FOR
COLUMN CO-ORDINATES
IS SHOWN
IN SECTION 4.6

℄ OF COLUMNS

OVERALL

COLUMN
CO-ORDINATES

REINFORCED CONCRETE CONSTRUCTION DIMENSIONING – PLAN

TO DETAILS I.E. WINDOWS, DOORS, CHIMNEYS ETC

MAIN WALL BREAKS

COLUMN
CO-ORDINATES

℄ OR FACE OF COLUMNS

OVERALL

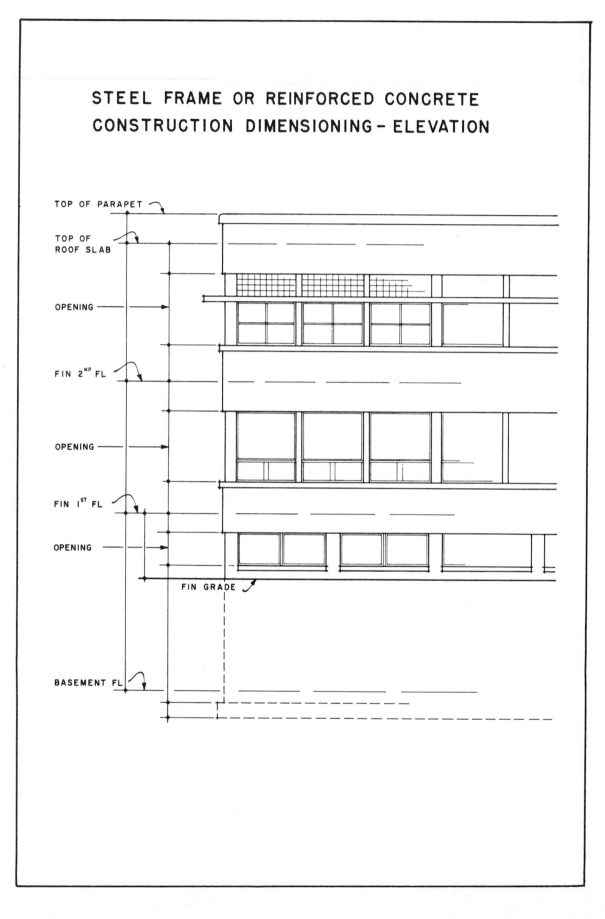

ELECTRICAL

CEIL   WALL

○   ─○   OUTLET   ○_{PC}   CEILING OUTLET WITH PULL CHAIN

Ⓕ   ─Ⓕ   FAN OUTLET

Ⓧ   ─Ⓧ   EXIT LIGHT OUTLET

Ⓒ   Ⓒ   CLOCK OUTLET. SPECIFY VOLTAGE

⊖   DUPLEX CONVENIENCE OUTLET

⊖_{1,3}   I = SINGLE,   3 = TRIPLEX, ETC.

⊖_R   RANGE OUTLET

⊖_S   SWITCH & CONVENIENCE OUTLET

MOTOR OUTLET

SPECIAL PURPOSE OUTLET

[A]   FLUORESCENT FIXTURE. LETTER DESIGNATES TYPE

[◯]   FLUORESCENT FIXTURE WITH OUTLET BOX

S   SINGLE POLE SWITCH

S_2   DOUBLE POLE SWITCH
S_3   THREE WAY   "
S_4   FOUR WAY   "

S_P   SWITCH & PILOT LAMP

▶   OUTSIDE TELEPHONE

▷   INTERCONNECTING TELEPHONE

▬   LIGHTING PANEL

▨   POWER PANEL

OVERHEAD
} HOME RUN TO PANEL BOARD. INDICATE NO. OF
CIRCUITS BY NO. OF ARROWS
UNDERFLOOR

CABLE DESIGNATION   DIA ↘   ⤻ NO. OF WIRES
                              → 3/4 - 5 - 12 ← GAUGE OF WIRE

PLUMBING

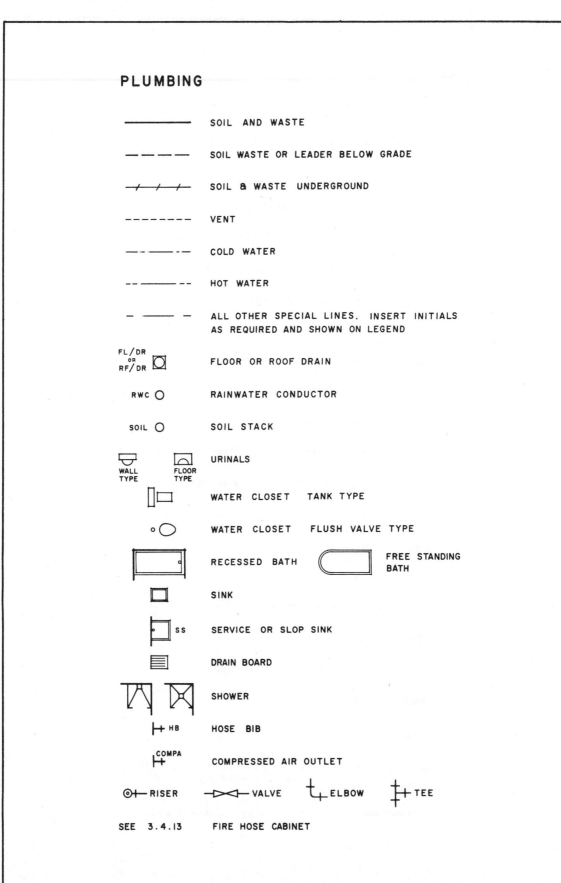

———————————   SOIL AND WASTE

— — — — —   SOIL WASTE OR LEADER BELOW GRADE

—/—/—/—   SOIL & WASTE UNDERGROUND

— — — — — — — —   VENT

— — — — — — —   COLD WATER

— — ——— — —   HOT WATER

— ——— —   ALL OTHER SPECIAL LINES. INSERT INITIALS
            AS REQUIRED AND SHOWN ON LEGEND

FL/DR
  OR   FLOOR OR ROOF DRAIN
RF/DR

RWC   RAINWATER CONDUCTOR

SOIL   SOIL STACK

URINALS
WALL TYPE   FLOOR TYPE

WATER CLOSET   TANK TYPE

WATER CLOSET   FLUSH VALVE TYPE

RECESSED BATH   FREE STANDING BATH

SINK

SS   SERVICE OR SLOP SINK

DRAIN BOARD

SHOWER

HB   HOSE BIB

COMPA   COMPRESSED AIR OUTLET

RISER   VALVE   ELBOW   TEE

SEE 3.4.13   FIRE HOSE CABINET

HEATING

—#——#——#—	HIGH-PRESSURE STEAM
—/———/———/—	MEDIUM-PRESSURE STEAM
———————	LOW-PRESSURE STEAM
—#—#—#—	HIGH-PRESSURE RETURN
—/—/—/—	MEDIUM-PRESSURE RETURN
———————	LOW-PRESSURE RETURN
—·—·—·—	AIR RELIEF LINE
———————	BOILER BLOW OFF
———————	HOT WATER HEATING SUPPLY
———————	HOT WATER HEATING RETURN
—▷◁—	GATE VALVE

⊙— RISER   ELBOW   TEE

⊏════⊐	RAD OR CONVECTOR
	FINNED TUBE
	BASEBOARD CONVECTOR
12 x 20	DUCT { 1ST DIM. APPLIES TO SIDE SHOWN / 2ND DIM. TO SIDE NOT SHOWN

DUCT { 1ST DIM. APPLIES TO SIDE SHOWN / 2ND DIM. TO SIDE NOT SHOWN

⊠	DUCT SECTION (SUPPLY)
⊠	DUCT SECTION (EXHAUST OR RETURN)
	CEILING REGISTER OR GRILLE
◀—	DUCT & DIRECTION OF FLOW

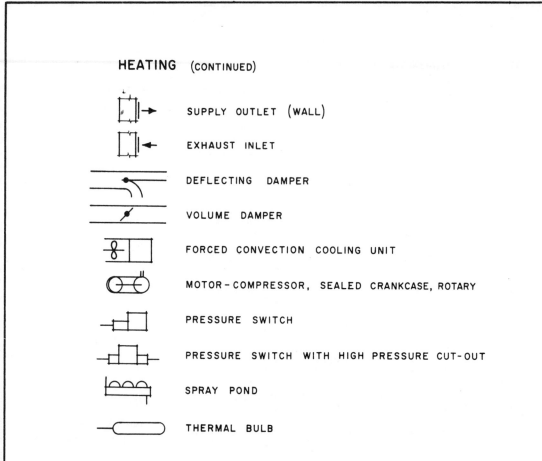

HEATING (CONTINUED)

SUPPLY OUTLET (WALL)

EXHAUST INLET

DEFLECTING DAMPER

VOLUME DAMPER

FORCED CONVECTION COOLING UNIT

MOTOR-COMPRESSOR, SEALED CRANKCASE, ROTARY

PRESSURE SWITCH

PRESSURE SWITCH WITH HIGH PRESSURE CUT-OUT

SPRAY POND

THERMAL BULB

STANDARD MODULAR WORKING DRAWINGS

THIS APPENDIX PRESENTS A BRIEF SUMMARY OF THE CONCEPT OF MODULAR DRAWING AND DIMENSIONING. FOR FURTHER INFORMATION ON THIS SUBJECT REFER TO THE FOLLOWING:

1   CANADIAN STANDARDS ASSOCIATION SPECIFICATION A 31 ON MODULAR CO-ORDINATION IN BUILDING.

2   MODULAR DRAFTING MANUAL, NRC 6344, PUBLISHED BY THE DIVISION OF BUILDING RESEARCH, NATIONAL RESEARCH COUNCIL.

THE POSITION, AND USUALLY THE SIZE OF BUILDING COMPONENTS IN THESE DRAWINGS ARE CONTROLLED BY THE STANDARD MODULAR GRID OF 4 INCHES. THE LARGE SCALE DRAWINGS SHOW IN CLEAR DETAIL THE FITTING OF COMPONENTS TO THE GRID.

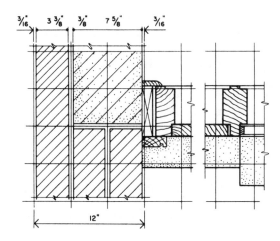

LARGE SCALE DETAIL

DIMENSION LINES TAKEN TO THE GRID ALWAYS HAVE AN *ARROWHEAD* ——————→

DIMENSION LINES TAKEN TO POINTS NOT ON THE GRID ALWAYS HAVE A *DOT* ——————●

THE DIMENSIONS OF THE MANUFACTURED COMPONENTS SHOWN IN THE LARGE SCALE DETAIL AS 3⅝" AND 7⅝" ARE THE *MANUFACTURE DIMENSIONS*

THE *MODULAR DIMENSIONS* OF THESE COMPONENTS ARE CONSIDERED AS 4" AND 8"

THE ³⁄₁₆" DIMENSION IS THE CONSTANT RELATIONSHIP OF THESE PARTICULAR COMPONENTS TO THE CONTROLLING GRID

THE SMALL SCALE PLAN AND ELEVATION DRAWINGS SHOW ASSEMBLY OF COMPONENTS. MOST OF THE DIMENSIONS GIVEN ARE MODULAR DIMENSIONS RUNNING TO CONTROLLING GRID LINES AND ARE IN MULTIPLES OF 4 INCHES.

BY APPLYING THE ARROWHEAD CONVENTION IT IS CLEAR WHICH DIMENSIONS ARE TO THE GRID LINES

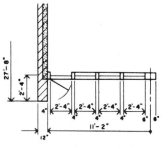

SMALL SCALE ASSEMBLY

1.35 NOTICES TO BIDDERS: SPECIFICATIONS

The first few pages of specifications are usually taken up with instructions for contractors in their submission of a tender. These notices should be very carefully studied by the estimator, and he should be in a position not only to take off the quantities of materials and labor required, but also to be able to advise his company upon their financial obligations, assuming they get the contract.

It is considered that where an architect issues about ten sets of drawings and specifications, a fair price may be expected from about six bids. In some cases, architects will call for invitation bids. In some cases, architects will call for invitation bids from about five or six reputable contractors. It is a compliment to the ability, experience, and sincerity of any firm to be included in such invitations.

1.36 EXAMPLES OF TENDERS

Sealed tenders addressed to _____ Architect, endorsed on the outside of the mailing container with the name of the work to which it refers, will be received for the building on the dates as published in the daily papers. Names of papers: _____ . Tenders shall be accompanied with *a certified check* in the sum of 5 percent of the tender as a guarantee of good faith that the successful tenderer will enter into contract.

Note: A certified check is one that is endorsed by the bank and guarantees that there are sufficient funds in the tenderer's account and that those funds will not be used for any other purpose until after the contract has been let.

Tenders will be opened at _____ time, _____ date, _____ place when tenderers may hear them read and declared.
Note: Tenders close at the time of the time zone in which they are called.
Note: Tenders for all public works must be opened in public.

All checks of unsuccessful tenderers will be returned.

1.37 DRAWINGS AND SPECIFICATIONS AND GENERAL CONDITIONS

The contractor and all subcontractors shall comply with all local, state, provincial, or national government rules, regulations, and ordinances. They will prepare and file all necessary documents or information, and pay for and obtain all licenses, permits, and certificates of inspection as may be specified or required.

Wherever municipal bylaws or state or provincial legislation require higher standards than those set forth herein, such higher standards shall govern.

Note: Assume there is some peculiar subsoil condition in a certain area. It is fair to assume that the local authority, knowing the conditions, may require special precautions to be taken, which are more stringent than those set forth in the specifications. In all cases, the highest building standards shall prevail.

1.38 EXAMINATION OF THE SITE

All contractors submitting bids for this work shall first visit the site.
Note: It is not sufficient to examine a small-building site which may be covered with snow, without testing the depth of the snow. Neither is it sufficient to examine the site without considering the water supply, which may have to be hauled by tanker, and so on.

Plans and specifications may be obtained from the office of the architect by depositing fifty dollars, which will be refunded upon return of the documents.

Bench Marks. The owner will establish the lot lines, restrictions, and a bench mark; all other grades, lines, and levels shall be established and maintained by the contractor.

1.39 EXCAVATION AND BACKFILL

Excavation to a greater than authorized depth shall be made good with extra construction at the contractor's expense as determined by the architect.

Footing excavations carried below the required depth shall be built up with the same concrete designed-mix reinforcing as for the footings above it.

1.40 PRECEDENCE OF DOCUMENTS

Specifications will take precedence over drawings and dimensioned figures. Detail drawings shall take precedence over scaled measurements from drawings.

1.41 DISPUTES

In case of disputes concerning the intention or meaning of the drawings and specifications, such disputes shall be decided by the architect. The intention and meaning of the drawings and specifications are to be taken as a whole. The contractor shall not avail himself of any unintentional error or omission, should any exist. The work shown on the drawings, if not fully described in the specifications, or vice versa, which is reasonably implied and is evidently necessary for the completion of each branch of the work, is to be done by the contractor as though both were specified. The contractor, however, shall have the right to appeal to the owner for arbitration on any matter when he feels that the architect's decision is not in accordance with the intention of the drawings and specifications.

QUESTIONS

1. Define a) collateral security; b) promissory note.

2. List six sources of credit for builders and contractors.

3. What is the function of the Small Business Administration in the United States or the Industrial Development Bank in Canada?

4. Define equity financing.

5. Define a bid depository.

6. List three advantages for the holder of a builder's license.

7. List three advantages of affiliating with a local and national construction association.

8. Assuming that you did not know the owner of a parcel of land in your area, list six steps that you could take in making a written offer to purchase such land from the owner.

9. Define the following:
   (a) searching a title
   (b) first and second mortgage
   (c) mechanic's lien
   (d) easement on a parcel of land
   (e) power of attorney on a parcel of land
   (f) writ or judgment on the title to property
   (g) caveat
   (h) abstract of title

10. Briefly define the functions of a city planning department.

11. List three city departments that you should consult before making application for a building permit for a residential project in your area.

12. Why is it sometimes very important to have a licensed surveyor establish the extremities of a parcel of land before you erect a building on it?

13. Define a) a datum line; b) bench mark.

14. List five agencies through which you could find the cost of labor in your area.

15. List three agencies from which you could get an indication of the availability of labor in your area.

16. Define demurrage.

17. Assuming that you are to build a school in your area, list six agencies (excluding subtrades) from which you could find the different building material costs.

18. Name two sources from which you could obtain the names and addresses of the manufacturers of doors and windows in your area.

19. Define the initials F.O.B. on a manufacturer's quotation for reinforcing steel.

20. List twelve subtrades.

21. Define the function of the Workmen's Compensation Board.

22. Define standard contract documents and support the view that they should be standard bidding procedure for all bidding contracts.

THE CONSTRUCTION COMPANY CHAIN OF COMMAND AND PROGRESS PAYMENTS

This chapter deals with the organization of the construction company, with particular emphasis on the role of the estimator. It also provides a guide list of overhead expenses.

The days of the one-man company are virtually over. It is now generally agreed that it is outside the ken of most men to have sufficient knowledge in all fields of building construction to be able to operate without the aid of technologists.

The smaller the contractor the more resourceful and knowledgeable he must be. Larger companies employ specialists for every operation both in the office and in the field. The world is becoming filled with specialists, and more and more men are required who have organizing ability. The construction industry is a rich field for capable, talented men. There are many important posts to be filled which offer ample rewards and great satisfaction.

2.1 BIDS AND ESTIMATES

Let it be accepted that all competitive bids are made against the same plans and specifications; theoretically, therefore, assuming that each bidding company is buying in the same market and paying the same union rates of labor, all the bids for the project, except an allowance for the organizing ability of the firms, should be the same. In the final analysis, each company is wagering that their efficiency is better than that of any other company. Remember that it is usually the lowest bidder who gets the job. *If your bid is 20 percent lower than anyone else's, then you have cause to reflect as to how efficient you really are.*

Estimates are made against the following:

(a) materials to be bought in the best market;

(b) staff to be employed because of their efficiency;

(c) subcontractors to be engaged for their abilities and competitive bids;

(d) company organizing ability (including financing);

(e) assessed time for completion date;

(f) the morale of the work crews—*the morale of the crews will never be greater than the morale of the man in charge.*

The successful contractor assembles the right things at the right times in the right places to be erected by the right people at the right prices. This is organization.

The general contractor's chain of command is shown in Table 2-1. This organization chart will be discussed at length in the following pages.

Table 2.1
CHAIN OF COMMAND

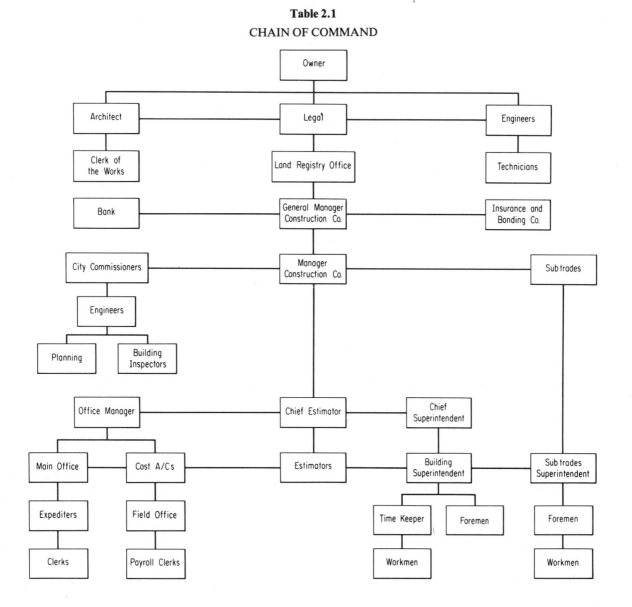

2.2 THE OWNER

The owner originates the idea for a structure and engages an architect and sometimes an engineer to make preliminary drawings and specifications for a building to meet his requirements within a stated cost. An approximate estimate of cost is made and the drawings and specifications are trimmed to the price that the owner is prepared to pay. Final drawings and specifications are made and the architect calls for tenders by competitive bid.

2.3 GENERAL MANAGER

He is responsible for financial and legal matters and for the policy and administration of the company. He makes final decisions. He meets architects, engineers, surveyors, lawyers, bankers, owners of proposed buildings, local authority departmental heads such as city engineers, and general managers of all subtrades.

He calls conferences of his staff as often as necessary. He delegates duties to men of his choosing and expects them to carry out the assignments with a minimum of guidance.

It has been said that the art of administration is to get people to do what you want them to do, when you want them to do it, and to have them like doing it.

2.4 MANAGER

He is responsible to the general manager for the more detailed administration of the business. He acts for the general manager when necessary. His duties parallel, in large measure, those of his senior. He watches very closely the progress schedules of every job under construction, and while not usurping the powers of his senior, he makes important decisions within his province without guidance. (For progress schedules see Chap. 12.)

His energies are directed to production and coordinating the offices of the chief accountant, chief estimator, and chief superintendent. As occasion demands, he will call a conference of his department heads and make reports to the general manager. Decisions regarding policy would be referred to the general manager.

2.5 THE ARCHITECT

He is responsible for interpreting into physical being the building that the owner wants at the specified price. Either the architect or his own inspector (sometimes called the clerk of the works) will carefully inspect the building during construction to ensure that all the materials and workmanship are being implemented as directed by the plans and specifications.

The architect issues progress payment certificates to the contractor at stipulated times upon satisfactory completion of different phases of the construction of the building.

An inspection certificate is a recommendation to the owner, by the architect, that the contractor is now entitled to a progress payment in the amount shown on the certificate. The owner may then instruct the bank to release to the contractor the amount stated on the certificate. *Some specifications call for evidence that all materials and labor used in the erection of the work certified as satisfactorily completed have indeed been paid for by the contractor before the certificate will be issued.* It is usual for the architect to withhold 10 to 15 percent of the value of the work performed. This represents the contractor's profit and maybe more. When the building is completed, the architect then makes a final holdback for the period of time stated in the contract.

The following progress payment schedule shows the manner of payments for an $80,000 residence with a 10 percent holdback at six stages of progress, and the final 10 percent holdback of the final payment for a further six months, until the contractor has made the final check and made good all necessary adjustments due to settling of the building or for any other legitimate reason.

2.6 THE ARCHITECT'S PROGRESS CERTIFICATES

PROGRESS PAYMENTS			
Date	Value of Work Completed (cumulative)	Value of Certificates Issued (individual)	Value of 10% Holdback (cumulative)
1	2	3	4
April	$ 9,600	$ 8,640	$ 960
May	24,000	12,960	2,400
June	40,000	14,400	4,000
July	51,200	10,080	5,120
August	72,000	18,720	7,200
September	80,000	7,200	8,000
	Total	$ 72,000	

At the end of September the total value of work completed was $80,000 (column 2). The total value of certificates issued was $72,000 (column 3) and the total value of holdback payments was $8,000 (column 4).

Let us examine the position with the issue of the July payment certificate of $10,080 (column 3).

Step 1: The value of work completed as of July was $51,200 (column 2).

Step 2: The value of all previous certificates issued was:

April	$ 8,640
May	12,960
June	14,400 (see column 3)
	$36;000

Step 3: The value of the 10 percent holdback as of July was 10 percent of $51,2000 = $5,120 (column 4).

Step 4: The value of the July certificate was equal to the *total value* of all work completed as of July, which was $51,200 as in Step 1, less the cumulative value of certificates already issued—$36,000 as in Step 2—and also less the new holdback on the July work completed, amounting to $51,200 as at Step 3.

Step 5: The value of the July certificate was:

$$\$51,200 - (\$36,000 + \$5,120) = \$10,080.$$

Problem

Pattern your work after the example shown and analyze the position with the payment certificate of $18,720 at the end of August.

2.7 THE ENGINEERS

They are responsible for the field work such as layout and detail drawings and for checking the work as it proceeds. They are consultants to the architect on structures.

2.8 THE TECHNICIAN

A technician in the construction industry is a junior engineer who has been given special training to stand between the professional men, such as architects and engineers, and tradesmen. Many technical schools offer courses leading to degrees in this field. The curriculum may include the following subjects: English, structures laboratory, mechanics, drafting and shop drawings, mathematics, estimating, contracts and specifications, mechanical systems, and surveying.

The technician has an intimate knowledge of materials. As an example, a bricklayer has a knowledge of bricks and mortar, but the technician knows the chemical properties of bricks and mortar and how to test for factors determining the strength in psi (pounds per square inch), decibel readings, fire ratings, and so on. Such men find ready employment in the construction industry as job sponsors, estimators, materials men, intricate detail draftsmen, grade strikers for city departments, government housing agents, and building inspectors; some may later decide to operate their own contracting businesses.

2.9 ATTORNEYS AND LAWYERS

They are responsible for expressing in written legal terms the liabilities and rewards of all contracting parties and for advising on legal matters. They also represent disputing parties in courts of law.

2.10 THE ESTIMATOR

He is responsible for estimating the quantities and cost of labor and materials for proposed projects, either new or renovations. The beginning estimator must be careful not to underestimate the time required to accomplish certain phases of work. *He must also estimate the time it will take him to prepare an estimate so that he may meet the deadline for placing a tender.*

Many estimators are promoted from the ranks of tradesmen. They tend to forget their early ambitious efficiency when they were quicker than the average tradesman and made fewer mistakes. They may also forget that during their construction careers they had a greater interest in their work, read more textbooks and trade magazines, and attended more technical courses and lectures. It is a very common error for such men (in the early stages) to underestimate the time required for the average tradesman to do a unit of work.

The estimator should be:

(a) a mature man of sound judgment with a wide experience in the field and preferably with a basic trade—this would give him the feel of the physical job;

(b) able to express himself freely, adequately, and decisively in a few words, either written or spoken;

(c) able to read drawings and to make rapid, clearcut, freehand sketches;

(d) able to handle numbers with facility (many offices will not permit their estimators to use a slide rule because of the risk of error in the placing of the decimal point—they use computing machines because these figures may be checked at any time);

(e) methodical and neat, with good handwriting or lettering;

(f) of a disciplined mind, knowing the order of building operations and taking off the quantities in that order—some larger organizations take off quantities by trades or operations such as foundations, framing (carcassing), finishing, and so on;

(g) up to date with the national and local building codes and amendments;

(h) versatile in knowing where to look for information that he requires, in technical libraries, his own library, or technical publications filed in the A.I.A. (American Institute of Architects) file.

(i) up to date with materials, methods, and designs;

(j) knowledgeable about job conditions for all seasons of the year;

(k) familiar with the financial columns of the daily press (he should know the trends of business generally and of the building industry in particular; able to make reports on financial matters relating to each project; and able to interpret the progress payment schedule against the company's financial position);

(l) well informed about the trade union movement and about new negotiations that might affect the materials costs, freight rates, delays through strikes, wages, and fringe benefits;

(m) able to assess what a man or machine can do in a given time;

(n) able to thoroughly organize his department and maintain a good reference library and A.I.A. file;

(o) on top of morale at all times and aware that the morale of his own staff will be patterned on his own.

He should try to organize field trips through the builders' exchange to visit such places as:

State or Provincial Government

Land Registry Office

Department of Labor

Apprenticeship Board

Department of Public Works

Workmen's Compensation Board

City

Planning Department

Engineers Department

Building Inspector's Office

Engineers: Electrical

Commercial

Telephone Company

Gas Company

Concrete manufacturers

Lime products manufacturers

Brick products manufacturers, and so on

It is good for the estimator, his staff, and all senior officials of any construction company to know personally the persons in official positions with whom they will continually be coming into contact. The author has always found such people to be most helpful and cooperative. Their job is to help and they are only too pleased to do so when asked.

2.11 THE JUNIOR ESTIMATOR

Some building contractors appoint one of their keen junior men to be a job sponsor. His role is to double up with the estimator in helping to take off all the materials from the plans and specifications. He then goes onto the job and doubles up with and acts as the right-hand man of the superintendent.

This method of training has much to commend it; furthermore, the job sponsor certainly earns his salary. It is a full-time job from the moment a set of plans and specifications are received in the office until the final handover of the completed building to the owner.

Since the success of any contractor is dependent upon the successful bidding by the estimating department, the training of the junior estimators is of paramount importance.

2.12 THE BUILDING SUPERINTENDENT

He is responsible for running the job. His job knowledge should complement that of the estimator. Many estimators have been promoted from the ranks to superintendent. The building superintendent should at all times keep in close touch with the estimators. He orders materials and coordinates delivery. He watches very closely the progress schedule and has the subtrades move on and off the job on scheduled time. (For progress schedules see Sec. 12.6.)

He should be a forthright man who either commands or demands respect from everyone with whom he comes in contact. He meets with the office staff, the architect and his inspector, city building inspectors, subcontractors, his foremen, and his own company tradesmen. He attracts to himself a loyal staff with whom he should be lucid and fair.

2.13 THE TIMEKEEPER

He works directly on the job under the superintendent. He checks the time of arrival and departure of men and materials and makes out a daily progress report

DAILY PROGRESS REPORT

THE MANSFIELD ENGINEERING & CONSTRUCTION CO. LTD.

JOB No.

FOREMAN'S SIGNATURE   DATE

JOB NAME   AND LOCATION

WEATHER CONDITIONS   LOW TEMP.   HIGH TEMP.

MEN ON JOB	No.	TOTAL HOURS WORKED	SUB TRADES ON JOB
Carpenters			
Laborers			
Others			

WORK ACCOMPLISHED AND REMARKS

CHANGE ORDERS OR EXTRAS AS REQUESTED OR APPROVED

EQUIPMENT ON JOB	EQUIPMENT ON RENTAL

(see page 45). He keeps a daily diary in which he records the weather, the main happenings of the day, accidents, visits to the site by important people, and a short note on the work being done that day. The diary can be a most important document, especially in case of some later dispute that may go to arbitration.

Either he or the foreman (sometimes the workmen themselves) will fill out a daily time sheet. This affects only the general contractor's men. On this sheet is recorded *the unit number of the work* each man is engaged on, the length of time he is engaged on it and, most important, the date (also see Sec. 12.11). This is a very important document, and it is only from the information therein that the estimating and cost-accounting departments can find out the time required for workmen to accomplish certain units of work.

If men are required to fill out a time sheet, they should be told the reason why. This information enables the contractor to estimate intelligently the cost of each unit of labor for a proposed new building, and thus keep his staff, including the men who make out the time sheets, efficiently employed.

There are no estimates of labor time required for building operations like your own. They should be built up and constantly revised.

2.14 THE GENERAL CONTRACTOR'S FOREMAN

He is directly responsible to the job superintendent and is delegated those duties that he and his crew can complete at one time.

He must know:

(a) his own trade very thoroughly;

(b) how to read drawings and specifications with complete understanding;

(c) how his own trade ties in with all others, and how other trades tie in with each other;

(d) how to make out a *neatly written or lettered* daily progress report, which will be passed through the job superintendent to the general contractor's office, where it will be recorded against the progress schedule (see Sec. 12.6). The daily progress report is a very important document and may even be used as evidence in the unfortunate event of court action in connection with the work;

(e) how to fill out a weekly time card describing the work done on each day and the amount of time spent on each task;

THE MANSFIELD ENGINEERING & CONSTRUCTION CO. LTD.
PROGRESS PAYMENTS SCHEDULE

Project _Residence: Mr. H. Ross._ Date _15 April_

Estimated Cost _$34,000.00_ Actual Cost _$33,800.00_

Starting Date _1 May_ Completion Date _1 September_

Unit Completion date	Unit	Actual Cost to date	Payments 85% to date	Holdback 15% to date
22 May	Sub. Floor completed	$4,800.00	$4,080.00	$720.00
19 June	Roof completed	$10,300.00	4,675.00	825.00
1 Aug.	Rough utilities completed	$17,900.00	6,460.00	1,140.00
26 Aug.	Plastering completed	$26,400.00	7,225.00	1,275.00
30 Aug.	Painting & Decorating cpt	30,600.00	3,570.00	630.00
30 Aug.	Landscaping	33,800.00	2,720.00	480.00
	Totals	33,800.00	28,730.00	5,070.00

(f) how to handle all the men on the job, not only the general contractor's men but also all other associated tradesmen and their foremen.

In the whole field of general contracting, the central theme is to select men who can organize others to get along with each other and achieve desired results at the right time and price. There is nothing more important in life than the ability either to wisely and intelligently lead or to accept a disciplined, wise, and intelligent leadership.

2.15 THE SUBCONTRACTORS

Subcontractors usually bid (on any specific job) to several contractors, except that on some large projects their bids have to be placed directly to the architect, engineers, or owners.

Many general contractors invite every subcontractor in the area to bid on work for which they are tendering; they then give the work to the bondable subcontractor submitting the lowest bid.

Subcontractors are not bound to give bids in the same amount to each general contractor. Where a subcontractor has formerly had problems with a general contractor he might increase his bid, or if he has a full work program he might submit a high bid—this in effect is a policy bid to keep his company name before the general contractor(s). (For a guide list of subcontractors, see Sec. 9.7.)

2.16 WORKMEN

They are responsible to their foremen. They are employed for their craftsmanship and should feel a reasonable sense of security in their jobs. Their loyalties should be rewarded with as regular employment as the business of the firm will permit.

QUESTIONS

1. List five main points against which estimates are made.

2. Write a short account (fifty words or less) of the role of the following persons in a large building contractor's employ: general manager; manager; architect; engineer; attorney and lawyer; estimator; technician; job sponsor; building superintendent; general contractor's foreman; subcontractor's foreman; timekeeper; workmen.

3. List ten desirable qualities for one who would wish to become a successful estimator.

4. Complete the following progress payments schedule.

PROGRESS PAYMENTS SCHEDULE			
Date	Value of Work Completed	Value of Certificate Issued	Total Value of 10% Holdback
April	$ 8,000	$7,200	$800
May	14,000		
June	20,000		
July	32,000		
August	48,000		
September	60,000		
Totals	$60,000	$54,000	$6,000

3

BONDS, CONTRACTS, AND INSURANCE

This chapter is designed to aid in the understanding of some of the main types of contracts and bonds that a successful tenderer would immediately enter into on being awarded a contract. A contract is a legal agreement between two or more parties and is also the memorandum of the agreement of the signatories. A standard printed short form is usually used. It is very directly worded and may have to be interpreted in court in the event of some major breach of contract. (See the last pages of this chapter.)

The actual cost of making an estimate is considerable in itself. This cost would depend upon the complexity and location of the project. Most people would agree that the cost of making an estimate would be between one-quarter and one-half of 1 percent of the cost of the building. Assume that a job is to cost $200,000. The estimate for such a project could cost $400 to $600 and not be considered excessive. The smaller the job the higher will be the relative cost for estimating.

What happens to the unsuccessful bidder? The cost of the estimate would have to be absorbed in the general overhead expenses of the company. From this it may be seen that no company can remain in business for very long without securing a reasonable average of successful bids.

There are a great many different types of bonds into which a contractor may have to enter, and these cost money. The estimator must always be aware that he is not just estimating the physical parts of a job, but he must also consider the many less obvious contractual and financial elements.

3.1 BONDS

It is necessary for the estimator not only to be aware of the different types of bonds that a building contractor may have to enter into, but also to be very much aware that the cost of these bonds must form part of his estimated bid. These costs would appear under "job overhead expenses" on the general estimate sheets.

3.2 TENDER OR BID BOND

This type of bond (sometimes called a proposal bond or guarantee bond) under the terms of the contract may have to be furnished by the contractor along with his tender. This is a legal document giving a guarantee by the bonding company that if the contractor stated on the face of the bond is awarded the contract and refuses to enter into contract they will indemnify the owners to the amount enfaced on the certificate. The amount is usually about 5 percent of the tender, but for public works the amount may be 10 percent. A letter from the bonding company is sometimes acceptable surety and constitutes a bond. (See the specimen bid bond and a typical letter from a bonding company headed "Consent of Surety" on the following pages.)

Some specifications call for a certified check in an amount equal to 5 percent (or maybe 10 percent for public works) instead of a bid bond. A further clause in the specifications usually states that, upon signing the contract, the successful tenderer will immediately furnish a performance bond.

3.3 PERFORMANCE BOND

There are several classes of performance bonds corresponding to the different classes of work involved, and the different rates may be obtained from the bonding company.

After having been awarded the contract, the successful tenderer pledges himself (through the bonding com-

pany) to faithfully perform the contract in accordance with the intentions of the drawings and specifications for the work. This may also cover a maintenance period after the building has been handed over to the owner.

The usual requirement for a performance bond is for a 100 percent coverage. The general contractor in turn will require a performance bond from his subcontractors. In this manner, the cost is spread against all the individual risks. (See the specimen performance bond on the following pages.)

3.4 LABOR AND MATERIAL PAYMENT BOND

The object of this bond is to protect the labor and material suppliers against nonpayment for work performed and materials supplied, respectively. A claimant under this type of bond must have a direct contract with the principal named in the bond; this will normally be the general contractor, but could be a subcontractor, depending on the specification requirements. (See the specimen labor and material payment bond on the following pages.)

3.5 LIENS

Liens may be registered against property for nonpayment of wages to mechanics (called a mechanic's lien) or by merchants who have delivered materials to a project and cannot recover payment for such services or materials from the contractor. There is a limited time during which such liens may be registered after the service or delivery of goods. *The contractor should make inquiries at the land registry office in the area to determine the rules and regulations governing liens for each particular area.*

When liens have been registered against the owner's property, the owner is often embarrassed. In effect, it means that enfaced on the title deeds of the property are registered claims of persons stating the amount of indebtedness the property owes them.

Sometimes the notice to tenderers in the preface to specifications will call for a lien bond to cover such emergencies.

3.6 LIEN BOND

This type of bond covers a situation where there is a claim registered against the property by any person or persons saying that they have rendered service or supplied goods to the property and have been unable to collect their just dues from the contractor. Such claims are written onto the title deeds with relevant dates, stating the amount of money claimed that was not met by the contractor at the time of handover of the building to the owner. The lien bond covers such situations, and the bonding company, in case of default by the contractor, will indemnify the lien holders and render the title deeds to the owner clear of such encumbrances. (See the specimen labor and material payment bond.)

3.7 FIRE INSURANCE

The contractor, unless otherwise stipulated, shall maintain and pay for fire insurance in the joint names of the owner and the contractor, totaling not less than 80 percent of the total value of the work done and material delivered on the site. Any loss under such policies shall be payable to both parties as their interest shall appear. Duplicates of such policies shall be furnished to the other party.

3.8 BONDS

The material on pages 51-78 has been supplied by the American Institute of Architects and is reproduced with their permission.

3.9 STIPULATED-SUM CONTRACT

This is often called a *lump-sum contract,* which is an agreed and contracted price between the owner and the contractor for the delivery of a completed building (to the architect's satisfaction) on an agreed date for an agreed price.

The contractor's price includes:

(a) his complete service and business organization in the erection and superintendence of the building;
(b) estimating the amount of materials needed;
(c) estimating the cost of the materials needed;
(d) estimating the cost of the labor and the organizing of all trades for the erection of the building.

In general, items (a), (b), and (c) are equal for all contractors when bidding. They all have to throw their office and field organizations behind the job. They can all estimate the amount and cost of materials required. It is indeed quite remarkable how close estimators usually are in estimating the amount of materials required. They all have the opportunity to purchase the same materials in the same markets and pay the same freight rates. In the final analysis, each company is estimating that their own organizing ability, staff skills, and company morale are better than the next.

In modern buildings, the general contractor's financial responsibility is probably only about 15 to 25 per-

cent of the total cost of the building. The remainder is largely taken up by the subcontractors.

On many jobs the general contractor is responsible for the complete building, including the cost of all subtrades. On some projects the subcontractors for plumbing, heating, electric wiring, and so on contract directly to the owner or architect. In this case it is important for the general contractor to read carefully the terms and conditions of contract: he must know who is responsible for cutting away and making good after subtrades. As an example, assume that a new building had been taken into use and it was discovered that a warm-air register had not been provided at a specified place. To rectify this may require the services of a tinsmith, carpenter, bricklayer, electrician, plumber, and painter; furthermore, the work may have to be done during a weekend, when premium rates must be paid to tradesmen. Who is responsible for the charges? Be careful! Estimating is more than the taking-off of quantities and the pricing of labor and materials. You must always read the terms and conditions of contract.

It is the contractor's responsibility, with all contracts, to draw up a progress schedule to the satisfaction of the architect (see Chap. 12). This is a very important document and is a timetable of what stage of progress the building will be in from starting to completion.

With the stipulated-sum contract the owner knows the cost of the building before the actual work starts.

3.10 COST PLUS A PERCENTAGE

In this type of contract, the builder is paid a fixed percentage of the actual cost of the building plus an additional compensation for his overhead expenses, services, and profit.

Under this system, the contract requires a company of known integrity, because it is patently obvious that unless the contractor keeps the job moving and observes every prudent economy in the building, the total cost would be higher and his percentage of profit greater. It is quite common for invitation bids to be called for this type of contract.

A cost-plus job soon becomes known to the workmen, and, unless they are kept under fair but keen discipline, their output may be considerably lower than on a stipulated-sum contract.

Some of the advantages of this type of contract include:

(a) in emergency cases, especially a national emergency, where work must start immediately regardless of cost;
(b) the method lends itself to getting things done more quickly;

THE AMERICAN INSTITUTE OF ARCHITECTS

AIA Document A201

General Conditions of the Contract for Construction

THIS DOCUMENT HAS IMPORTANT LEGAL CONSEQUENCES; CONSULTATION WITH AN ATTORNEY IS ENCOURAGED WITH RESPECT TO ITS MODIFICATION

1976 EDITION
TABLE OF ARTICLES

1. CONTRACT DOCUMENTS

2. ARCHITECT

3. OWNER

4. CONTRACTOR

5. SUBCONTRACTORS

6. WORK BY OWNER OR BY SEPARATE CONTRACTORS

7. MISCELLANEOUS PROVISIONS

8. TIME

9. PAYMENTS AND COMPLETION

10. PROTECTION OF PERSONS AND PROPERTY

11. INSURANCE

12. CHANGES IN THE WORK

13. UNCOVERING AND CORRECTION OF WORK

14. TERMINATION OF THE CONTRACT

This document has been approved and endorsed by The Associated General Contractors of America.

Copyright 1911, 1915, 1918, 1925, 1937, 1951, 1958, 1961, 1963, 1966, 1967, 1970, © 1976 by The American Institute of Architects, 1735 New York Avenue, N.W., Washington, D. C. 20006. Reproduction of the material herein or substantial quotation of its provisions without permission of the AIA violates the copyright laws of the United States and will be subject to legal prosecution.

INDEX

Acceptance of Defective or Non-Conforming Work . .6.2.2, **13.3**
Acceptance of Work 5.4.2, 9.5.5, 9.8.1, 9.9.1, 9.9.3
Access to Work 2.2.5, 6.2.1
Accident Prevention 2.2.4, 10
Acts and Omissions 2.2.4, 4.18.3, 7.4, 7.6.2, 8.3.1, 10.2.5
Additional Costs, Claims for 12.3
Administration of the Contract **2.2,** 4.3.3
All Risk Insurance 11.3.1
Allowances **4.8**
Applications for Payment 2.2.6, 9.2, **9.3,** 9.4,
        9.5.3, 9.6.1, 9.8.2, 9.9.1, 9.9.3, 14.2.2
Approvals . . . . .2.2.14, 3.4, 4.3.3, 4.5, 4.12.4 through
        4.12.6, 4.12.8, 4.18.3, 7.7, 9.3.2
Arbitration . . . . .2.2.7 through 2.2.13, 2.2.19, 6.2.5,
        7.9, 8.3.1, 11.3.7, 11.3.8
ARCHITECT **2**
Architect, **Definition** of **2.1**
Architect, Extent of Authority 2.2, 3.4, 4.12.8, 5.2, 6.3, 7.7.2,
        8.1.3, 8.3.1, 9.2, 9.3.1, 9.4, 9.5.3, 9.6, 9.8, 9.9.1, 9.9.3, 12.1.1,
        12.1.4, 12.3.1, 12.4.1, 13.1, 13.2.1, 13.2.5, 14.2
Architect, Limitations of Authority and Responsibility 2.2.2
        through 2.2.4, 2.2.10 through 2.2.14, 2.2.17, 2.2.18,
        4.3.3, 4.12.6, 5.2.1, 9.4.2, 9.5.4, 9.5.5, 12.4
Architect's Additional Services . .3.4, 7.7.2, 13.2.1, 13.2.5, 14.2.2
Architect's Approvals 2.2.14, 3.4, 4.5, 4.12.6, 4.12.8, 4.18.3
Architect's Authority to Reject Work 2.2.13, 4.5, 13.1.2, 13.2
Architect's Copyright 1.3
Architect's Decisions . . . . .2.2.7 through 2.2.13, 6.3, 7.7.2,
        7.9.1, 8.3.1, 9.2, 9.4, 9.6.1, 9.8.1, 12.1.4, 12.3.1
Architect's Inspections 2.2.13, 2.2.16, 9.8.1, 9.9.1
Architect's Instructions 2.2.13, 2.2.15, 7.7.2, 12.4, 13.1
Architect's Interpretations . . . . .2.2.7 through 2.2.10, 12.3.2
Architect's On-Site Observations 2.2.3, 2.2.5, 2.2.6, 2.2.17,
        7.7.1, 7.7.4, 9.4.2, 9.6.1, 9.9.1
Architect's Project Representative 2.2.17, 2.2.18
Architect's Relationship with Contractor 1.1.2, 2.2.4, 2.2.5,
        2.2.10, 2.2.13, 4.3.3, 4.5, 4.7.3, 4.12.6, 4.18, 11.3.6
Architect's Relationship with
        Subcontractors 1.1.2, 2.2.13, 9.5.3, 9.5.4
Architect's Representations 9.4.2, 9.6.1, 9.9.1
Artistic Effect 1.2.3, 2.2.11, 2.2.12, 7.9.1
Attorneys' Fees .4.18.1, 6.2.5, 9.9.2
Award of Separate Contracts 6.1.1
Award of Subcontracts and Other Contracts for
        Portions of the Work **5.2**
Bonds, Lien 9.9.2
Bonds, Performance, Labor and Material Payment 7.5, 9.9.3
Building Permit 4.7
Certificate of Substantial Completion 9.8.1
Certificates of Inspection, Testing or Approval 7.7.3
Certificates of Insurance 9.3.2, 11.1.4
Certificates for Payment 2.2.6, 2.2.16, **9.4,** 9.5.1, 9.5.5, 9.6.1,
        9.7.1, 9.8.2, 9.9.1, 9.9.3, 12.1.4, 14.2.2
Change Orders 1.1.1, 2.2.15, 3.4, 4.8.2.3, 5.2.3, 7.7.2,
        8.3.1, 9.7, 9.9.3, 11.3.1, 11.3.5, 11.3.7,
        12.1, 13.1.2, 13.2.5, 13.3.1
Change Orders, Definition of 12.1.1
CHANGES IN THE WORK 2.2.15, 4.1.1, **12**
Claims for Additional Cost or Time 8.3.2, 8.3.3, 12.2.1, **12.3**
Claims for Damages 6.1.1, 6.2.5, **7.4,** 8.3, 9.6.1.1

Cleaning Up **4.15,** 6.3
Commencement of the Work, Conditions Relating to. .3.2.1, 4.2,
        4.7.1, 4.10, 5.2.1, 6.2.2, 7.5, 9.2, 11.1.4, 11.3.4
Commencement of the Work, Definition of 8.1.2
Communications .2.2.2, 3.2.6, 4.9.1, **4.16**
Completion,
        Conditions Relating to . . .2.2.16, 4.11, 4.15, 9.4.2, 9.9, 13.2.2
COMPLETION, PAYMENTS AND **9**
Completion, Substantial 2.2.16, 8.1.1, 8.1.3, 8.2.2, 9.8, 13.2.2
Compliance with Laws 1.3, 2.1.1, 4.6, 4.7, 4.13,
        7.1, 7.7, 10.2.2, 14
Concealed Conditions **12.2**
Consent,
        Written . . .2.2.18, 4.14.2, 7.2, 7.6.2, 9.8.1, 9.9.2, 9.9.3, 11.3.9
Contract, Definition of **1.1.2**
Contract Administration 2.2, 4.3.3
Contract Award and Execution, Conditions
        Relating to 4.7.1, 4.10, 5.2, 7.5, 9.2, 11.1.4, 11.3.4
CONTRACT DOCUMENTS **1**
Contract Documents,
        Copies Furnished and Use of 1.3, 3.2.5, 5.3
Contract Documents, Definition of **1.1.1**
Contract Sum, Definition of **9.1.1**
Contract Termination **14**
Contract Time, Definition of 8.1.1
CONTRACTOR **4**
Contractor, **Definition** of **4.1,** 6.1.2
Contractor's Employees 4.3.2, 4.4.2, 4.8.1, 4.9, 4.18, 10.2.1
        through 10.2.4, 10.2.6, 10.3, 11.1.1
Contractor's Liability Insurance **11.1**
Contractor's Relationship with
        Separate Contractors and Owner's Forces 3.2.7, 6
Contractor's Relationship with
        Subcontractors 1.2.4, 5.2, 5.3, 9.5.2, 11.3.3, 11.3.6
Contractor's Relationship with the Architect 1.1.2, 2.2.4,
        2.2.5, 2.2.10, 2.2.13, 4.3.3, 4.5, 4.7.3, 4.12.6, 4.18, 11.3.6
Contractor's Representations 1.2.2, 4.5, 4.12.5, 9.3.3
Contractor's Responsibility for
        Those Performing the Work 4.3.2, 4.18, 10
Contractor's Review of Contract Documents 1.2.2, 4.2, 4.7.3
Contractor's Right to Stop the Work .9.7
Contractor's Right to Terminate the Contract 14.1
Contractor's Submittals 2.2.14, 4.10, 4.12, 5.2.1,
        5.2.3, 9.2, 9.3.1, 9.8.1, 9.9.2, 9.9.3
Contractor's Superintendent 4.9, 10.2.6
Contractor's Supervision and
        Construction Procedures 1.2.4, 2.2.4, 4.3, 4.4, 10
Contractual Liability Insurance 11.1.3
Coordination and
        Correlation 1.2.2, 1.2.4, 4.3.1, 4.10.1, 4.12.5, 6.1.3, 6.2.1
Copies Furnished of Drawings and Specifications . .1.3, 3.2.5, 5.3
Correction of Work .3.3, 3.4, 10.2.5, **13.2**
Cost, Definition of 12.1.4
Costs 3.4, 4.8.2, 4.15.2, 5.2.3, 6.1.1, 6.2.3, 6.2.5, 6.3, 7.7.1,
        7.7.2, 9.7, 11.3.1, 11.3.5, 12.1.3, 12.1.4, 12.3, 13.1.2, 13.2, 14
Cutting and Patching of Work **4.14,** 6.2
Damage to the Work 6.2.4, 6.2.5, 9.6.1.5, 9.8.1,
        10.2.1.2, 10.3, 11.3, 13.2.6
Damages, Claims for 6.1.1, 6.2.5, 7.4, 8.3.4, 9.6.1.2
Damages for Delay 6.1.1, 8.3.4, 9.7
Day, Definition of 8.1.4

Decisions of the Architect ......2.2.9 through 2.2.12, 6.3, 7.7.2,
     7.9.1, 8.3.1, 9.2, 9.4, 9.6.1, 9.8.1, 12.1.4, 12.3.1, 14.2.1
Defective or Non-Conforming Work, Acceptance, Rejection
    and Correction of 2.2.3, 2.2.13, 3.3, 3.4, 4.5, 6.2.2, 6.2.3,
                    9.6.1.1, 9.9.4.2, 13
Definitions ....**1.1, 2.1, 3.1, 4.1,** 4.12.1 through 4.12.3, **5.1,**
              6.1.2, **8.1,** 9.1.1, 12.1.1, 12.1.4
Delays and Extensions of Time **8.3**
Disputes 2.2.9, 2.2.12, 2.2.19, 6.2.5, 6.3, 7.9.1
Documents and Samples at the Site **4.11**
Drawings and Specifications, Use and
    Ownership of 1.1.1, 1.3, 3.2.5, 5.3
Emergencies **10.3**
Employees, Contractor's 4.3.2, 4.4.2, 4.8.1, 4.9, 4.18, 10.2.1
              through 10.2.4, 10.2.6, 10.3, 11.1.1
Equipment, Labor, Materials and 1.1.3, 4.4, 4.5, 4.12, 4.13,
        4.15.1, 6.2.1, 9.3.2, 9.3.3, 11.3, 13.2.2, 13.2.5, 14
Execution and Progress of the Work 1.1.3, 1.2.3, 2.2.3, 2.2.4,
        2.2.8, 4.2, 4.4.1, 4.5, 6.2.2, 7.9.3, 8.2,
           8.3, 9.6.1, 10.2.3, 10.2.4, 14.2
Execution, Correlation and Intent of the
    Contract Documents **1.2,** 4.7.1
Extensions of Time 8.3, 12.1
Failure of Payment by Owner **9.7,** 14.1
Failure of Payment of Subcontractors ..9.5.2, 9.6.1.3, 9.9.2, 14.2.1
Final Completion and Final Payment ..2.2.12, 2.2.16, **9.9,** 13.3.1
Financial Arrangements, Owner's 3.2.1
Fire and Extended Coverage Insurance 11.3.1
Governing Law **7.1**
Guarantees (See Warranty
    and Warranties) 2.2.16, 4.5, 9.3.3, 9.8.1, 9.9.4, 13.2.2
Indemnification 4.17, **4.18,** 6.2.5, 9.9.2
Identification of Contract Documents 1.2.1
Identification of Subcontractors and Suppliers 5.2.1
Information and
    Services Required of the Owner **3.2,** 6, 9, 11.2, 11.3
Inspections 2.2.13, 2.2.16, 4.3.3, 7.7, 9.8.1, 9.9.1
Instructions to Bidders 1.1.1, 7.5
Instructions to the
    Contractor 2.2.2, 3.2.6, 4.8.1, 7.7.2, 12.1.2, 12.1.4
INSURANCE 9.8.1, **11**
Insurance, Contractor's Liability 11.1
Insurance, Loss of Use 11.4
Insurance, Owner's Liability 11.2
Insurance, Property 11.3
Insurance, Stored Materials 9.3.2, 11.3.1
Insurance Companies, Consent to Partial Occupancy 11.3.9
Insurance Companies, Settlement With 11.3.8
Intent of
    the Contract Documents ...1.2.3, 2.2.10, 2.2.13, 2.2.14, 12.4
Interest **7.8**
Interpretations, Written 1.1.1, 2.2.7, 2.2.8, 2.2.10, 12.4
Labor and Materials, Equipment 1.1.3, **4.4,** 4.5, 4.12, 4.13,
        4.15.1, 6.2.1, 9.3.2, 9.3.3, 11.3, 13.2.2, 13.2.5, 14
Labor and Material Payment Bond 7.5
Labor Disputes 8.3.1
Laws and Regulations 1.3, 2.1, 4.6, 4.7, 4.13, 7.1,
                      7.7, 10.2.2, 14
Liens 9.3.3, 9.9.2, 9.9.4.1
Limitations of Authority 2.2.2, 2.2.17, 2.2.18, 11.3.8, 12.4.1
Limitations of Liability 2.2.10, 2.2.13, 2.2.14, 3.3, 4.2, 4.7.3,

        4.12.6, 4.17, 4.18.3, 6.2.2, 7.6.2, 9.4.2,
        9.9.4, 9.9.5, 10.2.5, 11.1.2, 11.3.6
Limitations of Time, General 2.2.8, 2.2.14, 3.2.4, 4.2, 4.7.3,
       4.12.4, 4.15, 5.2.1, 5.2.3, 7.4, 7.7, 8.2, 9.5.2, 9.6,
       9.8, 9.9, 11.3.4, 12.1.4, 12.4, 13.2.1, 13.2.2, 13.2.5
Limitations of Time, Specific 2.2.8, 2.2.12, 3.2.1, 3.4,
    4.10, 5.3, 6.2.2, 7.9.2, 8.2, 8.3.2, 8.3.3, 9.2, 9.3.1, 9.4.1, 9.5.1,
      9.7, 11.1.4, 11.3.1, 11.3.8, 11.3.9, 12.2, 12.3, 13.2.2,
              13.2.5, 13.2.7, 14.1, 14.2.1
Limitations, Statutes of 7.9.2, 13.2.2, 13.2.7
Loss of Use Insurance **11.4**
Materials, Labor, Equipment and 1.1.3, 4.4, 4.5, 4.12, 4.13,
    4.15.1, 6.2.1, 9.3.2, 9.3.3, 11.3.1, 13.2.2, 13.2.5, 14
Materials Suppliers 4.12.1, 5.2.1, 9.3.3
Means, Methods, Techniques, Sequences and
    Procedures of Construction 2.2.4, 4.3.1, 9.4.2
Minor Changes in the Work 1.1.1, 2.2.15, **12.4**
MISCELLANEOUS PROVISIONS **7**
Modifications, Definition of 1.1.1
Modifications to the Contract 1.1.1, 1.1.2, 2.2.2, 2.2.18,
                    4.7.3, 7.9.3, 12
Mutual Responsibility **6.2**
Non-Conforming Work, Acceptance of Defective or 13.3.1
Notice, Written 2.2.8, 2.2.12, 3.4, 4.2, 4.7.3, 4.7.4, 4.9,
    4.12.6, 4.12.7, 4.17, 5.2.1, 7.3, 7.4, 7.7, 7.9.2, 8.1.2, 8.3.2,
    8.3.3, 9.4.1, 9.6.1, 9.7, 9.9.1, 9.9.5, 10.2.6, 11.1.4, 11.3.1,
    11.3.4, 11.3.5, 11.3.7, 11.3.8, 12.2, 12.3, 13.2.2, 13.2.5, 14
Notices, Permits, Fees and 4.7, 10.2.2
Notice of Testing and Inspections 7.7
Notice to Proceed 8.1.2
Observations, Architect's On-Site 2.2.3, 7.7.1, 7.7.4, 9.4.2
Observations, Contractor's 1.2.2, 4.2.1, 4.7.3
Occupancy 8.1.3, 9.5.5, 11.3.9
On-Site Inspections by the Architect 2.2.3, 2.2.16, 9.4.2,
                      9.8.1, 9.9.1
On-Site Observations by the Architect 2.2.3, 2.2.6, 2.2.17,
          7.7.1, 7.7.4, 9.4.2, 9.6.1, 9.9.1
Orders, Written 3.3, 4.9, 12.1.4, 12.4.1, 13.1
OWNER **3**
Owner, **Definition** of **3.1**
Owner, Information and Services Required of the 3.2, 6.1.3,
                  6.2, 9, 11.2, 11.3
Owner's Authority 2.2.16, 4.8.1, 7.7.2, 9.3.1, 9.3.2,
         9.8.1, 11.3.8, 12.1.2, 12.1.4
Owner's Financial Capability 3.2.1
Owner's Liability Insurance **11.2**
Owner's Relationship with Subcontractors 1.1.2, 9.5.4
Owner's Right to Carry Out the Work **3.4,** 13.2.4
Owner's Right to Clean Up 4.15.2, **6.3**
Owner's Right to Perform Work and to Award
    Separate Contracts **6.1**
Owner's Right to Terminate the Contract 14.2
Owner's Right to Stop the Work **3.3**
Ownership and Use of Documents 1.1.1, **1.3,** 3.2.5, 5.2.3
Patching of Work, Cutting and 4.14, 6.2.2
Patents, Royalties and 4.17.1
Payment Bond, Labor and Material 7.5
Payment, Applications for 2.2.6, 9.2, 9.3, 9.4, 9.5.3,
          9.6.1, 9.8.2, 9.9.1, 9.9.3, 14.2.2
Payment, Certificates for 2.2.6, 2.2.16, 9.4, 9.5.1,
      9.5.5, 9.6.1, 9.7.1, 9.8.2, 9.9.1, 9.9.3, 12.1.4, 14.2.2

Payment, Failure of 9.5.2, 9.6.1.3, 9.7, 9.9.2, 14
Payment, Final 2.2.12, 2.2.16, 9.9, 13.3.1
Payments, Progress 7.8, 7.9.3, 9.5.5, 9.8.2, 9.9.3, 12.1.4
PAYMENTS AND COMPLETION **9**
Payments to Subcontractors 9.5.2, 9.5.3, 9.5.4,
9.6.1.3, 11.3.3, 14.2.1
Payments Withheld **9.6**
Performance Bond and Labor and Material Payment Bond ..**7.5**
Permits, Fees and Notices 3.2.3, **4.7,** 4.13
PERSONS AND PROPERTY, PROTECTION OF **10**
Product Data, Definition of 4.12.2
Product Data, Shop Drawings, Samples and ...2.2.14, 4.2.1, 4.12
Progress and Completion 2.2.3, 7.9.3, **8.2**
Progress Payments 7.8, 7.9.3, 9.5.5, 9.8.2, 9.9.3, 12.1.4
Progress Schedule **4.10**
Project, Definition of **1.1.4**
Project Representative 2.2.17
Property Insurance **11.3**
PROTECTION OF PERSONS AND PROPERTY **10**
Regulations and Laws 1.3, 2.1.1, 4.6, 4.7, 4.13, 7.1, 10.2.2, 14
Rejection of Work 2.2.13, 4.5.1, 13.2
Releases of Waivers and Liens 9.9.2, 9.9.4
Representations 1.2.2, 4.5, 4.12.5, 9.4.2, 9.6.1, 9.9.1
Representatives 2.1, 2.2.2, 2.2.17,
2.2.18, 3.1, 4.1, 4.9, 5.1, 9.3.3
Responsibility for Those Performing the Work 2.2.4, 4.3.2,
6.1.3, 6.2, 9.8.1
Retainage 9.3.1, 9.5.2, 9.8.2, 9.9.2, 9.9.3
Review of Contract Documents
by the Contractor 1.2.2, **4.2,** 4.7.3
Reviews of Contractor's Submittals by
Owner and Architect 2.2.14, 4.10, 4.12, 5.2.1, 5.2.3, 9.2
Rights and Remedies 1.1.2, 2.2.12, 2.2.13, 3.3, 3.4, 5.3, 6.1,
6.3, **7.6,** 7.9, 8.3.1, 9.6.1, 9.7, 10.3, 12.1.2, 12.2, 13.2.2, 14
Royalties and Patents **4.17**
Safety of Persons and Property **10.2**
Safety Precautions and Programs 2.2.4, **10.1**
Samples, Definition of 4.12.3
Samples, Shop Drawings, Product Data and 2.2.14, 4.2, 4.12
Samples at the Site, Documents and 4.11
Schedule of Values **9.2**
Schedule, Progress 4.10
Separate Contracts and Contractors 4.14.2, 6, 11.3.6, 13.1.2
Shop Drawings, Definition of 4.12.1
Shop Drawings, Product Data and Samples ... 2.2.14, 4.2, **4.12**
Site, Use of 4.13, 6.2.1
Site Visits, Architect's 2.2.3, 2.2.5, 2.2.6, 2.2.17,
7.7.1, 7.7.4, 9.4.2, 9.6.1, 9.9.1
Site Inspections 1.2.2, 2.2.3, 2.2.16, 7.7, 9.8.1, 9.9.1
Special Inspection and Testing 2.2.13, 7.7
Specifications 1.1.1, 1.2.4, 1.3
Statutes of Limitations 7.9.2, 13.2.2, 13.2.7
Stopping the Work 3.3, 9.7.1, 10.3, 14.1
Stored Materials 6.2.1, 9.3.2, 10.2.1.2, 11.3.1, 13.2.5

SUBCONTRACTORS **5**
Subcontractors, **Definition** of **5.1**
Subcontractors, Work by 1.2.4, 2.2.4, 4.3.1, 4.3.2
Subcontractual Relations **5.3**
Submittals 1.3, 4.10, 4.12, 5.2.1, 5.2.3, 9.2,
9.3.1, 9.8.1, 9.9.2, 9.9.3
Subrogation, Waiver of 11.3.6
Substantial Completion 2.2.16, 8.1.1, 8.1.3, 8.2.2, **9.8,** 13.2.2
Substantial Completion, Definition of 8.1.3
Substitution of Subcontractors 5.2.3, 5.2.4
Substitution of the Architect 2.2.19
Substitutions of Materials 4.5, 12.1.4
Sub-subcontractors, Definition of 5.1.2
Subsurface Conditions 12.2.1
Successors and Assigns **7.2**
Supervision and Construction Procedures .1.2.4, 2.2.4, **4.3,** 4.4, 10
Superintendent, Contractor's **4.9,** 10.2.6
Surety, Consent of 9.9.2, 9.9.3
Surveys 3.2.2, 4.18.3
Taxes **4.6**
Termination by the Contractor **14.1**
Termination by the Owner **14.2**
Termination of the Architect 2.2.19
TERMINATION OF THE CONTRACT **14**
Tests 2.2.13, 4.3.3, **7.7,** 9.4.2
Time **8**
Time, **Definition** of **8.1**
Time, Delays and Extensions of 8.3, 12.1, 12.3, 13.2.7
Time Limits, Specific 2.2.8, 2.2.12, 3.2.1, 3.4,
4.10, 5.3, 6.2.2, 7.9.2, 8.2, 8.3.2, 8.3.3, 9.2, 9.3.1,
9.4.1, 9.5.1, 9.7, 11.1.4, 11.3.1, 11.3.8, 11.3.9,
12.2, 12.3, 13.2.2, 13.2.5, 13.2.7, 14.1, 14.2.1
Title to Work 9.3.2, 9.3.3
UNCOVERING AND CORRECTION OF WORK **13**
Uncovering of Work **13.1**
Unforseen Conditions 8.3, 12.2
Unit Prices 12.1.3, 12.1.5
Use of Documents 1.1.1, 1.3, 3.2.5, 5.3
Use of Site **4.13,** 6.2.1
Values, Schedule of 9.2
Waiver of Claims by the Contractor 7.6.2, 8.3.2, 9.9.5, 11.3.6
Waiver of Claims by the Owner 7.6.2, 9.9.4, 11.3.6, 11.4.1
Waiver of Liens 9.9.2
Warranty and Warranties 2.2.16, **4.5,** 9.3.3, 9.8.1, 9.9.4, 13.2.2
Weather Delays 8.3.1
Work, Definition of **1.1.3**
Work by Owner or by Separate Contractors **6**
Written Consent 2.2.18, 4.14.2, 7.2, 7.6.2, 9.8.1, 9.9.3, 9.9.4
Written Interpretations 1.1.1, 1.2.4, 2.2.8, 12.3.2
Written Notice 2.2.8, 2.2.12, 3.4, 4.2, 4.7.3, 4.7.4, 4.9, 4.12.6,
4.12.7, 4.17, 5.2.1, **7.3,** 7.4, 7.7, 9.2, 8.1.2, 8.3.2, 8.3.3,
9.4.1, 9.6.1, 9.7, 9.9.1, 9.9.5, 10.2.6, 11.1.4, 11.3.1, 11.3.4,
11.3.5, 11.3.7, 11.3.8, 12.2, 12.3, 13.2.2, 13.2.5, 14
Written Orders 3.3, 4.9, 12.1.4, 12.4.1, 13.1

AIA DOCUMENT A201 • GENERAL CONDITIONS OF THE CONTRACT FOR CONSTRUCTION • THIRTEENTH EDITION • AUGUST 1976
AIA® • © 1976 • THE AMERICAN INSTITUTE OF ARCHITECTS, 1735 NEW YORK AVENUE, N.W., WASHINGTON, D.C. 20006

> # GENERAL CONDITIONS OF THE CONTRACT FOR CONSTRUCTION

ARTICLE 1

CONTRACT DOCUMENTS

1.1   DEFINITIONS

1.1.1   THE CONTRACT DOCUMENTS

The Contract Documents consist of the Owner-Contractor Agreement, the Conditions of the Contract (General, Supplementary and other Conditions), the Drawings, the Specifications, and all Addenda issued prior to and all Modifications issued after execution of the Contract. A Modification is (1) a written amendment to the Contract signed by both parties, (2) a Change Order, (3) a written interpretation issued by the Architect pursuant to Subparagraph 2.2.8, or (4) a written order for a minor change in the Work issued by the Architect pursuant to Paragraph 12.4. The Contract Documents do not include Bidding Documents such as the Advertisement or Invitation to Bid, the Instructions to Bidders, sample forms, the Contractor's Bid or portions of Addenda relating to any of these, or any other documents, unless specifically enumerated in the Owner-Contractor Agreement.

1.1.2   THE CONTRACT

The Contract Documents form the Contract for Construction. This Contract represents the entire and integrated agreement between the parties hereto and supersedes all prior negotiations, representations, or agreements, either written or oral. The Contract may be amended or modified only by a Modification as defined in Subparagraph 1.1.1. The Contract Documents shall not be construed to create any contractual relationship of any kind between the Architect and the Contractor, but the Architect shall be entitled to performance of obligations intended for his benefit, and to enforcement thereof. Nothing contained in the Contract Documents shall create any contractual relationship between the Owner or the Architect and any Subcontractor or Sub-subcontractor.

1.1.3   THE WORK

The Work comprises the completed construction required by the Contract Documents and includes all labor necessary to produce such construction, and all materials and equipment incorporated or to be incorporated in such construction.

1.1.4   THE PROJECT

The Project is the total construction of which the Work performed under the Contract Documents may be the whole or a part.

1.2   EXECUTION, CORRELATION AND INTENT

1.2.1 The Contract Documents shall be signed in not less than triplicate by the Owner and Contractor. If either the Owner or the Contractor or both do not sign the Conditions of the Contract, Drawings, Specifications, or any of the other Contract Documents, the Architect shall identify such Documents.

1.2.2 By executing the Contract, the Contractor represents that he has visited the site, familiarized himself with the local conditions under which the Work is to be performed, and correlated his observations with the requirements of the Contract Documents.

1.2.3 The intent of the Contract Documents is to include all items necessary for the proper execution and completion of the Work. The Contract Documents are complementary, and what is required by any one shall be as binding as if required by all. Work not covered in the Contract Documents will not be required unless it is consistent therewith and is reasonably inferable therefrom as being necessary to produce the intended results. Words and abbreviations which have well-known technical or trade meanings are used in the Contract Documents in accordance with such recognized meanings.

1.2.4 The organization of the Specifications into divisions, sections and articles, and the arrangement of Drawings shall not control the Contractor in dividing the Work among Subcontractors or in establishing the extent of Work to be performed by any trade.

1.3   OWNERSHIP AND USE OF DOCUMENTS

1.3.1 All Drawings, Specifications and copies thereof furnished by the Architect are and shall remain his property. They are to be used only with respect to this Project and are not to be used on any other project. With the exception of one contract set for each party to the Contract, such documents are to be returned or suitably accounted for to the Architect on request at the completion of the Work. Submission or distribution to meet official regulatory requirements or for other purposes in connection with the Project is not to be construed as publication in derogation of the Architect's common law copyright or other reserved rights.

ARTICLE 2

ARCHITECT

2.1   DEFINITION

2.1.1 The Architect is the person lawfully licensed to practice architecture, or an entity lawfully practicing architecture identified as such in the Owner-Contractor Agreement, and is referred to throughout the Contract Documents as if singular in number and masculine in gender. The term Architect means the Architect or his authorized representative.

2.2   ADMINISTRATION OF THE CONTRACT

2.2.1 The Architect will provide administration of the Contract as hereinafter described.

2.2.2 The Architect will be the Owner's representative during construction and until final payment is due. The Architect will advise and consult with the Owner. The Owner's instructions to the Contractor shall be forwarded through the Architect. The Architect will have authority to act on behalf of the Owner only to the extent provided in the Contract Documents, unless otherwise modified by written instrument in accordance with Subparagraph 2.2.18.

2.2.3 The Architect will visit the site at intervals appropriate to the stage of construction to familiarize himself generally with the progress and quality of the Work and to determine in general if the Work is proceeding in accordance with the Contract Documents. However, the Architect will not be required to make exhaustive or continuous on-site inspections to check the quality or quantity of the Work. On the basis of his on-site observations as an architect, he will keep the Owner informed of the progress of the Work, and will endeavor to guard the Owner against defects and deficiencies in the Work of the Contractor.

2.2.4 The Architect will not be responsible for and will not have control or charge of construction means, methods, techniques, sequences or procedures, or for safety precautions and programs in connection with the Work, and he will not be responsible for the Contractor's failure to carry out the Work in accordance with the Contract Documents. The Architect will not be responsible for or have control or charge over the acts or omissions of the Contractor, Subcontractors, or any of their agents or employees, or any other persons performing any of the Work.

2.2.5 The Architect shall at all times have access to the Work wherever it is in preparation and progress. The Contractor shall provide facilities for such access so the Architect may perform his functions under the Contract Documents.

2.2.6 Based on the Architect's observations and an evaluation of the Contractor's Applications for Payment, the Architect will determine the amounts owing to the Contractor and will issue Certificates for Payment in such amounts, as provided in Paragraph 9.4.

2.2.7 The Architect will be the interpreter of the requirements of the Contract Documents and the judge of the performance thereunder by both the Owner and Contractor.

2.2.8 The Architect will render interpretations necessary for the proper execution or progress of the Work, with reasonable promptness and in accordance with any time limit agreed upon. Either party to the Contract may make written request to the Architect for such interpretations.

2.2.9 Claims, disputes and other matters in question between the Contractor and the Owner relating to the execution or progress of the Work or the interpretation of the Contract Documents shall be referred initially to the Architect for decision which he will render in writing within a reasonable time.

2.2.10 All interpretations and decisions of the Architect shall be consistent with the intent of and reasonably inferable from the Contract Documents and will be in writing or in the form of drawings. In his capacity as interpreter and judge, he will endeavor to secure faithful performance by both the Owner and the Contractor, will not

show partiality to either, and will not be liable for the result of any interpretation or decision rendered in good faith in such capacity.

2.2.11 The Architect's decisions in matters relating to artistic effect will be final if consistent with the intent of the Contract Documents.

2.2.12 Any claim, dispute or other matter in question between the Contractor and the Owner referred to the Architect, except those relating to artistic effect as provided in Subparagraph 2.2.11 and except those which have been waived by the making or acceptance of final payment as provided in Subparagraphs 9.9.4 and 9.9.5, shall be subject to arbitration upon the written demand of either party. However, no demand for arbitration of any such claim, dispute or other matter may be made until the earlier of (1) the date on which the Architect has rendered a written decision, or (2) the tenth day after the parties have presented their evidence to the Architect or have been given a reasonable opportunity to do so, if the Architect has not rendered his written decision by that date. When such a written decision of the Architect states (1) that the decision is final but subject to appeal, and (2) that any demand for arbitration of a claim, dispute or other matter covered by such decision must be made within thirty days after the date on which the party making the demand receives the written decision, failure to demand arbitration within said thirty days' period will result in the Architect's decision becoming final and binding upon the Owner and the Contractor. If the Architect renders a decision after arbitration proceedings have been initiated, such decision may be entered as evidence but will not supersede any arbitration proceedings unless the decision is acceptable to all parties concerned.

2.2.13 The Architect will have authority to reject Work which does not conform to the Contract Documents. Whenever, in his opinion, he considers it necessary or advisable for the implementation of the intent of the Contract Documents, he will have authority to require special inspection or testing of the Work in accordance with Subparagraph 7.7.2 whether or not such Work be then fabricated, installed or completed. However, neither the Architect's authority to act under this Subparagraph 2.2.13, nor any decision made by him in good faith either to exercise or not to exercise such authority, shall give rise to any duty or responsibility of the Architect to the Contractor, any Subcontractor, any of their agents or employees, or any other person performing any of the Work.

2.2.14 The Architect will review and approve or take other appropriate action upon Contractor's submittals such as Shop Drawings, Product Data and Samples, but only for conformance with the design concept of the Work and with the information given in the Contract Documents. Such action shall be taken with reasonable promptness so as to cause no delay. The Architect's approval of a specific item shall not indicate approval of an assembly of which the item is a component.

2.2.15 The Architect will prepare Change Orders in accordance with Article 12, and will have authority to order minor changes in the Work as provided in Subparagraph 12.4.1.

2.2.16 The Architect will conduct inspections to determine the dates of Substantial Completion and final completion, will receive and forward to the Owner for the Owner's review written warranties and related documents required by the Contract and assembled by the Contractor, and will issue a final Certificate for Payment upon compliance with the requirements of Paragraph 9.9.

2.2.17 If the Owner and Architect agree, the Architect will provide one or more Project Representatives to assist the Architect in carrying out his responsibilities at the site. The duties, responsibilities and limitations of authority of any such Project Representative shall be as set forth in an exhibit to be incorporated in the Contract Documents.

2.2.18 The duties, responsibilities and limitations of authority of the Architect as the Owner's representative during construction as set forth in the Contract Documents will not be modified or extended without written consent of the Owner, the Contractor and the Architect.

2.2.19 In case of the termination of the employment of the Architect, the Owner shall appoint an architect against whom the Contractor makes no reasonable objection whose status under the Contract Documents shall be that of the former architect. Any dispute in connection with such appointment shall be subject to arbitration.

ARTICLE 3

OWNER

3.1 DEFINITION

3.1.1 The Owner is the person or entity identified as such in the Owner-Contractor Agreement and is referred to throughout the Contract Documents as if singular in number and masculine in gender. The term Owner means the Owner or his authorized representative.

3.2 INFORMATION AND SERVICES REQUIRED OF THE OWNER

3.2.1 The Owner shall, at the request of the Contractor, at the time of execution of the Owner-Contractor Agreement, furnish to the Contractor reasonable evidence that he has made financial arrangements to fulfill his obligations under the Contract. Unless such reasonable evidence is furnished, the Contractor is not required to execute the Owner-Contractor Agreement or to commence the Work.

3.2.2 The Owner shall furnish all surveys describing the physical characteristics, legal limitations and utility locations for the site of the Project, and a legal description of the site.

3.2.3 Except as provided in Subparagraph 4.7.1, the Owner shall secure and pay for necessary approvals, easements, assessments and charges required for the construction, use or occupancy of permanent structures or for permanent changes in existing facilities.

3.2.4 Information or services under the Owner's control shall be furnished by the Owner with reasonable promptness to avoid delay in the orderly progress of the Work.

3.2.5 Unless otherwise provided in the Contract Documents, the Contractor will be furnished, free of charge, all copies of Drawings and Specifications reasonably necessary for the execution of the Work.

3.2.6 The Owner shall forward all instructions to the Contractor through the Architect.

3.2.7 The foregoing are in addition to other duties and responsibilities of the Owner enumerated herein and especially those in respect to Work by Owner or by Separate Contractors, Payments and Completion, and Insurance in Articles 6, 9 and 11 respectively.

3.3 OWNER'S RIGHT TO STOP THE WORK

3.3.1 If the Contractor fails to correct defective Work as required by Paragraph 13.2 or persistently fails to carry out the Work in accordance with the Contract Documents, the Owner, by a written order signed personally or by an agent specifically so empowered by the Owner in writing, may order the Contractor to stop the Work, or any portion thereof, until the cause for such order has been eliminated; however, this right of the Owner to stop the Work shall not give rise to any duty on the part of the Owner to exercise this right for the benefit of the Contractor or any other person or entity, except to the extent required by Subparagraph 6.1.3.

3.4 OWNER'S RIGHT TO CARRY OUT THE WORK

3.4.1 If the Contractor defaults or neglects to carry out the Work in accordance with the Contract Documents and fails within seven days after receipt of written notice from the Owner to commence and continue correction of such default or neglect with diligence and promptness, the Owner may, after seven days following receipt by the Contractor of an additional written notice and without prejudice to any other remedy he may have, make good such deficiencies. In such case an appropriate Change Order shall be issued deducting from the payments then or thereafter due the Contractor the cost of correcting such deficiencies, including compensation for the Architect's additional services made necessary by such default, neglect or failure. Such action by the Owner and the amount charged to the Contractor are both subject to the prior approval of the Architect. If the payments then or thereafter due the Contractor are not sufficient to cover such amount, the Contractor shall pay the difference to the Owner.

ARTICLE 4

CONTRACTOR

4.1 DEFINITION

4.1.1 The Contractor is the person or entity identified as such in the Owner-Contractor Agreement and is referred to throughout the Contract Documents as if singular in number and masculine in gender. The term Contractor means the Contractor or his authorized representative.

4.2 REVIEW OF CONTRACT DOCUMENTS

4.2.1 The Contractor shall carefully study and compare the Contract Documents and shall at once report to the Architect any error, inconsistency or omission he may discover. The Contractor shall not be liable to the Owner or the Architect for any damage resulting from any such errors, inconsistencies or omissions in the Contract Documents. The Contractor shall perform no portion of the Work at any time without Contract Documents or, where required, approved Shop Drawings, Product Data or Samples for such portion of the Work.

4.3   SUPERVISION AND CONSTRUCTION PROCEDURES

4.3.1 The Contractor shall supervise and direct the Work, using his best skill and attention. He shall be solely responsible for all construction means, methods, techniques, sequences and procedures and for coordinating all portions of the Work under the Contract.

4.3.2 The Contractor shall be responsible to the Owner for the acts and omissions of his employees, Subcontractors and their agents and employees, and other persons performing any of the Work under a contract with the Contractor.

4.3.3 The Contractor shall not be relieved from his obligations to perform the Work in accordance with the Contract Documents either by the activities or duties of the Architect in his administration of the Contract, or by inspections, tests or approvals required or performed under Paragraph 7.7 by persons other than the Contractor.

4.4   LABOR AND MATERIALS

4.4.1 Unless otherwise provided in the Contract Documents, the Contractor shall provide and pay for all labor, materials, equipment, tools, construction equipment and machinery, water, heat, utilities, transportation, and other facilities and services necessary for the proper execution and completion of the Work, whether temporary or permanent and whether or not incorporated or to be incorporated in the Work.

4.4.2 The Contractor shall at all times enforce strict discipline and good order among his employees and shall not employ on the Work any unfit person or anyone not skilled in the task assigned to him.

4.5   WARRANTY

4.5.1 The Contractor warrants to the Owner and the Architect that all materials and equipment furnished under this Contract will be new unless otherwise specified, and that all Work will be of good quality, free from faults and defects and in conformance with the Contract Documents. All Work not conforming to these requirements, including substitutions not properly approved and authorized, may be considered defective. If required by the Architect, the Contractor shall furnish satisfactory evidence as to the kind and quality of materials and equipment. This warranty is not limited by the provisions of Paragraph 13.2.

4.6   TAXES

4.6.1 The Contractor shall pay all sales, consumer, use and other similar taxes for the Work or portions thereof provided by the Contractor which are legally enacted at the time bids are received, whether or not yet effective.

4.7   PERMITS, FEES AND NOTICES

4.7.1 Unless otherwise provided in the Contract Documents, the Contractor shall secure and pay for the building permit and for all other permits and governmental fees, licenses and inspections necessary for the proper execution and completion of the Work which are customarily secured after execution of the Contract and which are legally required at the time the bids are received.

4.7.2 The Contractor shall give all notices and comply with all laws, ordinances, rules, regulations and lawful orders of any public authority bearing on the performance of the Work.

4.7.3 It is not the responsibility of the Contractor to make certain that the Contract Documents are in accordance with applicable laws, statutes, building codes and regulations. If the Contractor observes that any of the Contract Documents are at variance therewith in any respect, he shall promptly notify the Architect in writing, and any necessary changes shall be accomplished by appropriate Modification.

4.7.4 If the Contractor performs any Work knowing it to be contrary to such laws, ordinances, rules and regulations, and without such notice to the Architect, he shall assume full responsibility therefor and shall bear all costs attributable thereto.

4.8   ALLOWANCES

4.8.1 The Contractor shall include in the Contract Sum all allowances stated in the Contract Documents. Items covered by these allowances shall be supplied for such amounts and by such persons as the Owner may direct, but the Contractor will not be required to employ persons against whom he makes a reasonable objection.

4.8.2 Unless otherwise provided in the Contract Documents:

  .1 these allowances shall cover the cost to the Contractor, less any applicable trade discount, of the materials and equipment required by the allowance delivered at the site, and all applicable taxes;

  .2 the Contractor's costs for unloading and handling on the site, labor, installation costs, overhead, profit and other expenses contemplated for the original allowance shall be included in the Contract Sum and not in the allowance;

  .3 whenever the cost is more than or less than the allowance, the Contract Sum shall be adjusted accordingly by Change Order, the amount of which will recognize changes, if any, in handling costs on the site, labor, installation costs, overhead, profit and other expenses.

4.9   SUPERINTENDENT

4.9.1 The Contractor shall employ a competent superintendent and necessary assistants who shall be in attendance at the Project site during the progress of the Work. The superintendent shall represent the Contractor and all communications given to the superintendent shall be as binding as if given to the Contractor. Important communications shall be confirmed in writing. Other communications shall be so confirmed on written request in each case.

4.10   PROGRESS SCHEDULE

4.10.1 The Contractor, immediately after being awarded the Contract, shall prepare and submit for the Owner's and Architect's information an estimated progress schedule for the Work. The progress schedule shall be related to the entire Project to the extent required by the Contract Documents, and shall provide for expeditious and practicable execution of the Work.

4.11   DOCUMENTS AND SAMPLES AT THE SITE

4.11.1 The Contractor shall maintain at the site for the Owner one record copy of all Drawings, Specifications, Addenda, Change Orders and other Modifications, in good order and marked currently to record all changes made during construction, and approved Shop Drawings, Product Data and Samples. These shall be available to the Architect and shall be delivered to him for the Owner upon completion of the Work.

4.12   SHOP DRAWINGS, PRODUCT DATA AND SAMPLES

4.12.1 Shop Drawings are drawings, diagrams, schedules and other data specially prepared for the Work by the Contractor or any Subcontractor, manufacturer, supplier or distributor to illustrate some portion of the Work.

4.12.2 Product Data are illustrations, standard schedules, performance charts, instructions, brochures, diagrams and other information furnished by the Contractor to illustrate a material, product or system for some portion of the Work.

4.12.3 Samples are physical examples which illustrate materials, equipment or workmanship and establish standards by which the Work will be judged.

4.12.4 The Contractor shall review, approve and submit, with reasonable promptness and in such sequence as to cause no delay in the Work or in the work of the Owner or any separate contractor, all Shop Drawings, Product Data and Samples required by the Contract Documents.

4.12.5 By approving and submitting Shop Drawings, Product Data and Samples, the Contractor represents that he has determined and verified all materials, field measurements, and field construction criteria related thereto, or will do so, and that he has checked and coordinated the information contained within such submittals with the requirements of the Work and of the Contract Documents.

4.12.6 The Contractor shall not be relieved of responsibility for any deviation from the requirements of the Contract Documents by the Architect's approval of Shop Drawings, Product Data or Samples under Subparagraph 2.2.14 unless the Contractor has specifically informed the Architect in writing of such deviation at the time of submission and the Architect has given written approval to the specific deviation. The Contractor shall not be relieved from responsibility for errors or omissions in the Shop Drawings, Product Data or Samples by the Architect's approval thereof.

4.12.7 The Contractor shall direct specific attention, in writing or on resubmitted Shop Drawings, Product Data or Samples, to revisions other than those requested by the Architect on previous submittals.

4.12.8 No portion of the Work requiring submission of a Shop Drawing, Product Data or Sample shall be commenced until the submittal has been approved by the Architect as provided in Subparagraph 2.2.14. All such

portions of the Work shall be in accordance with approved submittals.

4.13   USE OF SITE

4.13.1 The Contractor shall confine operations at the site to areas permitted by law, ordinances, permits and the Contract Documents and shall not unreasonably encumber the site with any materials or equipment.

4.14   CUTTING AND PATCHING OF WORK

4.14.1 The Contractor shall be responsible for all cutting, fitting or patching that may be required to complete the Work or to make its several parts fit together properly.

4.14.2 The Contractor shall not damage or endanger any portion of the Work or the work of the Owner or any separate contractors by cutting, patching or otherwise altering any work, or by excavation. The Contractor shall not cut or otherwise alter the work of the Owner or any separate contractor except with the written consent of the Owner and of such separate contractor. The Contractor shall not unreasonably withhold from the Owner or any separate contractor his consent to cutting or otherwise altering the Work.

4.15   CLEANING UP

4.15.1 The Contractor at all times shall keep the premises free from accumulation of waste materials or rubbish caused by his operations. At the completion of the Work he shall remove all his waste materials and rubbish from and about the Project as well as all his tools, construction equipment, machinery and surplus materials.

4.15.2 If the Contractor fails to clean up at the completion of the Work, the Owner may do so as provided in Paragraph 3.4 and the cost thereof shall be charged to the Contractor.

4.16   COMMUNICATIONS

4.16.1 The Contractor shall forward all communications to the Owner through the Architect.

4.17   ROYALTIES AND PATENTS

4.17.1 The Contractor shall pay all royalties and license fees. He shall defend all suits or claims for infringement of any patent rights and shall save the Owner harmless from loss on account thereof, except that the Owner shall be responsible for all such loss when a particular design, process or the product of a particular manufacturer or manufacturers is specified, but if the Contractor has reason to believe that the design, process or product specified is an infringement of a patent, he shall be responsible for such loss unless he promptly gives such information to the Architect.

4.18   INDEMNIFICATION

4.18.1 To the fullest extent permitted by law, the Contractor shall indemnify and hold harmless the Owner and the Architect and their agents and employees from and against all claims, damages, losses and expenses, including but not limited to attorneys' fees, arising out of or resulting from the performance of the Work, provided that any such claim, damage, loss or expense (1) is attributable to bodily injury, sickness, disease or death, or to injury to or destruction of tangible property (other than the Work itself) including the loss of use resulting therefrom, and (2) is caused in whole or in part by any negligent act or omission of the Contractor, any Subcontractor, anyone directly or indirectly employed by any of them or anyone for whose acts any of them may be liable, regardless of whether or not it is caused in part by a party indemnified hereunder. Such obligation shall not be construed to negate, abridge, or otherwise reduce any other right or obligation of indemnity which would otherwise exist as to any party or person described in this Paragraph 4.18.

4.18.2 In any and all claims against the Owner or the Architect or any of their agents or employees by any employee of the Contractor, any Subcontractor, anyone directly or indirectly employed by any of them or anyone for whose acts any of them may be liable, the indemnification obligation under this Paragraph 4.18 shall not be limited in any way by any limitation on the amount or type of damages, compensation or benefits payable by or for the Contractor or any Subcontractor under workers' or workmen's compensation acts, disability benefit acts or other employee benefit acts.

4.18.3 The obligations of the Contractor under this Paragraph 4.18 shall not extend to the liability of the Architect, his agents or employees, arising out of (1) the preparation or approval of maps, drawings, opinions, reports, surveys, change orders, designs or specifications, or (2) the giving of or the failure to give directions or instructions by the Architect, his agents or employees provided such giving or failure to give is the primary cause of the injury or damage.

ARTICLE 5

SUBCONTRACTORS

5.1 DEFINITION

5.1.1 A Subcontractor is a person or entity who has a direct contract with the Contractor to perform any of the Work at the site. The term Subcontractor is referred to throughout the Contract Documents as if singular in number and masculine in gender and means a Subcontractor or his authorized representative. The term Subcontractor does not include any separate contractor or his subcontractors.

5.1.2 A Sub-subcontractor is a person or entity who has a direct or indirect contract with a Subcontractor to perform any of the Work at the site. The term Sub-subcontractor is referred to throughout the Contract Documents as if singular in number and masculine in gender and means a Sub-subcontractor or an authorized representative thereof.

5.2 AWARD OF SUBCONTRACTS AND OTHER CONTRACTS FOR PORTIONS OF THE WORK

5.2.1 Unless otherwise required by the Contract Documents or the Bidding Documents, the Contractor, as soon as practicable after the award of the Contract, shall furnish to the Owner and the Architect in writing the names of the persons or entities (including those who are to furnish materials or equipment fabricated to a special design) proposed for each of the principal portions of the Work. The Architect will promptly reply to the Contractor in writing stating whether or not the Owner or the Architect, after due investigation, has reasonable objection to any

such proposed person or entity. Failure of the Owner or Architect to reply promptly shall constitute notice of no reasonable objection.

5.2.2 The Contractor shall not contract with any such proposed person or entity to whom the Owner or the Architect has made reasonable objection under the provisions of Subparagraph 5.2.1. The Contractor shall not be required to contract with anyone to whom he has a reasonable objection.

5.2.3 If the Owner or the Architect has reasonable objection to any such proposed person or entity, the Contractor shall submit a substitute to whom the Owner or the Architect has no reasonable objection, and the Contract Sum shall be increased or decreased by the difference in cost occasioned by such substitution and an appropriate Change Order shall be issued; however, no increase in the Contract Sum shall be allowed for any such substitution unless the Contractor has acted promptly and responsively in submitting names as required by Subparagraph 5.2.1.

5.2.4 The Contractor shall make no substitution for any Subcontractor, person or entity previously selected if the Owner or Architect makes reasonable objection to such substitution.

5.3 SUBCONTRACTUAL RELATIONS

5.3.1 By an appropriate agreement, written where legally required for validity, the Contractor shall require each Subcontractor, to the extent of the Work to be performed by the Subcontractor, to be bound to the Contractor by the terms of the Contract Documents, and to assume toward the Contractor all the obligations and responsibilities which the Contractor, by these Documents, assumes toward the Owner and the Architect. Said agreement shall preserve and protect the rights of the Owner and the Architect under the Contract Documents with respect to the Work to be performed by the Subcontractor so that the subcontracting thereof will not prejudice such rights, and shall allow to the Subcontractor, unless specifically provided otherwise in the Contractor-Subcontractor agreement, the benefit of all rights, remedies and redress against the Contractor that the Contractor, by these Documents, has against the Owner. Where appropriate, the Contractor shall require each Subcontractor to enter into similar agreements with his Sub-subcontractors. The Contractor shall make available to each proposed Subcontractor, prior to the execution of the Subcontract, copies of the Contract Documents to which the Subcontractor will be bound by this Paragraph 5.3, and identify to the Subcontractor any terms and conditions of the proposed Subcontract which may be at variance with the Contract Documents. Each Subcontractor shall similarly make copies of such Documents available to his Sub-subcontractors.

ARTICLE 6

WORK BY OWNER OR BY SEPARATE CONTRACTORS

6.1 OWNER'S RIGHT TO PERFORM WORK AND TO AWARD SEPARATE CONTRACTS

6.1.1 The Owner reserves the right to perform work related to the Project with his own forces, and to award separate contracts in connection with other portions of the Project or other work on the site under these or similar Conditions of the Contract. If the Contractor claims that delay or additional cost is involved because of such action by the Owner, he shall make such claim as provided elsewhere in the Contract Documents.

6.1.2 When separate contracts are awarded for different portions of the Project or other work on the site, the term Contractor in the Contract Documents in each case shall mean the Contractor who executes each separate Owner-Contractor Agreement.

6.1.3 The Owner will provide for the coordination of the work of his own forces and of each separate contractor with the Work of the Contractor, who shall cooperate therewith as provided in Paragraph 6.2.

6.2 MUTUAL RESPONSIBILITY

6.2.1 The Contractor shall afford the Owner and separate contractors reasonable opportunity for the introduction and storage of their materials and equipment and the execution of their work, and shall connect and coordinate his Work with theirs as required by the Contract Documents.

6.2.2 If any part of the Contractor's Work depends for proper execution or results upon the work of the Owner or any separate contractor, the Contractor shall, prior to proceeding with the Work, promptly report to the Architect any apparent discrepancies or defects in such other work that render it unsuitable for such proper execution and results. Failure of the Contractor so to report shall constitute an acceptance of the Owner's or separate contractors' work as fit and proper to receive his Work, except as to defects which may subsequently become apparent in such work by others.

6.2.3 Any costs caused by defective or ill-timed work shall be borne by the party responsible therefor.

6.2.4 Should the Contractor wrongfully cause damage to the work or property of the Owner, or to other work on the site, the Contractor shall promptly remedy such damage as provided in Subparagraph 10.2.5.

6.2.5 Should the Contractor wrongfully cause damage to the work or property of any separate contractor, the Contractor shall upon due notice promptly attempt to settle with such other contractor by agreement, or otherwise to resolve the dispute. If such separate contractor sues or initiates an arbitration proceeding against the Owner on account of any damage alleged to have been caused by the Contractor, the Owner shall notify the Contractor who shall defend such proceedings at the Owner's expense, and if any judgment or award against the Owner arises therefrom the Contractor shall pay or satisfy it and shall reimburse the Owner for all attorneys' fees and court or arbitration costs which the Owner has incurred.

6.3 OWNER'S RIGHT TO CLEAN UP

6.3.1 If a dispute arises between the Contractor and separate contractors as to their responsibility for cleaning up as required by Paragraph 4.15, the Owner may clean up

and charge the cost thereof to the contractors responsible therefor as the Architect shall determine to be just.

ARTICLE 7

MISCELLANEOUS PROVISIONS

7.1 GOVERNING LAW

7.1.1 The Contract shall be governed by the law of the place where the Project is located.

7.2 SUCCESSORS AND ASSIGNS

7.2.1 The Owner and the Contractor each binds himself, his partners, successors, assigns and legal representatives to the other party hereto and to the partners, successors, assigns and legal representatives of such other party in respect to all covenants, agreements and obligations contained in the Contract Documents. Neither party to the Contract shall assign the Contract or sublet it as a whole without the written consent of the other, nor shall the Contractor assign any moneys due or to become due to him hereunder, without the previous written consent of the Owner.

7.3 WRITTEN NOTICE

7.3.1 Written notice shall be deemed to have been duly served if delivered in person to the individual or member of the firm or entity or to an officer of the corporation for whom it was intended, or if delivered at or sent by registered or certified mail to the last business address known to him who gives the notice.

7.4 CLAIMS FOR DAMAGES

7.4.1 Should either party to the Contract suffer injury or damage to person or property because of any act or omission of the other party or of any of his employees, agents or others for whose acts he is legally liable, claim shall be made in writing to such other party within a reasonable time after the first observance of such injury or damage.

**7.5 PERFORMANCE BOND AND LABOR AND
MATERIAL PAYMENT BOND**

7.5.1 The Owner shall have the right to require the Contractor to furnish bonds covering the faithful performance of the Contract and the payment of all obligations arising thereunder if and as required in the Bidding Documents or in the Contract Documents.

7.6 RIGHTS AND REMEDIES

7.6.1 The duties and obligations imposed by the Contract Documents and the rights and remedies available thereunder shall be in addition to and not a limitation of any duties, obligations, rights and remedies otherwise imposed or available by law.

7.6.2 No action or failure to act by the Owner, Architect or Contractor shall constitute a waiver of any right or duty afforded any of them under the Contract, nor shall any such action or failure to act constitute an approval of or acquiescence in any breach thereunder, except as may be specifically agreed in writing.

7.7 TESTS

7.7.1 If the Contract Documents, laws, ordinances, rules, regulations or orders of any public authority having jurisdiction require any portion of the Work to be inspected, tested or approved, the Contractor shall give the Architect timely notice of its readiness so the Architect may observe such inspection, testing or approval. The Contractor shall bear all costs of such inspections, tests or approvals conducted by public authorities. Unless otherwise provided, the Owner shall bear all costs of other inspections, tests or approvals.

7.7.2 If the Architect determines that any Work requires special inspection, testing, or approval which Subparagraph 7.7.1 does not include, he will, upon written authorization from the Owner, instruct the Contractor to order such special inspection, testing or approval, and the Contractor shall give notice as provided in Subparagraph 7.7.1. If such special inspection or testing reveals a failure of the Work to comply with the requirements of the Contract Documents, the Contractor shall bear all costs thereof, including compensation for the Architect's additional services made necessary by such failure; otherwise the Owner shall bear such costs, and an appropriate Change Order shall be issued.

7.7.3 Required certificates of inspection, testing or approval shall be secured by the Contractor and promptly delivered by him to the Architect.

7.7.4 If the Architect is to observe the inspections, tests or approvals required by the Contract Documents, he will do so promptly and, where practicable, at the source of supply.

7.8 INTEREST

7.8.1 Payments due and unpaid under the Contract Documents shall bear interest from the date payment is due at such rate as the parties may agree upon in writing or, in the absence thereof, at the legal rate prevailing at the place of the Project.

7.9 ARBITRATION

7.9.1 All claims, disputes and other matters in question between the Contractor and the Owner arising out of, or relating to, the Contract Documents or the breach thereof, except as provided in Subparagraph 2.2.11 with respect to the Architect's decisions on matters relating to artistic effect, and except for claims which have been waived by the making or acceptance of final payment as provided by Subparagraphs 9.9.4 and 9.9.5, shall be decided by arbitration in accordance with the Construction Industry Arbitration Rules of the American Arbitration Association then obtaining unless the parties mutually agree otherwise. No arbitration arising out of or relating to the Contract Documents shall include, by consolidation, joinder or in any other manner, the Architect, his employees or consultants except by written consent containing a specific reference to the Owner-Contractor Agreement and signed by the Architect, the Owner, the Contractor and any other person sought to be joined. No arbitration shall include by consolidation, joinder or in any other manner, parties other than the Owner, the Contractor and any other persons substantially involved in a common question of fact or law, whose presence is required if complete relief is to be accorded in the arbitration. No person other than the Owner or Contractor shall be included as an original third party or additional third party to an arbitration whose interest or responsibility is insubstantial. Any consent to arbitration involving an additional person or persons shall not constitute consent to arbitration of any dispute not described therein or with any person not named or described therein. The foregoing agreement to arbitrate and any other agreement to arbitrate with an additional person or persons duly consented to by the parties to the Owner-Contractor Agreement shall be specifically enforceable under the prevailing arbitration law. The award rendered by the arbitrators shall be final, and judgment may be entered upon it in accordance with applicable law in any court having jurisdiction thereof.

7.9.2 Notice of the demand for arbitration shall be filed in writing with the other party to the Owner-Contractor Agreement and with the American Arbitration Association, and a copy shall be filed with the Architect. The demand for arbitration shall be made within the time limits specified in Subparagraph 2.2.12 where applicable, and in all other cases within a reasonable time after the claim, dispute or other matter in question has arisen, and in no event shall it be made after the date when institution of legal or equitable proceedings based on such claim, dispute or other matter in question would be barred by the applicable statute of limitations.

7.9.3 Unless otherwise agreed in writing, the Contractor shall carry on the Work and maintain its progress during any arbitration proceedings, and the Owner shall continue to make payments to the Contractor in accordance with the Contract Documents.

ARTICLE 8

TIME

8.1 DEFINITIONS

8.1.1 Unless otherwise provided, the Contract Time is the period of time allotted in the Contract Documents for Substantial Completion of the Work as defined in Subparagraph 8.1.3, including authorized adjustments thereto.

8.1.2 The date of commencement of the Work is the date established in a notice to proceed. If there is no notice to proceed, it shall be the date of the Owner-Contractor Agreement or such other date as may be established therein.

8.1.3 The Date of Substantial Completion of the Work or designated portion thereof is the Date certified by the Architect when construction is sufficiently complete, in accordance with the Contract Documents, so the Owner can occupy or utilize the Work or designated portion thereof for the use for which it is intended.

8.1.4 The term day as used in the Contract Documents shall mean calendar day unless otherwise specifically designated.

8.2 PROGRESS AND COMPLETION

8.2.1 All time limits stated in the Contract Documents are of the essence of the Contract.

AIA DOCUMENT A201 • GENERAL CONDITIONS OF THE CONTRACT FOR CONSTRUCTION • THIRTEENTH EDITION • AUGUST 1976
AIA® • © 1976 • THE AMERICAN INSTITUTE OF ARCHITECTS, 1735 NEW YORK AVENUE, N.W., WASHINGTON, D.C. 20006

8.2.2 The Contractor shall begin the Work on the date of commencement as defined in Subparagraph 8.1.2. He shall carry the Work forward expeditiously with adequate forces and shall achieve Substantial Completion within the Contract Time.

8.3   DELAYS AND EXTENSIONS OF TIME

8.3.1 If the Contractor is delayed at any time in the progress of the Work by any act or neglect of the Owner or the Architect, or by any employee of either, or by any separate contractor employed by the Owner, or by changes ordered in the Work, or by labor disputes, fire, unusual delay in transportation, adverse weather conditions not reasonably anticipatable, unavoidable casualties, or any causes beyond the Contractor's control, or by delay authorized by the Owner pending arbitration, or by any other cause which the Architect determines may justify the delay, then the Contract Time shall be extended by Change Order for such reasonable time as the Architect may determine.

8.3.2 Any claim for extension of time shall be made in writing to the Architect not more than twenty days after the commencement of the delay; otherwise it shall be waived. In the case of a continuing delay only one claim is necessary. The Contractor shall provide an estimate of the probable effect of such delay on the progress of the Work.

8.3.3 If no agreement is made stating the dates upon which interpretations as provided in Subparagraph 2.2.8 shall be furnished, then no claim for delay shall be allowed on account of failure to furnish such interpretations until fifteen days after written request is made for them, and not then unless such claim is reasonable.

8.3.4 This Paragraph 8.3 does not exclude the recovery of damages for delay by either party under other provisions of the Contract Documents.

ARTICLE 9

PAYMENTS AND COMPLETION

9.1   CONTRACT SUM

9.1.1 The Contract Sum is stated in the Owner-Contractor Agreement and, including authorized adjustments thereto, is the total amount payable by the Owner to the Contractor for the performance of the Work under the Contract Documents.

9.2   SCHEDULE OF VALUES

9.2.1 Before the first Application for Payment, the Contractor shall submit to the Architect a schedule of values allocated to the various portions of the Work, prepared in such form and supported by such data to substantiate its accuracy as the Architect may require. This schedule, unless objected to by the Architect, shall be used only as a basis for the Contractor's Applications for Payment.

9.3   APPLICATIONS FOR PAYMENT

9.3.1 At least ten days before the date for each progress payment established in the Owner-Contractor Agreement, the Contractor shall submit to the Architect an itemized Application for Payment, notarized if required, supported by such data substantiating the Contractor's right to payment as the Owner or the Architect may require, and reflecting retainage, if any, as provided elsewhere in the Contract Documents.

9.3.2 Unless otherwise provided in the Contract Documents, payments will be made on account of materials or equipment not incorporated in the Work but delivered and suitably stored at the site and, if approved in advance by the Owner, payments may similarly be made for materials or equipment suitably stored at some other location agreed upon in writing. Payments for materials or equipment stored on or off the site shall be conditioned upon submission by the Contractor of bills of sale or such other procedures satisfactory to the Owner to establish the Owner's title to such materials or equipment or otherwise protect the Owner's interest, including applicable insurance and transportation to the site for those materials and equipment stored off the site.

9.3.3 The Contractor warrants that title to all Work, materials and equipment covered by an Application for Payment will pass to the Owner either by incorporation in the construction or upon the receipt of payment by the Contractor, whichever occurs first, free and clear of all liens, claims, security interests or encumbrances, hereinafter referred to in this Article 9 as "liens"; and that no Work, materials or equipment covered by an Application for Payment will have been acquired by the Contractor, or by any other person performing Work at the site or furnishing materials and equipment for the Project, subject to an agreement under which an interest therein or an encumbrance thereon is retained by the seller or otherwise imposed by the Contractor or such other person.

9.4   CERTIFICATES FOR PAYMENT

9.4.1 The Architect will, within seven days after the receipt of the Contractor's Application for Payment, either issue a Certificate for Payment to the Owner, with a copy to the Contractor, for such amount as the Architect determines is properly due, or notify the Contractor in writing his reasons for withholding a Certificate as provided in Subparagraph 9.6.1.

9.4.2 The issuance of a Certificate for Payment will constitute a representation by the Architect to the Owner, based on his observations at the site as provided in Subparagraph 2.2.3 and the data comprising the Application for Payment, that the Work has progressed to the point indicated; that, to the best of his knowledge, information and belief, the quality of the Work is in accordance with the Contract Documents (subject to an evaluation of the Work for conformance with the Contract Documents upon Substantial Completion, to the results of any subsequent tests required by or performed under the Contract Documents, to minor deviations from the Contract Documents correctable prior to completion, and to any specific qualifications stated in his Certificate); and that the Contractor is entitled to payment in the amount certified. However, by issuing a Certificate for Payment, the Architect shall not thereby be deemed to represent that he has made exhaustive or continuous on-site inspections to check the quality or quantity of the Work or that he has reviewed the construction means, methods, techniques, sequences or procedures, or that he has made any examination to ascertain how or for what purpose the Contractor has used the moneys previously paid on account of the Contract Sum.

9.5 PROGRESS PAYMENTS

9.5.1 After the Architect has issued a Certificate for Payment, the Owner shall make payment in the manner and within the time provided in the Contract Documents.

9.5.2 The Contractor shall promptly pay each Subcontractor, upon receipt of payment from the Owner, out of the amount paid to the Contractor on account of such Subcontractor's Work, the amount to which said Subcontractor is entitled, reflecting the percentage actually retained, if any, from payments to the Contractor on account of such Subcontractor's Work. The Contractor shall, by an appropriate agreement with each Subcontractor, require each Subcontractor to make payments to his Subsubcontractors in similar manner.

9.5.3 The Architect may, on request and at his discretion, furnish to any Subcontractor, if practicable, information regarding the percentages of completion or the amounts applied for by the Contractor and the action taken thereon by the Architect on account of Work done by such Subcontractor.

9.5.4 Neither the Owner nor the Architect shall have any obligation to pay or to see to the payment of any moneys to any Subcontractor except as may otherwise be required by law.

9.5.5 No Certificate for a progress payment, nor any progress payment, nor any partial or entire use or occupancy of the Project by the Owner, shall constitute an acceptance of any Work not in accordance with the Contract Documents.

9.6 PAYMENTS WITHHELD

9.6.1 The Architect may decline to certify payment and may withhold his Certificate in whole or in part, to the extent necessary reasonably to protect the Owner, if in his opinion he is unable to make representations to the Owner as provided in Subparagraph 9.4.2. If the Architect is unable to make representations to the Owner as provided in Subparagraph 9.4.2 and to certify payment in the amount of the Application, he will notify the Contractor as provided in Subparagraph 9.4.1. If the Contractor and the Architect cannot agree on a revised amount, the Architect will promptly issue a Certificate for Payment for the amount for which he is able to make such representations to the Owner. The Architect may also decline to certify payment or, because of subsequently discovered evidence or subsequent observations, he may nullify the whole or any part of any Certificate for Payment previously issued, to such extent as may be necessary in his opinion to protect the Owner from loss because of:

.1 defective work not remedied,

.2 third party claims filed or reasonable evidence indicating probable filing of such claims,

.3 failure of the Contractor to make payments properly to Subcontractors or for labor, materials or equipment,

.4 reasonable evidence that the Work cannot be completed for the unpaid balance of the Contract Sum,

.5 damage to the Owner or another contractor,

.6 reasonable evidence that the Work will not be completed within the Contract Time, or

.7 persistent failure to carry out the Work in accordance with the Contract Documents.

9.6.2 When the above grounds in Subparagraph 9.6.1 are removed, payment shall be made for amounts withheld because of them.

9.7 FAILURE OF PAYMENT

9.7.1 If the Architect does not issue a Certificate for Payment, through no fault of the Contractor, within seven days after receipt of the Contractor's Application for Payment, or if the Owner does not pay the Contractor within seven days after the date established in the Contract Documents any amount certified by the Architect or awarded by arbitration, then the Contractor may, upon seven additional days' written notice to the Owner and the Architect, stop the Work until payment of the amount owing has been received. The Contract Sum shall be increased by the amount of the Contractor's reasonable costs of shut-down, delay and start-up, which shall be effected by appropriate Change Order in accordance with Paragraph 12.3.

9.8 SUBSTANTIAL COMPLETION

9.8.1 When the Contractor considers that the Work, or a designated portion thereof which is acceptable to the Owner, is substantially complete as defined in Subparagraph 8.1.3, the Contractor shall prepare for submission to the Architect a list of items to be completed or corrected. The failure to include any items on such list does not alter the responsibility of the Contractor to complete all Work in accordance with the Contract Documents. When the Architect on the basis of an inspection determines that the Work or designated portion thereof is substantially complete, he will then prepare a Certificate of Substantial Completion which shall establish the Date of Substantial Completion, shall state the responsibilities of the Owner and the Contractor for security, maintenance, heat, utilities, damage to the Work, and insurance, and shall fix the time within which the Contractor shall complete the items listed therein. Warranties required by the Contract Documents shall commence on the Date of Substantial Completion of the Work or designated portion thereof unless otherwise provided in the Certificate of Substantial Completion. The Certificate of Substantial Completion shall be submitted to the Owner and the Contractor for their written acceptance of the responsibilities assigned to them in such Certificate.

9.8.2 Upon Substantial Completion of the Work or designated portion thereof and upon application by the Contractor and certification by the Architect, the Owner shall make payment, reflecting adjustment in retainage, if any, for such Work or portion thereof, as provided in the Contract Documents.

9.9 FINAL COMPLETION AND FINAL PAYMENT

9.9.1 Upon receipt of written notice that the Work is ready for final inspection and acceptance and upon receipt of a final Application for Payment, the Architect will

AIA DOCUMENT A201 • GENERAL CONDITIONS OF THE CONTRACT FOR CONSTRUCTION • THIRTEENTH EDITION • AUGUST 1976
AIA® • © 1976 • THE AMERICAN INSTITUTE OF ARCHITECTS, 1735 NEW YORK AVENUE, N.W., WASHINGTON, D.C. 20006

promptly make such inspection and, when he finds the Work acceptable under the Contract Documents and the Contract fully performed, he will promptly issue a final Certificate for Payment stating that to the best of his knowledge, information and belief, and on the basis of his observations and inspections, the Work has been completed in accordance with the terms and conditions of the Contract Documents and that the entire balance found to be due the Contractor, and noted in said final Certificate, is due and payable. The Architect's final Certificate for Payment will constitute a further representation that the conditions precedent to the Contractor's being entitled to final payment as set forth in Subparagraph 9.9.2 have been fulfilled.

9.9.2 Neither the final payment nor the remaining retained percentage shall become due until the Contractor submits to the Architect (1) an affidavit that all payrolls, bills for materials and equipment, and other indebtedness connected with the Work for which the Owner or his property might in any way be responsible, have been paid or otherwise satisfied, (2) consent of surety, if any, to final payment and (3), if required by the Owner, other data establishing payment or satisfaction of all such obligations, such as receipts, releases and waivers of liens arising out of the Contract, to the extent and in such form as may be designated by the Owner. If any Subcontractor refuses to furnish a release or waiver required by the Owner, the Contractor may furnish a bond satisfactory to the Owner to indemnify him against any such lien. If any such lien remains unsatisfied after all payments are made, the Contractor shall refund to the Owner all moneys that the latter may be compelled to pay in discharging such lien, including all costs and reasonable attorneys' fees.

9.9.3 If, after Substantial Completion of the Work, final completion thereof is materially delayed through no fault of the Contractor or by the issuance of Change Orders affecting final completion, and the Architect so confirms, the Owner shall, upon application by the Contractor and certification by the Architect, and without terminating the Contract, make payment of the balance due for that portion of the Work fully completed and accepted. If the remaining balance for Work not fully completed or corrected is less than the retainage stipulated in the Contract Documents, and if bonds have been furnished as provided in Paragraph 7.5, the written consent of the surety to the payment of the balance due for that portion of the Work fully completed and accepted shall be submitted by the Contractor to the Architect prior to certification of such payment. Such payment shall be made under the terms and conditions governing final payment, except that it shall not constitute a waiver of claims.

9.9.4 The making of final payment shall constitute a waiver of all claims by the Owner except those arising from:
  .1 unsettled liens,
  .2 faulty or defective Work appearing after Substantial Completion,
  .3 failure of the Work to comply with the requirements of the Contract Documents, or
  .4 terms of any special warranties required by the Contract Documents.

9.9.5 The acceptance of final payment shall constitute a waiver of all claims by the Contractor except those previously made in writing and identified by the Contractor as unsettled at the time of the final Application for Payment.

ARTICLE 10
PROTECTION OF PERSONS AND PROPERTY

10.1 SAFETY PRECAUTIONS AND PROGRAMS

10.1.1 The Contractor shall be responsible for initiating, maintaining and supervising all safety precautions and programs in connection with the Work.

10.2 SAFETY OF PERSONS AND PROPERTY

10.2.1 The Contractor shall take all reasonable precautions for the safety of, and shall provide all reasonable protection to prevent damage, injury or loss to:
  .1 all employees on the Work and all other persons who may be affected thereby;
  .2 all the Work and all materials and equipment to be incorporated therein, whether in storage on or off the site, under the care, custody or control of the Contractor or any of his Subcontractors or Sub-subcontractors; and
  .3 other property at the site or adjacent thereto, including trees, shrubs, lawns, walks, pavements, roadways, structures and utilities not designated for removal, relocation or replacement in the course of construction.

10.2.2 The Contractor shall give all notices and comply with all applicable laws, ordinances, rules, regulations and lawful orders of any public authority bearing on the safety of persons or property or their protection from damage, injury or loss.

10.2.3 The Contractor shall erect and maintain, as required by existing conditions and progress of the Work, all reasonable safeguards for safety and protection, including posting danger signs and other warnings against hazards, promulgating safety regulations and notifying owners and users of adjacent utilities.

10.2.4 When the use or storage of explosives or other hazardous materials or equipment is necessary for the execution of the Work, the Contractor shall exercise the utmost care and shall carry on such activities under the supervision of properly qualified personnel.

10.2.5 The Contractor shall promptly remedy all damage or loss (other than damage or loss insured under Paragraph 11.3) to any property referred to in Clauses 10.2.1.2 and 10.2.1.3 caused in whole or in part by the Contractor, any Subcontractor, any Sub-subcontractor, or anyone directly or indirectly employed by any of them, or by anyone for whose acts any of them may be liable and for which the Contractor is responsible under Clauses 10.2.1.2 and 10.2.1.3, except damage or loss attributable to the acts or omissions of the Owner or Architect or anyone directly or indirectly employed by either of them, or by anyone for whose acts either of them may be liable, and not attributable to the fault or negligence of the Contractor. The foregoing obligations of the Contractor are in addition to his obligations under Paragraph 4.18.

10.2.6 The Contractor shall designate a responsible member of his organization at the site whose duty shall be the prevention of accidents. This person shall be the Contractor's superintendent unless otherwise designated by the Contractor in writing to the Owner and the Architect.

10.2.7 The Contractor shall not load or permit any part of the Work to be loaded so as to endanger its safety.

10.3 EMERGENCIES

10.3.1 In any emergency affecting the safety of persons or property, the Contractor shall act, at his discretion, to prevent threatened damage, injury or loss. Any additional compensation or extension of time claimed by the Contractor on account of emergency work shall be determined as provided in Article 12 for Changes in the Work.

ARTICLE 11

INSURANCE

11.1 CONTRACTOR'S LIABILITY INSURANCE

11.1.1 The Contractor shall purchase and maintain such insurance as will protect him from claims set forth below which may arise out of or result from the Contractor's operations under the Contract, whether such operations be by himself or by any Subcontractor or by anyone directly or indirectly employed by any of them, or by anyone for whose acts any of them may be liable:

.1 claims under workers' or workmen's compensation, disability benefit and other similar employee benefit acts;

.2 claims for damages because of bodily injury, occupational sickness or disease, or death of his employees;

.3 claims for damages because of bodily injury, sickness or disease, or death of any person other than his employees;

.4 claims for damages insured by usual personal injury liability coverage which are sustained (1) by any person as a result of an offense directly or indirectly related to the employment of such person by the Contractor, or (2) by any other person;

.5 claims for damages, other than to the Work itself, because of injury to or destruction of tangible property, including loss of use resulting therefrom; and

.6 claims for damages because of bodily injury or death of any person or property damage arising out of the ownership, maintenance or use of any motor vehicle.

11.1.2 The insurance required by Subparagraph 11.1.1 shall be written for not less than any limits of liability specified in the Contract Documents, or required by law, whichever is greater.

11.1.3 The insurance required by Subparagraph 11.1.1 shall include contractual liability insurance applicable to the Contractor's obligations under Paragraph 4.18.

11.1.4 Certificates of Insurance acceptable to the Owner shall be filed with the Owner prior to commencement of the Work. These Certificates shall contain a provision that coverages afforded under the policies will not be cancelled until at least thirty days' prior written notice has been given to the Owner.

11.2 OWNER'S LIABILITY INSURANCE

11.2.1 The Owner shall be responsible for purchasing and maintaining his own liability insurance and, at his option, may purchase and maintain such insurance as will protect him against claims which may arise from operations under the Contract.

11.3 PROPERTY INSURANCE

11.3.1 Unless otherwise provided, the Owner shall purchase and maintain property insurance upon the entire Work at the site to the full insurable value thereof. This insurance shall include the interests of the Owner, the Contractor, Subcontractors and Sub-subcontractors in the Work and shall insure against the perils of fire and extended coverage and shall include "all risk" insurance for physical loss or damage including, without duplication of coverage, theft, vandalism and malicious mischief. If the Owner does not intend to purchase such insurance for the full insurable value of the entire Work, he shall inform the Contractor in writing prior to commencement of the Work. The Contractor may then effect insurance which will protect the interests of himself, his Subcontractors and the Sub-subcontractors in the Work, and by appropriate Change Order the cost thereof shall be charged to the Owner. If the Contractor is damaged by failure of the Owner to purchase or maintain such insurance and to so notify the Contractor, then the Owner shall bear all reasonable costs properly attributable thereto. If not covered under the all risk insurance or otherwise provided in the Contract Documents, the Contractor shall effect and maintain similar property insurance on portions of the Work stored off the site or in transit when such portions of the Work are to be included in an Application for Payment under Subparagraph 9.3.2.

11.3.2 The Owner shall purchase and maintain such boiler and machinery insurance as may be required by the Contract Documents or by law. This insurance shall include the interests of the Owner, the Contractor, Subcontractors and Sub-subcontractors in the Work.

11.3.3 Any loss insured under Subparagraph 11.3.1 is to be adjusted with the Owner and made payable to the Owner as trustee for the insureds, as their interests may appear, subject to the requirements of any applicable mortgagee clause and of Subparagraph 11.3.8. The Contractor shall pay each Subcontractor a just share of any insurance moneys received by the Contractor, and by appropriate agreement, written where legally required for validity, shall require each Subcontractor to make payments to his Sub-subcontractors in similar manner.

11.3.4 The Owner shall file a copy of all policies with the Contractor before an exposure to loss may occur.

11.3.5 If the Contractor requests in writing that insurance for risks other than those described in Subparagraphs 11.3.1 and 11.3.2 or other special hazards be included in the property insurance policy, the Owner shall, if possible, include such insurance, and the cost thereof shall be charged to the Contractor by appropriate Change Order.

11.3.6 The Owner and Contractor waive all rights against (1) each other and the Subcontractors, Sub-subcontractors, agents and employees each of the other, and (2) the Architect and separate contractors, if any, and their subcontractors, sub-subcontractors, agents and employees, for damages caused by fire or other perils to the extent covered by insurance obtained pursuant to this Paragraph 11.3 or any other property insurance applicable to the Work, except such rights as they may have to the proceeds of such insurance held by the Owner as trustee. The foregoing waiver afforded the Architect, his agents and employees shall not extend to the liability imposed by Subparagraph 4.18.3. The Owner or the Contractor, as appropriate, shall require of the Architect, separate contractors, Subcontractors and Sub-subcontractors by appropriate agreements, written where legally required for validity, similar waivers each in favor of all other parties enumerated in this Subparagraph 11.3.6.

11.3.7 If required in writing by any party in interest, the Owner as trustee shall, upon the occurrence of an insured loss, give bond for the proper performance of his duties. He shall deposit in a separate account any money so received, and he shall distribute it in accordance with such agreement as the parties in interest may reach, or in accordance with an award by arbitration in which case the procedure shall be as provided in Paragraph 7.9. If after such loss no other special agreement is made, replacement of damaged work shall be covered by an appropriate Change Order.

11.3.8 The Owner as trustee shall have power to adjust and settle any loss with the insurers unless one of the parties in interest shall object in writing within five days after the occurrence of loss to the Owner's exercise of this power, and if such objection be made, arbitrators shall be chosen as provided in Paragraph 7.9. The Owner as trustee shall, in that case, make settlement with the insurers in accordance with the directions of such arbitrators. If distribution of the insurance proceeds by arbitration is required, the arbitrators will direct such distribution.

11.3.9 If the Owner finds it necessary to occupy or use a portion or portions of the Work prior to Substantial Completion thereof, such occupancy or use shall not commence prior to a time mutually agreed to by the Owner and Contractor and to which the insurance company or companies providing the property insurance have consented by endorsement to the policy or policies. This insurance shall not be cancelled or lapsed on account of such partial occupancy or use. Consent of the Contractor and of the insurance company or companies to such occupancy or use shall not be unreasonably withheld.

11.4   LOSS OF USE INSURANCE

11.4.1 The Owner, at his option, may purchase and maintain such insurance as will insure him against loss of use of his property due to fire or other hazards, however caused. The Owner waives all rights of action against the Contractor for loss of use of his property, including consequential losses due to fire or other hazards however caused, to the extent covered by insurance under this Paragraph 11.4.

ARTICLE 12

CHANGES IN THE WORK

12.1   CHANGE ORDERS

12.1.1 A Change Order is a written order to the Contractor signed by the Owner and the Architect, issued after execution of the Contract, authorizing a change in the Work or an adjustment in the Contract Sum or the Contract Time. The Contract Sum and the Contract Time may be changed only by Change Order. A Change Order signed by the Contractor indicates his agreement therewith, including the adjustment in the Contract Sum or the Contract Time.

12.1.2 The Owner, without invalidating the Contract, may order changes in the Work within the general scope of the Contract consisting of additions, deletions or other revisions, the Contract Sum and the Contract Time being adjusted accordingly. All such changes in the Work shall be authorized by Change Order, and shall be performed under the applicable conditions of the Contract Documents.

12.1.3 The cost or credit to the Owner resulting from a change in the Work shall be determined in one or more of the following ways:

   .1 by mutual acceptance of a lump sum properly itemized and supported by sufficient substantiating data to permit evaluation;

   .2 by unit prices stated in the Contract Documents or subsequently agreed upon;

   .3 by cost to be determined in a manner agreed upon by the parties and a mutually acceptable fixed or percentage fee; or

   .4 by the method provided in Subparagraph 12.1.4.

12.1.4 If none of the methods set forth in Clauses 12.1.3.1, 12.1.3.2 or 12.1.3.3 is agreed upon, the Contractor, provided he receives a written order signed by the Owner, shall promptly proceed with the Work involved. The cost of such Work shall then be determined by the Architect on the basis of the reasonable expenditures and savings of those performing the Work attributable to the change, including, in the case of an increase in the Contract Sum, a reasonable allowance for overhead and profit. In such case, and also under Clauses 12.1.3.3 and 12.1.3.4 above, the Contractor shall keep and present, in such form as the Architect may prescribe, an itemized accounting together with appropriate supporting data for inclusion in a Change Order. Unless otherwise provided in the Contract Documents, cost shall be limited to the following: cost of materials, including sales tax and cost of delivery; cost of labor, including social security, old age and unemployment insurance, and fringe benefits required by agreement or custom; workers' or workmen's compensation insurance; bond premiums; rental value of equipment and machinery; and the additional costs of supervision and field office personnel directly attributable to the change. Pending final determination of cost to the Owner, payments on account shall be made on the Architect's Certificate for Payment. The amount of credit to be allowed by the Contractor to the Owner for any deletion or change which results in a net decrease in the Contract Sum will be the amount of the actual net cost as confirmed by the Architect. When both additions and credits covering related Work or substitutions are involved in any one change, the allowance for overhead and profit shall be figured on the basis of the net increase, if any, with respect to that change.

12.1.5 If unit prices are stated in the Contract Documents or subsequently agreed upon, and if the quantities originally contemplated are so changed in a proposed Change Order that application of the agreed unit prices to the quantities of Work proposed will cause substantial inequity to the Owner or the Contractor, the applicable unit prices shall be equitably adjusted.

12.2   CONCEALED CONDITIONS

12.2.1 Should concealed conditions encountered in the performance of the Work below the surface of the ground or should concealed or unknown conditions in an existing structure be at variance with the conditions indicated by the Contract Documents, or should unknown physical conditions below the surface of the ground or should concealed or unknown conditions in an existing structure of an unusual nature, differing materially from those ordinarily encountered and generally recognized as inherent in work of the character provided for in this Contract, be encountered, the Contract Sum shall be equitably adjusted by Change Order upon claim by either party made within twenty days after the first observance of the conditions.

12.3   CLAIMS FOR ADDITIONAL COST

12.3.1 If the Contractor wishes to make a claim for an increase in the Contract Sum, he shall give the Architect written notice thereof within twenty days after the occurrence of the event giving rise to such claim. This notice shall be given by the Contractor before proceeding to execute the Work, except in an emergency endangering life or property in which case the Contractor shall proceed in accordance with Paragraph 10.3. No such claim shall be valid unless so made. If the Owner and the Contractor cannot agree on the amount of the adjustment in the Contract Sum, it shall be determined by the Architect. Any change in the Contract Sum resulting from such claim shall be authorized by Change Order.

12.3.2 If the Contractor claims that additional cost is involved because of, but not limited to, (1) any written interpretation pursuant to Subparagraph 2.2.8, (2) any order by the Owner to stop the Work pursuant to Paragraph 3.3 where the Contractor was not at fault, (3) any written order for a minor change in the Work issued pursuant to Paragraph 12.4, or (4) failure of payment by the Owner pursuant to Paragraph 9.7, the Contractor shall make such claim as provided in Subparagraph 12.3.1.

12.4   MINOR CHANGES IN THE WORK

12.4.1 The Architect will have authority to order minor changes in the Work not involving an adjustment in the Contract Sum or an extension of the Contract Time and not inconsistent with the intent of the Contract Documents. Such changes shall be effected by written order, and shall be binding on the Owner and the Contractor.

The Contractor shall carry out such written orders promptly.

ARTICLE 13

UNCOVERING AND CORRECTION OF WORK

13.1   UNCOVERING OF WORK

13.1.1 If any portion of the Work should be covered contrary to the request of the Architect or to requirements specifically expressed in the Contract Documents, it must, if required in writing by the Architect, be uncovered for his observation and shall be replaced at the Contractor's expense.

13.1.2 If any other portion of the Work has been covered which the Architect has not specifically requested to observe prior to being covered, the Architect may request to see such Work and it shall be uncovered by the Contractor. If such Work be found in accordance with the Contract Documents, the cost of uncovering and replacement shall, by appropriate Change Order, be charged to the Owner. If such Work be found not in accordance with the Contract Documents, the Contractor shall pay such costs unless it be found that this condition was caused by the Owner or a separate contractor as provided in Article 6, in which event the Owner shall be responsible for the payment of such costs.

13.2   CORRECTION OF WORK

13.2.1 The Contractor shall promptly correct all Work rejected by the Architect as defective or as failing to conform to the Contract Documents whether observed before or after Substantial Completion and whether or not fabricated, installed or completed. The Contractor shall bear all costs of correcting such rejected Work, including compensation for the Architect's additional services made necessary thereby.

13.2.2 If, within one year after the Date of Substantial Completion of the Work or designated portion thereof or within one year after acceptance by the Owner of designated equipment or within such longer period of time as may be prescribed by law or by the terms of any applicable special warranty required by the Contract Documents, any of the Work is found to be defective or not in accordance with the Contract Documents, the Contractor shall correct it promptly after receipt of a written notice from the Owner to do so unless the Owner has previously given the Contractor a written acceptance of such condition. This obligation shall survive termination of the Contract. The Owner shall give such notice promptly after discovery of the condition.

13.2.3 The Contractor shall remove from the site all portions of the Work which are defective or non-conforming and which have not been corrected under Subparagraphs 4.5.1, 13.2.1 and 13.2.2, unless removal is waived by the Owner.

13.2.4 If the Contractor fails to correct defective or non-conforming Work as provided in Subparagraphs 4.5.1, 13.2.1 and 13.2.2, the Owner may correct it in accordance with Paragraph 3.4.

AIA DOCUMENT A201 • GENERAL CONDITIONS OF THE CONTRACT FOR CONSTRUCTION • THIRTEENTH EDITION • AUGUST 1976
18   A201-1976   AIA® • © 1976 • THE AMERICAN INSTITUTE OF ARCHITECTS, 1735 NEW YORK AVENUE, N.W., WASHINGTON, D.C. 20006

13.2.5 If the Contractor does not proceed with the correction of such defective or non-conforming Work within a reasonable time fixed by written notice from the Architect, the Owner may remove it and may store the materials or equipment at the expense of the Contractor. If the Contractor does not pay the cost of such removal and storage within ten days thereafter, the Owner may upon ten additional days' written notice sell such Work at auction or at private sale and shall account for the net proceeds thereof, after deducting all the costs that should have been borne by the Contractor, including compensation for the Architect's additional services made necessary thereby. If such proceeds of sale do not cover all costs which the Contractor should have borne, the difference shall be charged to the Contractor and an appropriate Change Order shall be issued. If the payments then or thereafter due the Contractor are not sufficient to cover such amount, the Contractor shall pay the difference to the Owner.

13.2.6 The Contractor shall bear the cost of making good all work of the Owner or separate contractors destroyed or damaged by such correction or removal.

13.2.7 Nothing contained in this Paragraph 13.2 shall be construed to establish a period of limitation with respect to any other obligation which the Contractor might have under the Contract Documents, including Paragraph 4.5 hereof. The establishment of the time period of one year after the Date of Substantial Completion or such longer period of time as may be prescribed by law or by the terms of any warranty required by the Contract Documents relates only to the specific obligation of the Contractor to correct the Work, and has no relationship to the time within which his obligation to comply with the Contract Documents may be sought to be enforced, nor to the time within which proceedings may be commenced to establish the Contractor's liability with respect to his obligations other than specifically to correct the Work.

13.3 ACCEPTANCE OF DEFECTIVE OR NON-CONFORMING WORK

13.3.1 If the Owner prefers to accept defective or non-conforming Work, he may do so instead of requiring its removal and correction, in which case a Change Order will be issued to reflect a reduction in the Contract Sum where appropriate and equitable. Such adjustment shall be effected whether or not final payment has been made.

ARTICLE 14

TERMINATION OF THE CONTRACT

14.1 TERMINATION BY THE CONTRACTOR

14.1.1 If the Work is stopped for a period of thirty days under an order of any court or other public authority having jurisdiction, or as a result of an act of government, such as a declaration of a national emergency making materials unavailable, through no act or fault of the Contractor or a Subcontractor or their agents or employees or any other persons performing any of the Work under a contract with the Contractor, or if the Work should be stopped for a period of thirty days by the Contractor because the Architect has not issued a Certificate for Payment as provided in Paragraph 9.7 or because the Owner has not made payment thereon as provided in Paragraph 9.7, then the Contractor may, upon seven additional days' written notice to the Owner and the Architect, terminate the Contract and recover from the Owner payment for all Work executed and for any proven loss sustained upon any materials, equipment, tools, construction equipment and machinery, including reasonable profit and damages.

14.2 TERMINATION BY THE OWNER

14.2.1 If the Contractor is adjudged a bankrupt, or if he makes a general assignment for the benefit of his creditors, or if a receiver is appointed on account of his insolvency, or if he persistently or repeatedly refuses or fails, except in cases for which extension of time is provided, to supply enough properly skilled workmen or proper materials, or if he fails to make prompt payment to Subcontractors or for materials or labor, or persistently disregards laws, ordinances, rules, regulations or orders of any public authority having jurisdiction, or otherwise is guilty of a substantial violation of a provision of the Contract Documents, then the Owner, upon certification by the Architect that sufficient cause exists to justify such action, may, without prejudice to any right or remedy and after giving the Contractor and his surety, if any, seven days' written notice, terminate the employment of the Contractor and take possession of the site and of all materials, equipment, tools, construction equipment and machinery thereon owned by the Contractor and may finish the Work by whatever method he may deem expedient. In such case the Contractor shall not be entitled to receive any further payment until the Work is finished.

14.2.2 If the unpaid balance of the Contract Sum exceeds the costs of finishing the Work, including compensation for the Architect's additional services made necessary thereby, such excess shall be paid to the Contractor. If such costs exceed the unpaid balance, the Contractor shall pay the difference to the Owner. The amount to be paid to the Contractor or to the Owner, as the case may be, shall be certified by the Architect, upon application, in the manner provided in Paragraph 9.4, and this obligation for payment shall survive the termination of the Contract.

AIA DOCUMENT A201 • GENERAL CONDITIONS OF THE CONTRACT FOR CONSTRUCTION • THIRTEENTH EDITION • AUGUST 1976
AIA® • © 1976 • THE AMERICAN INSTITUTE OF ARCHITECTS, 1735 NEW YORK AVENUE, N.W., WASHINGTON, D.C. 20006 **A201-1976** **19**

THE AMERICAN INSTITUTE OF ARCHITECTS

· *AIA Document A101*

Standard Form of Agreement Between Owner and Contractor

where the basis of payment is a

STIPULATED SUM

1977 EDITION

THIS DOCUMENT HAS IMPORTANT LEGAL CONSEQUENCES; CONSULTATION WITH AN ATTORNEY IS ENCOURAGED WITH RESPECT TO ITS COMPLETION OR MODIFICATION

Use only with the 1976 Edition of AIA Document A201, General Conditions of the Contract for Construction.

This document has been approved and endorsed by The Associated General Contractors of America.

AGREEMENT

made as of the   day of   in the year of Nineteen
Hundred and

BETWEEN the Owner:

and the Contractor:

The Project:

The Architect:

The Owner and the Contractor agree as set forth below.

Copyright 1915, 1918, 1925, 1937, 1951, 1958, 1961, 1963, 1967, 1974, © 1977 by the American Institute of Architects, 1735 New York Avenue, N.W., Washington, D. C. 20006. Reproduction of the material herein or substantial quotation of its provisions without permission of the AIA violates the copyright laws of the United States and will be subject to legal prosecution.

ARTICLE 1

THE CONTRACT DOCUMENTS

The Contract Documents consist of this Agreement, the Conditions of the Contract (General, Supplementary and other Conditions), the Drawings, the Specifications, all Addenda issued prior to and all Modifications issued after execution of this Agreement. These form the Contract, and all are as fully a part of the Contract as if attached to this Agreement or repeated herein. An enumeration of the Contract Documents appears in Article 7.

ARTICLE 2

THE WORK

The Contractor shall perform all the Work required by the Contract Documents for
(Here insert the caption descriptive of the Work as used on other Contract Documents.)

ARTICLE 3

TIME OF COMMENCEMENT AND SUBSTANTIAL COMPLETION

The Work to be performed under this Contract shall be commenced

and, subject to authorized adjustments, Substantial Completion shall be achieved not later than

(Here insert any special provisions for liquidated damages relating to failure to complete on time.)

ARTICLE 4

CONTRACT SUM

The Owner shall pay the Contractor in current funds for the performance of the Work, subject to additions and deductions by Change Order as provided in the Contract Documents, the Contract Sum of

The Contract Sum is determined as follows:
(State here the base bid or other lump sum amount, accepted alternates, and unit prices, as applicable.)

ARTICLE 5

PROGRESS PAYMENTS

Based upon Applications for Payment submitted to the Architect by the Contractor and Certificates for Payment issued by the Architect, the Owner shall make progress payments on account of the Contract Sum to the Contractor as provided in the Contract Documents for the period ending the   day of the month as follows:

Not later than   days following the end of the period covered by the Application for Payment   percent (%) of the portion of the Contract Sum properly allocable to labor, materials and equipment incorporated in the Work and   percent (%) of the portion of the Contract Sum properly allocable to materials and equipment suitably stored at the site or at some other location agreed upon in writing, for the period covered by the Application for Payment, less the aggregate of previous payments made by the Owner; and upon Substantial Completion of the entire Work, a sum sufficient to increase the total payments to   percent (%) of the Contract Sum, less such amounts as the Architect shall determine for all incomplete Work and unsettled claims as provided in the Contract Documents.

(If not covered elsewhere in the Contract Documents, here insert any provision for limiting or reducing the amount retained after the Work reaches a certain stage of completion.)

Payments due and unpaid under the Contract Documents shall bear interest from the date payment is due at the rate entered below, or in the absence thereof, at the legal rate prevailing at the place of the Project.
(Here insert any rate of interest agreed upon.)

(Usury laws and requirements under the Federal Truth in Lending Act, similar state and local consumer credit laws and other regulations at the Owner's and Contractor's principal places of business, the location of the Project and elsewhere may affect the validity of this provision. Specific legal advice should be obtained with respect to deletion, modification, or other requirements such as written disclosures or waivers.)

AIA DOCUMENT A101 • OWNER-CONTRACTOR AGREEMENT • ELEVENTH EDITION • JUNE 1977 • AIA®
©1977 • THE AMERICAN INSTITUTE OF ARCHITECTS, 1735 NEW YORK AVE., N.W., WASHINGTON, D. C. 20006   **A101-1977   3**

ARTICLE 6

FINAL PAYMENT

Final payment, constituting the entire unpaid balance of the Contract Sum, shall be paid by the Owner to the Contractor when the Work has been completed, the Contract fully performed, and a final Certificate for Payment has been issued by the Architect.

ARTICLE 7

MISCELLANEOUS PROVISIONS

7.1   Terms used in this Agreement which are defined in the Conditions of the Contract shall have the meanings designated in those Conditions.

7.2   The Contract Documents, which constitute the entire agreement between the Owner and the Contractor, are listed in Article 1 and, except for Modifications issued after execution of this Agreement, are enumerated as follows:

(List below the Agreement, the Conditions of the Contract (General, Supplementary, and other Conditions), the Drawings, the Specifications, and any Addenda and accepted alternates, showing page or sheet numbers in all cases and dates where applicable.)

This Agreement entered into as of the day and year first written above.

OWNER   CONTRACTOR

_____   _____

_____   _____

_____   _____

THE AMERICAN INSTITUTE OF ARCHITECTS

AIA Document A310

Bid Bond

KNOW ALL MEN BY THESE PRESENTS, that we
(Here insert full name and address or legal title of Contractor)

as Principal, hereinafter called the Principal, and

(Here insert full name and address or legal title of Surety)

a corporation duly organized under the laws of the State of
as Surety, hereinafter called the Surety, are held and firmly bound unto

(Here insert full name and address or legal title of Owner)

as Obligee, hereinafter called the Obligee, in the sum of

 Dollars ($),

for the payment of which sum well and truly to be made, the said Principal and the said Surety, bind ourselves, our heirs, executors, administrators, successors and assigns, jointly and severally, firmly by these presents.

WHEREAS, the Principal has submitted a bid for

(Here insert full name, address and description of project)

NOW, THEREFORE, if the Obligee shall accept the bid of the Principal and the Principal shall enter into a Contract with the Obligee in accordance with the terms of such bid, and give such bond or bonds as may be specified in the bidding or Contract Documents with good and sufficient surety for the faithful performance of such Contract and for the prompt payment of labor and material furnished in the prosecution thereof, or in the event of the failure of the Principal to enter such Contract and give such bond or bonds, if the Principal shall pay to the Obligee the difference not to exceed the penalty hereof between the amount specified in said bid and such larger amount for which the Obligee may in good faith contract with another party to perform the Work covered by said bid, then this obligation shall be null and void, otherwise to remain in full force and effect.

Signed and sealed this   day of   19

(Witness)		(Principal)	(Seal)
		(Title)	
(Witness)		(Surety)	(Seal)
		(Title)	

AIA DOCUMENT A310 • BID BOND • AIA ® • FEBRUARY 1970 ED • THE AMERICAN
INSTITUTE OF ARCHITECTS, 1735 N.Y. AVE., N.W., WASHINGTON, D. C. 20006

1

THE AMERICAN INSTITUTE OF ARCHITECTS

AIA Document A311

Performance Bond

KNOW ALL MEN BY THESE PRESENTS: that

(Here insert full name and address or legal title of Contractor)

as Principal, hereinafter called Contractor, and,

(Here insert full name and address or legal title of Surety)

as Surety, hereinafter called Surety, are held and firmly bound unto

(Here insert full name and address or legal title of Owner)

as Obligee, hereinafter called Owner, in the amount of

Dollars ($),

for the payment whereof Contractor and Surety bind themselves, their heirs, executors, administrators, successors and assigns, jointly and severally, firmly by these presents.

WHEREAS,

Contractor has by written agreement dated   19   , entered into a contract with Owner for
(Here insert full name, address and description of project)

in accordance with Drawings and Specifications prepared by

(Here insert full name and address or legal title of Architect)

which contract is by reference made a part hereof, and is hereinafter referred to as the Contract.

PERFORMANCE BOND

NOW, THEREFORE, THE CONDITION OF THIS OBLIGATION is such that, if Contractor shall promptly and faithfully perform said Contract, then this obligation shall be null and void; otherwise it shall remain in full force and effect.

The Surety hereby waives notice of any alteration or extension of time made by the Owner.

Whenever Contractor shall be, and declared by Owner to be in default under the Contract, the Owner having performed Owner's obligations thereunder, the Surety may promptly remedy the default, or shall promptly

1) Complete the Contract in accordance with its terms and conditions, or

2) Obtain a bid or bids for completing the Contract in accordance with its terms and conditions, and upon determination by Surety of the lowest responsible bidder, or, if the Owner elects, upon determination by the Owner and the Surety jointly of the lowest responsible bidder, arrange for a contract between such bidder and Owner, and make available as Work progresses (even though there should be a default or a succession of defaults under the contract or contracts of completion arranged under this paragraph) sufficient funds to pay the cost of completion less the balance of the contract price; but not exceeding, including other costs and damages for which the Surety may be liable hereunder, the amount set forth in the first paragraph hereof. The term "balance of the contract price," as used in this paragraph, shall mean the total amount payable by Owner to Contractor under the Contract and any amendments thereto, less the amount properly paid by Owner to Contractor.

Any suit under this bond must be instituted before the expiration of two (2) years from the date on which final payment under the Contract falls due.

No right of action shall accrue on this bond to or for the use of any person or corporation other than the Owner named herein or the heirs, executors, administrators or successors of the Owner.

Signed and sealed this   day of   19

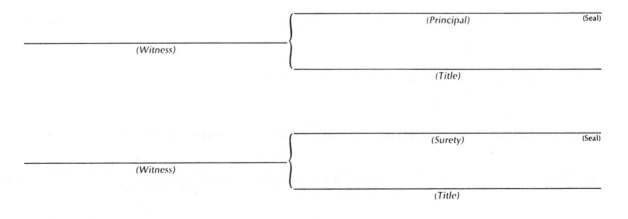

| | | (Principal) | (Seal) |
|------------------------| |--------------------|--------|
(Witness)		(Title)	
		(Surety)	(Seal)
(Witness)		(Title)	

AIA DOCUMENT A311 • PERFORMANCE BOND AND LABOR AND MATERIAL PAYMENT BOND • AIA ®
FEBRUARY 1970 ED. • THE AMERICAN INSTITUTE OF ARCHITECTS, 1735 N.Y. AVE., N.W., WASHINGTON, D. C. 20006

2

THE AMERICAN INSTITUTE OF ARCHITECTS

AIA Document A311

Labor and Material Payment Bond

THIS BOND IS ISSUED SIMULTANEOUSLY WITH PERFORMANCE BOND IN FAVOR OF THE
OWNER CONDITIONED ON THE FULL AND FAITHFUL PERFORMANCE OF THE CONTRACT

KNOW ALL MEN BY THESE PRESENTS: that

(Here insert full name and address or legal title of Contractor)

as Principal, hereinafter called Principal, and,

(Here insert full name and address or legal title of Surety)

as Surety, hereinafter called Surety, are held and firmly bound unto

(Here insert full name and address or legal title of Owner)

as Obligee, hereinafter called Owner, for the use and benefit of claimants as hereinbelow defined, in the

amount of

(Here insert a sum equal to at least one-half of the contract price)   Dollars ($),

for the payment whereof Principal and Surety bind themselves, their heirs, executors, administrators, successors and assigns, jointly and severally, firmly by these presents.

WHEREAS,

Principal has by written agreement dated   19   , entered into a contract with Owner for

(Here insert full name, address and description of project)

in accordance with Drawings and Specifications prepared by

(Here insert full name and address or legal title of Architect)

which contract is by reference made a part hereof, and is hereinafter referred to as the Contract.

LABOR AND MATERIAL PAYMENT BOND

NOW, THEREFORE, THE CONDITION OF THIS OBLIGATION is such that, if Principal shall promptly make payment to all claimants as hereinafter defined, for all labor and material used or reasonably required for use in the performance of the Contract, then this obligation shall be void; otherwise it shall remain in full force and effect, subject, however, to the following conditions:

1. A claimant is defined as one having a direct contract with the Principal or with a Subcontractor of the Principal for labor, material, or both, used or reasonably required for use in the performance of the Contract, labor and material being construed to include that part of water, gas, power, light, heat, oil, gasoline, telephone service or rental of equipment directly applicable to the Contract.

2. The above named Principal and Surety hereby jointly and severally agree with the Owner that every claimant as herein defined, who has not been paid in full before the expiration of a period of ninety (90) days after the date on which the last of such claimant's work or labor was done or performed, or materials were furnished by such claimant, may sue on this bond for the use of such claimant, prosecute the suit to final judgment for such sum or sums as may be justly due claimant, and have execution thereon. The Owner shall not be liable for the payment of any costs or expenses of any such suit.

3. No suit or action shall be commenced hereunder by any claimant:

a) Unless claimant, other than one having a direct contract with the Principal, shall have given written notice to any two of the following: the Principal, the Owner, or the Surety above named, within ninety (90) days after such claimant did or performed the last of the work or labor, or furnished the last of the materials for which said claim is made, stating with substantial

accuracy the amount claimed and the name of the party to whom the materials were furnished, or for whom the work or labor was done or performed. Such notice shall be served by mailing the same by registered mail or certified mail, postage prepaid, in an envelope addressed to the Principal, Owner or Surety, at any place where an office is regularly maintained for the transaction of business, or served in any manner in which legal process may be served in the state in which the aforesaid project is located, save that such service need not be made by a public officer.

b) After the expiration of one (1) year following the date on which Principal ceased Work on said Contract, it being understood, however, that if any limitation embodied in this bond is prohibited by any law controlling the construction hereof such limitation shall be deemed to be amended so as to be equal to the minimum period of limitation permitted by such law.

c) Other than in a state court of competent jurisdiction in and for the county or other political subdivision of the state in which the Project, or any part thereof, is situated, or in the United States District Court for the district in which the Project, or any part thereof, is situated, and not elsewhere.

4. The amount of this bond shall be reduced by and to the extent of any payment or payments made in good faith hereunder, inclusive of the payment by Surety of mechanics' liens which may be filed of record against said improvement, whether or not claim for the amount of such lien be presented under and against this bond.

Signed and sealed this   day of   19

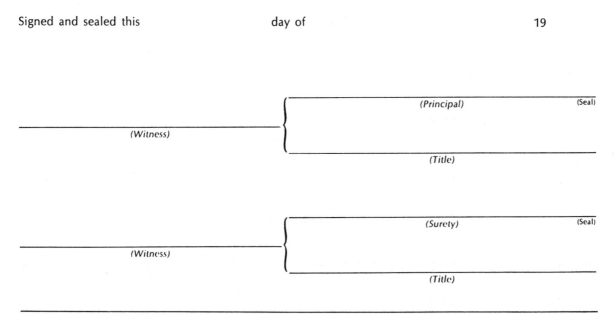

(Witness)	(Principal) (Seal)
	(Title)
(Witness)	(Surety) (Seal)
	(Title)

AIA DOCUMENT A311 • PERFORMANCE BOND AND LABOR AND MATERIAL PAYMENT BOND • AIA ®
FEBRUARY 1970 ED. • THE AMERICAN INSTITUTE OF ARCHITECTS, 1735 N.Y. AVE., N.W., WASHINGTON, D. C. 20006

4

(c) where there are unknown factors and contractors would be very reluctant to bid, for example, where work is subject to tides or flooding;

(d) where work must start even before the final drawings, specifications, and engineering services are completed.

3.11 COST PLUS A FIXED FEE

This type of contract is a refinement of the cost plus a percentage. Here the contractor would be paid for his overhead expenses plus a fixed contracted sum for his services and profit. In this manner, it is in the interest of the contractor to keep the job moving because the sooner he hands over the building to the owner, the sooner he will receive his final payment.

It is usual in this type of contract for the owner to make installments of the contractor's fee at agreed times against the progress of the building.

With this type of contract the owner knows very closely what the cost of the building will be, and the contractor knows what his profit will be. A great advantage comes to both parties when the contractor pushes the job through to an early conclusion. He has earned his fee in less time and the owner of the building has the use of it at an earlier date.

All bills are paid by the owner, which obviates any padding of the contractor's accounts.

3.12 CONTRACT FOR A COST PLUS A VARIABLE SUM

Under this form of contract, the builder undertakes to deliver to the owner, for a fixed sum on a fixed date, the completed building. With the variable premium, the contractor is rewarded or penalized for half the difference between the estimated cost and the actual cost. In addition, he would be rewarded or penalized for each day of early or late delivery of the building. Assume that an estimated $250,000 building actually cost $240,000. The difference of $10,000 would be shared by the owner and the contractor. The contractor's variable premium would be increased by $5,000 and the final cost to the owner would be $245,000 instead of $250,000. In contrast, assume that the building cost $260,000. The extra $10,000 would be absorbed by a deduction from the contractor's variable premium by an amount of $5,000 and the other $5,000 would be absorbed by the owner. In this case, the building would cost the owner $255,000. These are liquidated damages.

A further refinement of the variable sum paid to the contractor may be reckoned by the contractor being awarded a bonus for each day of delivery earlier than the agreed date. However, the contractor may be penalized for each day he is later than the contracted delivery date.

In all contracts, there is an imperative need for a really good progress schedule (see Chap. 12, p. 211).

3.13 QUANTITY ESTIMATE

This is a unit-price estimate. Assume a large project is to be built. The engineers or quantity surveyors will work out the number of units of work to be done. As an example:

(a) cut and haul 185,000 cu yd of earth (type of earth);

(b) supply and place 17,000 cu yd of concrete (type);

(c) supply and place 40,000 lb of reinforcing bars;

(d) supply and place 7,800 sq yd of 0'-6" concrete floor, and so on.

With this type of contract, all the tenderers are bidding to do the work against the same given quantities. For cut and haul, the unit price of one bidder might be $1.15 per cu yd. For concrete, his bid may be $9.81 per cu yd. Other bidders may be more or less per unit. In short, all the contractors are pitting their own company's abilities and equipment and labor force against the rest. Many large projects are estimated in this manner.

Where there are disparities between the original stated number of units and the actual, an adjustment is made to the contractor or owner. Mistakes do occur, since it is not in the affairs of humans to be infallible.

3.14 CONSTRUCTION BY DAY LABOR

With this type of building operation, the owner recruits his own labor force, pays for all labor, materials, subcontractors, rentals for equipment, and all other expenses such as permits, licenses, and so on. He may, through his architects, hire the service of an engineer. He would recruit the services of a superintendent. He may even hire the complete services of a building contractor. This method allows for great flexibility in alterations during construction, with the owner paying for everything. With this type of contract, there may be some conflict of views between the subcontractors on the issue of cutting away and making good, and attending on other trades.

3.15 CHECKLIST OF BONDS AND INSURANCES FOR CONTRACTORS

In recapitulation, the following material has been excerpted, by permission, from the second edition of *Insurance for Contractors* by Walter T. Derk, available for two dollars from the publisher, Fred S. James & Co., 1 North LaSalle Street, Chicago, Illinois.

Contract and Other Bonds

A very necessary adjunct to administration of a contractor's insurance program is the performance of similar services with regard to his contract bond requirements. One complements the other in that close working knowledge of work in progress, projects being bid or completed helps to enhance the close relationship which exists between the contractor and his insurance/bond counselor. Obviously it is a relationship depending upon confidence; in many respects it parallels a good banking connection.

Keeping a personal vow made when contemplating the outline of these pages, we shall not begin with the usual statement about the basic difference between insurance and bonding on the grounds that it is too basic. It does seem necessary to define some terms and briefly illustrate what a few particular contract bonds do, however:

Bid bonds: Given by the contractor to the owner, guaranteeing that, if awarded the contract, he will accept it and furnish final performance or payment bonds as required.
Performance bonds: Given by the contractor to the owner, guaranteeing that he will complete the contract as specified.
Labor and material payment bonds: Given by the contractor to the owner, guaranteeing that he will pay all labor and material bills arising out of the contract.
Maintenance bonds: Given by the contractor to the owner, guaranteeing to rectify defects in workmanship or materials for a specified time following completion. A one-year maintenance bond is normally included in the performance bond without additional charge.
Completion bonds: Given by the contractor to the owner and lending institution, guaranteeing that the work will be completed and that funds will be provided for that purpose.
Supply bonds: Given by manufacturer or supply distributor to owner, guaranteeing that the materials contracted for will be delivered as specified in the contract.
Subcontractor bonds: Given by subcontractor to contractor, guaranteeing performance of his contract and payment of all labor and material bills.

Such bonds are required by statute for Federal, state, and local government work and, of course, for a great deal of private construction. They are the best form of guarantee that construction will be finished as required and that all bills will be paid. Amounts of bond required may vary from 10 percent to 100 percent of the total price of the contract, but are normally 100 percent.

Any undue delay or outright failure to secure a required contract bond could cost the contractor the job, so performance of the bond agent is all-important to success of the contractor/counselor partnership. You can help by promptly supplying all financial information requested and, in general, keeping him posted about the status of your present and future work program. Those who are relatively new to the contracting business should strive to establish a strong working relationship with such an insurance/bond source.

Now, some definitions of other bonds you will encounter:

License or permit bonds: Given by the contractor/licensee to a public body, guaranteeing compliance with statutes or ordinances, sometimes holding the public body harmless.
Subdivision bonds: Given by the developer to a public body, guaranteeing construction of all necessary improvements and utilities; similar to a completion bond.
Union wage bonds: Given by the contractor to a union, guaranteeing that the contractor will pay union scale wages to employees and remit to the union any welfare funds withheld.
Self-Insurers' workmen's compensation bonds: Given by a self-insured employer to the state, guaranteeing payment of statutory benefits to injured employees.

Others falling into the broad categories of court bonds and fidelity bonds will not be dwelled upon here; need for the former will be made known to you when and if the time comes, while the best method for protection against employee dishonesty will be brought to your attention by the professional handling your combined insurance/surety account.

Coverage Checklist

While by no means suggesting that one should necessarily buy all of the following basic coverages and extensions or that doing so will guarantee a good insurance program, it may be well to list here most of what is commonly available as a checklist. You or your insurance representative can examine current or renewal policies to determine which of these are presently insured, how much additional premium needed extensions would cost, and which, if any, to buy.

General Liability

☐   1. Comprehensive policy form
☐   2. Property damage liability
☐   3. Elevator liability
☐   4. Independent contractor's protective liability
☐   5. Contractual liability—blanket coverage
☐   6. Completed operations—products liability
☐   7. Limits consistent with automobile liability
☐   8. Occurrence-basis bodily injury
☐   9. Occurrence-basis property damage
☐ 10. Property damage—explosion, collapse, or underground damage

☐ 11. Broad form property damage liability

☐ 12. Personal injury liability

☐ 13. Liquor law liability

☐ 14. Fire legal liability

☐ 15. Vendors' liability

☐ 16. Professional liability

☐ 17. Worldwide coverage

☐ 18. Watercraft liability

☐ 19. Aircraft liability

Automobile Liability

☐ 1. Comprehensive policy form

☐ 2. Occurrence-basis bodily injury

☐ 3. Occurrence-basis property damage

☐ 4. Medical payments

☐ 5. Uninsured motorists coverage

☐ 6. Use of other automobiles endorsement

☐ 7. Foreign coverage

☐ 8. Limits consistent with general liability

Automobile Physical Damage

☐ 1. Fleet automatic coverage

☐ 2. Fire

☐ 3. Theft

☐ 4. Combined additional coverage

☐ 5. Malicious mischief and vandalism

☐ 6. Comprehensive coverage

☐ 7. Collision

☐ 8. Towing

☐ 9. Foreign coverage

☐ 10. Leased equipment coverage

Workmen's Compensation and Employer's Liability

☐ 1. Increased limits—employers' liability

☐ 2. All states endorsement

☐ 3. Status of executive officers or partners

☐ 4. Longshoremen's and harbor workers' Jones Act, or federal employers' liability coverage

☐ 5. Additional medical coverage endorsement

☐ 6. Separate coverage as required in monopolistic fund states

☐ 7. Status of foreign employees

Umbrella Excess Liability

☐ 1. Accurate schedule of underlying primary policies

☐ 2. Employee benefit liability coverage

Contractors' Equipment Floater

☐ 1. "All-risk" perils

☐ 2. On-premises coverage

☐ 3. Material in transit

☐ 4. On-site coverage

General

☐ 1. Complete and accurate list of insured entities

Workmen's Compensation and Employers' Liability

A detailed analysis of the standard workmen's compensation and employers' liability policy form will not be made here, because it does little more than agree to provide compensation and medical benefits to injured employees in accordance with the provision of the applicable workmen's compensation law. The variations and technicalities arise from differences in the workmen's compensation acts of the various states, not from the policy itself.

The policy jacket, which more than anything else is a vehicle to establish effective and expiration dates, states covered and premium rates applicable to payroll, refers to Coverage A—workmen's compensation, which agrees to "pay promptly when due all compensation and other benefits required of the insured by the Workmen's Compensation Law." Coverage B—employers' liability provides common law defense of the employer for employee injuries, usually subject to a basic limit of liability of $25,000 "for all damages because of bodily injury by accident, including death at any time resulting therefrom, sustained by one or more employees in any one accident."

Policy Extensions: Increased Limits; Employers' Liability

In most states it is permissible to increase the basic limit of Coverage B from $25,000 per accident to $100,000 for approximately 2 percent additional premium.

QUESTIONS

1. How is the cost of an unsuccessful building contractor's bid absorbed by the company?

2. In ordinary language, define the following six types of building contracts:
   (a) stipulated sum
   (b) cost plus a percentage
   (c) cost plus a fixed fee
   (d) contract for a cost plus a variable sum
   (e) quantity estimate
   (f) construction by day labor

3. Define the following five types of building contractor's bonds:
   (a) tender or bid bond
   (b) performance bond
   (c) lien bond
   (d) contract bond
   (e) labor and materials bond

4. There are several main articles in *The Standard Form of Agreement between Contractor and Owner for Construction of Building.* In your own words, define the following articles:
   (a) "Scope of the Work"
   (b) "Time of Completion"
   (c) "Contract Sum"
   (d) "Progress Payments"
   (e) "Acceptance and Final Payments"
   (f) "The Contract Documents"

5. Does a building contractor's fire insurance policy cover materials on the ground not yet built into the structure?

6. If a building contractor wins the contract for a $350,000 project with work to commence on May 1 and the building to be handed over by October 1, how much fire insurance would be taken out on May 1?

NOTICE TO TENDERERS; GENERAL CONDITIONS; INDENTURE

This chapter is designed to introduce you to the notices to tenderers, general conditions of the contract, the type of indenture to be signed by the successful bidder, and finally the examinations of the plans and specifications by the estimator before any actual work is done on the estimate. Only a few of the most important examples that might apply for an ordinary commercial building are given.

4.1 NOTICE TO TENDERERS

The first few pages of specifications are usually taken up with instructions for contractors in their submission of a tender. These notices should be very carefully studied by the estimator and he should be in a position not only to take off the quantities of materials and labor required, but also to be able to advise his company upon its financial obligations, assuming it gets the contract.

It is considered that when an architect issues about ten sets of drawings and specifications, a fair price may be expected from about six bids. In some cases, architects will call for invitation bids from about five or six reputable contractors. It is a compliment to the ability, experience, and sincerity of any firm to be included in such invitations.

4.2 EXAMPLES OF TENDERS

Sealed tenders addressed to _____ Architect, and endorsed on the outside of the mailing container with the name of the work to which it refers will be received for the building on the dates as published in the daily papers. Names of papers _____ . Tenders shall be accompanied with a certified check in the sum of 5 percent of the tender as a guarantee of good faith that the successful tenderer will enter into contract.
Note: A certified or marked check is one that is endorsed by the bank and guarantees that there are sufficient funds in the tenderer's account and that those funds will not be used for any other purpose until after the contract has been let.

Tenders will be opened at _____ time _____ date _____ place when tenderers may hear them read and declared.
Note: Tenders close at the time of the time zone in which they are called.
Note: Tenders for all public works must be opened in public.

All checks of unsuccessful tenderers will be returned within four days of the opening of the tenders.

4.3 PLAN AND SPECIFICATIONS—GENERAL CONDITIONS

The contractor and all subcontractors shall comply with all local, state, provincial, or national government rules, regulations, and ordinances. They will prepare and file all necessary documents or information, pay for and obtain all licenses, permits, and certificates of inspection as may be specified or required.

Wherever municipal bylaws or state or provincial legislation require higher standards than those set forth herein, such higher standards shall govern.
Note: Assume there is some peculiar subsoil condition in a certain area. It is fair to assume that the local authority, knowing the conditions, may require special precautions to be taken which are more stringent than those set forth in the specifications. In all cases, the highest building standards shall prevail.

4.4 EXAMINATION OF THE SITE

All contractors submitting bids for this work shall first visit the site.
Note: It is not sufficient to examine a site for a small building which may be covered with snow without testing the depth of the snow. Neither is it sufficient to examine the site without considering the water supply, which may have to be hauled by tanker, or by drilling a well, and so on.

Plans and specifications may be obtained from the office of the architect by depositing a stated sum of dollars, which will be refunded upon return of the documents.

Bench Marks

The owner will establish the lot lines, restrictions, and a bench mark; all other grades, lines, and levels shall be established and maintained by the contractor.

Excavation and Backfill

Excavation to a greater than authorized depth shall be made good with extra construction at the contractor's expense as determined by the architect.

Footing excavations carried below the required depth shall be built up with the same concrete designed mix and reinforcing as for the footings about it.

Surplus excavated material shall be hauled from the site and legally dumped at the contractor's expense.

4.5 PRECEDENCE OF DOCUMENTS

Specifications will take precedence over drawings and dimensioned figures. Detail drawings shall take precedence over scaled measurements from drawings.

4.6 DISPUTES

In case of disputes concerning the intention or meaning of the drawings and specifications, such disuptes shall be decided by the architect. The intention and meaning of the drawings and specifications are to be taken as a whole. The contractor shall not avail himself of any unintentional error or omission, should any exist. The work shown on the drawings, if not fully described in the specifications, or vice versa, which is reasonably implied and is evidently necessary for the complete finish of each branch of the work, is to be done by the contractor as though both were specified. The contractor, however, shall have the right to appeal to the owner for arbitration on any matter when he feels that the architect's decision is not in accordance with the intention of the drawings and specifications.

4.7 ARBITRATION

When it is necessary to go to arbitration, a board shall be set up by each contending party, appointing one arbitrator of his own choosing, and then the two chosen arbitrators shall appoint a third arbitrator acceptable to both, who shall act as chairman. Whenever two members of the board are in agreement, such agreement shall be binding on both the contending parties. The fees for the board will be met by each party paying his own appointed arbitrator, and the fee of the chairman shall be shared equally by both contending parties. *Note: The elected arbitrators are usually reputable men of good character having a very wide experience in the field of construction. The law courts are the last place that the contending parties usually wish to go. They are slow and expensive, and judgments may be given by a court that does not have the wide experience that arbitrators have. The method of arbitration has stood the test of time by being cheaper, speedier, and usually satisfactory to all parties.*

There are other types of arbitration boards, but **no arbitration board is valid that attempts to deny either party access to the courts.**

4.8 BONDS

The successful bidder will enter into a performance bond and bid bond; he will also take out all necessary workmen's compensation coverage for the area of operations (see Chap. 3).

4.9 PROGRESS SCHEDULE AND DIARY

A construction schedule shall be submitted to the engineer within ten days of the notice of acceptance (see Sec. 12.6.) This schedule will be in a form satisfactory to the engineer and show a detailed breakdown of the work; it shall be revised and resubmitted with each progress claim.

A progress diary shall be maintained by the contractor from the date of commencement of the work. Recording shall show the progress of the work, daily weather conditions, excavation work, concrete placing and finishing, removal of forms, subtrade progress, and relation to construction schedule. This record shall

be open to inspection by the engineer at all reasonable times (see Sec. 12.6).

4.10 PROGRESS REPORTS AND PHOTOGRAPHS

Biweekly, the contractor shall submit to the architect a written report accompanied with two 8″ × 10″ glossy prints taken by a professional photographer. One copy of the report and one photograph will be retained by the engineer.

4.11 PAYMENTS

The contractor will, on the fifteenth of each month, submit to the architect a statement of all the work done and materials supplied onto the job, including wage sheets and receipted material accounts. The contractor will be allowed to add 10 percent to these costs to cover his overhead expenses and the contractor shall be paid 85 percent of the amount of his monthly estimate. The final payment of balance will become due and payable six months after occupation of the building by the owner.

4.12 TEMPORARY OFFICE AND STORE SHEDS

The contractor shall provide a field office with light and heat, and he shall also provide suitable storage sheds for the protection of all building materials that may be damaged by the weather.

4.13 TEMPORARY SERVICES

Temporary power, gas, water, sewer, and telephone shall be provided by the contractor at his own expense.

4.14 SUPERVISION

The contractor shall give his personal supervision to the work and shall have a competent superintendent constantly on the job during normal working hours who shall have authority to take instruction from the architect or his representative. Such superintendent shall not be taken away from the job except with the agreement of the architect.

4.15 CONTINGENCY SUM

All bidders shall include the sum of $3,000 as a contingency in their proposal.

Note: A contingency sum is a fixed fund upon which the architect may draw for the erection of the building to meet unforeseen expenses such as minor alterations, additions, and so on.

Any portion of such unused contingency fund shall be returned to the owner at the conclusion of the work.

4.16 PRIME COSTS

The contractor shall include in his tender all prime costs as stated in the specifications.

Note: A prime cost (P.C.) means an original cost of any item or items to be installed in a new building, such as special plumbing fixtures, hardware, and so on. In the case of hardware, the architect in consultation with the owner will make a list of requirements. The wholesalers of hardware are then invited to give a price for the complete list, the contract usually going to the lowest bidder. The P.C. sum is stated by the architect in the notices to tenderers and the contractor makes an allowance for the amount stated plus his profit.

When the actual amount of money spent by the architect on the P.C. item is either more or less than that stated in the specifications, an adjustment is made on the final payment to the contractor.

4.17 SUBCONTRACTOR

The owner will recognize the contractor only. No portion of the work is to be sublet without the approval of the owner, to whom the names of all subcontractors shall be submitted. On large jobs the subcontracts may be tendered separately from the general contract.

4.18 FIRE INSURANCE

The contractor shall insure the works and keep them insured against loss by fire until they are delivered up to the owner. It is usual for the contractors' fire insurance to lapse when the building is handed over to the owner.

4.19 LIENS

The contractor shall hand over the building at completion free from all liens and encumbrances or give a bond of guarantee freeing the owner from all responsibilities in the same (see Sec. 3.5).

4.20 CLEANUP

The contractor shall keep the premises clean at all times. He shall replace broken glass and fixtures and leave the building free from fingermarks and ready for the final inspection.

4.21 EXAMINING THE PLANS AND SPECIFICATIONS

On first receiving the drawings and specifications, the estimator should scan the drawings to get the feel and visualization of the floor plans, the elevations, and the typical wall sections. Next, he should read the specifications very thoroughly along with the drawings, and, most important, he should make very careful note of any new features or materials.

Where any discrepancies are noted between the drawings and the specifications, they should be brought to the attention of the architect. When such discrepancies are valid, the architect will immediately issue an addendum to all bidders. In this way, every contractor is estimating against the same things.

No estimator has any worries about traditional materials, features, and methods, but architectural concepts are never constant and neither are the manufacturers of building materials. In all cases of new materials, the estimator should try to see them, and, in consultation with his colleagues, he should arrive at an estimate of the time required for labor and machines to erect or install such new items. New methods and materials of construction are discussed freely at builders' exchanges (see Sec. 1.36.)

When the actual estimates are being taken off, the estimator must visualize every feature of the building, including things that could easily be missed during actual construction. All such items must be noted in writing.

Assume that a fire hydrant is to be set into a concrete wall and afterward framed with wood. Such an item would require a wood buck during construction. This should be a written note. The estimator should hand over all the written notes to the superintendent, who will in turn, as the building progresses, instruct his foremen accordingly. It would be a direct loss in time and money—and, not the least, in the morale of the work crews—to have to be taking jack hammers to tear completed work apart. *The morale of the crews should be given constant thought.* The estimator's and superintendent's knowledge should be complementary to each other, and they should be in close touch with each other during the erection of the building.

4.22 ALTERATIONS AND ADDITIONS

It is quite common for the owner to require some alterations from the original drawings during actual construction. This may be a case where a contractor will have an unavoidable delay in the handing over of the building. A suitable notation should be made in the job field diary by the timekeeper. In any case, no alterations should be made without written instructions from the architect. (See the typical daily extra work record sheet below.) A large alteration may delay completion date of the building and exclude the contractor from bidding on future jobs. The field diary is important for recording such events, which may become exhibits in future arbitrations.

4.23 EXTRAS

No extra work of any description should be started without written instructions and cost negotiations. A notation must be made in the field diary. These notations may be of great value in case of arbitration.

A change order is originated by the architect; this involves extra work for the contractor and the fee for such work is negotiable.

DAILY EXTRA WORK ORDER

Send To Head Office Daily

Contract No. _____

Job _____

Charge to _____

Date _____

Authorized by _____

Extra Order No. _____

Detail Supervision, Materials, Equipment Hire or Own

List Numbers of Men by Trades	$	cts	$	cts
Fringe Benefits				
$				
Materials:				
Hired Equipment:				
$				
Overhead Expenses % Commission %				
Own Equipment:				
TOTAL $				

Work done or charged to subcontractors.

Authorized by _____

Reported by _____

To be approved

Approved by _____

Superintendent _____

4.24 DELAYS

The estimator or office should immediately notify the architect in writing of any delay in operations, giving the date, cause, weather, and action taken. This should also be noted in the diary.

4.25 INSTRUCTIONS TO BIDDERS

The following pages are specimen instructions to bidders.

INSTRUCTIONS TO BIDDERS

Tenders shall be made on the blanks provided,   -1-
enclosed in an envelope, endorsed (
     (Here state the name of the work)

     ), sealed and addressed to the
Architect.   All blanks shall be completely filled in
giving all proposals asked and all figures shall be stated
in writing.   Failure to observe this ruling shall cause
rejection of the proposal.

Each tender shall be accompanied by a certified cheque   -2-
equal in amount to Ten (10%) per cent of the contract price   Tender
as evidence of good faith and to the effect that should the   Deposit
proposal be accepted the tenderer shall enter into a contract and furnish satisfactory deposit or bond.   The said
deposit or certified cheque to be forfeited if the tenderer
whose proposal is accepted shall fail to give satisfactory
guarantee deposit or bond or to execute a contract, or shall
be returned after guarantee deposit or bond is given and
contract executed.   Tender deposits or cheques of unsuccessful bidders shall be returned after the award has
been given and contract executed.

Persons or firms submitting tenders or proposals shall be actively engaged in the line of work required by the specification and shall be able to refer to work of a similar character performed by them.

- 3 -

Tenderers.

They shall be fully conversant with the general technical phraseology in the English language, of the lines of work covered by the drawings and specifications.

Before submitting a tender, bidders shall carefully examine the drawings and specifications, visit the building site or premises, and fully inform themselves as to all existing conditions and limitations.

- 4 -

Inspection of Sites.

Tenderers finding discrepancies in or omissions from the drawings, specifications, or other documents, or having any doubt as to the meaning or intent of any part thereof, should at once inform the Architect, who will send written instructions or explanations to all tenderers. Neither Owner nor the Architect shall be held responsible for oral instructions.

- 5 -

Discrepancies and Omissions.

The Contractor shall bring to the attention of the Architect any obvious omission and/or discrepancies failing which he shall become responsible for completing such work without added compensation.

Failure of the tenderer (should he become the successful bidder) to inform or bring to the attention of the Architect at the time of tendering, any such items as appear obvious, shall be assumed as indicating his willingness to accept the original intent and meaning as may be interpreted by the Architect, within the reason of standard practice, or

requirements necessary to obtain a complete work, or any other
such work, supply or installation, failing which would jeopardize
the structure or leave an unfinished or incomplete work in part
or in whole.

Addenda or corrections issued during the time of
tendering are to be covered in the proposal and shall become
part of the contract documents.

The Architect reserves the right to reject any or all
proposals.   Contracts when awarded shall be for each branch
or division of the work, or for the whole work as deemed for
the best interests by the Architect.

The tenderer whose proposal is accepted will be required
to furnish a marked cheque or bond or bonds or Guarantee Bond
for supply of proper materials, faithful performance and
maintenance of the work as provided for under article
"Guarantee Deposit or Bond".

A written request for approval of equivalents for
products and materials and/or associated alternative
subcontractors shall be submitted in order to reach this
office not later than ten days prior to closing date of
Tender.   All subcontractors must be noted in the list
submitted by the General Contractor with his tender.

All Contractors are to itemize, with their bids,
all contingency and P.C. Sums.

- 6 -

Awards.

- 7 -

Contract

deposit or

Guarantee

Bond.

- 8 -

Equivalents

and Sub-

contractors.

- 9 -

Contingency

& P.C. Sums.

If necessary, one (1) Addendum to the Specifications
and Drawings will be issued by the Architect, one (1) week
before closing date of tender.

This Addendum will contain all changes and clarifi-
cations to the Drawings and Specifications, as well as a
list of equivalent materials and products, as determined
by the Architect, or his representative.

No further Addenda will be issued after one week
Prior to Tendering and all Bidders shall refer to
"Interpretations of Specifications and Drawings" as covered
in the preface to Instructions to Bidders.

Tenderers shall make allowance for delay in mailing or
transportation, or preferably see his tender delivered
personally, that they reach the Architect within the stated
time, as no allowance can be made for extenuating circumstances.

The Mechanical and Electrical Subcontractors <u>shall</u> on
acceptance of tender (whether called for by the Owner or by the
General Contractor) and when completing agreement, render to
the General Contractor a surety bond for 50% of their contract to cover materials and performance for the faithful
completion of their contract.   They shall therefore allow
in their tender figure for the cost of such bond.   The
General Contractor in his turn shall not conclude any agreement with the Mechanical and Electrical Subcontractors until
he has received the covering surety bond.

- 10 -

Addenda

- 11 -

Tendering Date

- 12 -

Mechanical &
Electrical
Subcontractors
Bond

AGREEMENT

𝕿𝖍𝖎𝖘 𝕴𝖓𝖉𝖊𝖓𝖙𝖚𝖗𝖊,   made in triplicate this   — I —

day of   in the Year of Our Lord, One Thousand   Parties of
                                                                                     the Contract
Nine Hundred and

𝕭𝖊𝖙𝖜𝖊𝖊𝖓:

OF THE FIRST PART;

— and —

OF THE SECOND PART.   — 2 —
                        Project

WHEREAS the   is desirous of

and has caused drawings and specifications describing the work to be done
to be prepared by the Architectural Office, Department of Public Works.

WHEREAS the said drawings numbered:

and specifications have been signed by and on behalf of the parties hereto.

AND WHEREAS the contractor has agreed to execute upon and subject   — 3 —
to the conditions set forth in the General Conditions hereto with work shown   Amount
on the said drawings and described in the said specifications for the sum of

NOW IT IS HEREBY AGREED AS FOLLOWS:

In consideration of the sum of

or such other sums as shall become payable to be paid by
the Owner at the times and in the manner set forth herein
and the conditions set forth in the General Conditions
hereto, the Contractor shall upon and subject to the said
conditions execute and complete the work shown upon the
said drawings and described in the said specifications.

The Contractor shall be entitled, under the certificates to be issued by the Architect to the Contractor, to
monthly payments amounting to eighty-five (85%) per cent
or ninety (90%) per cent of the value of the work done up
to the date of the said certificates. The said payments
shall be made at the end of each and every month. The
final ten (10%) per cent or fifteen (15%) per cent, may be
retained until six months after the date of the final
certificates, or until all defects are made good, according
to the true intent and meaning thereof, whichever shall
happen last. The retention of any monies shall not be
subject to any charges.

The Contractor shall commence work on a date to be
fixed by the Architect by notice in writing, which date
may be forthwith after the receipt of such notice. He
shall begin the works immediately on the date set out
in the said notice, shall complete the same by the

day of   19   , subject nevertheless, to
the provisions for the extension of the time hereinafter
contained.   Subject to the provisions of the Delay Clause,
time shall be deemed to be of the essence of this contract.

The guarantee deposit of a marked cheque, bond or bonds   - 6 -
or surety bond in the amount of   Guarantee

Deposit or

deposited with the Architect by the Contractor is   Bond
incorporated into and forms a part of this Agreement.

The instructions to bidders and the General Conditions   - 7 -
hereinafter set out shall be read and construed as forming   Relation of
part of this agreement, and the parties hereto shall   Conditions
respectively abide by and submit themselves to the said
conditions and stipulations therein set out and perform
the agreement on their part respectively and in such
manner as may be required.

QUESTIONS

1. Define the following items that appear in specifications:
   (a) notices to tenderers
   (b) invitation bids
   (c) certified check
   (d) general conditions
   (e) examination of the site
   (f) bench marks
   (g) datum line
   (h) excavating to authorized depth and the consequences of excavating below such depth
   (i) precedence of documents
   (j) formation of an arbitration board
   (k) progress schedule as required by the architect
   (l) photographs that accompany progress reports
   (m) progress payments
   (n) temporary offices and store sheds
   (o) temporary services
   (p) supervision of the work
   (q) contingency sum as stipulated in the specifications
   (r) prime cost
   (s) fire insurance (contractors)
   (t) liens on property
   (u) examining the plans and specifications
   (v) cleanup
   (w) tender

2. Define authorized:
   (a) alterations
   (b) extras
   (c) delays

3. In instructions to bidders, define:
   (a) tender deposit
   (b) discrepancies and omissions
   (c) addenda
   (d) the use of alternative materials to those specified
   (e) equivalents for products and materials

5
OVERHEAD EXPENSES AND PROFIT

This chapter deals with the two types of overhead expenses and also discusses profit. It is a remarkable fact that many tenders are submitted with one or more of these items completely forgotten—with disastrous results for the contractor. More people go bankrupt in the building industry than in any other field of enterprise. **Don't forget to add profit.**

5.1 INDIVIDUAL JOB OVERHEAD EXPENSES

These are a direct charge to each specific job. These items will vary from job to job and from company to company depending on the nature of the business. A guide list of individual job expenses is shown on the following pages. This list should be checked very carefully and additions or omissions made to suit your own needs. The list is not exhaustive and it should be kept up to date. The charge for individual job overheads is made item-by-item on the first sheet of the general estimate sheets under the heading "Preliminaries."

5.2 GENERAL OVERHEAD EXPENSES

These constitute a first charge against every company irrespective of the amount of physical work being done at any one time. A proportionate charge for such expenses is made against every job whether large or small. A guide list of the general overhead expenses is shown on the following pages. It should be checked with your company overhead expenses and additions or omissions made to suit your needs. The charge for general overhead expenses is made on the last page of the general estimate sheets. It is made as a lump-sum charge.

5.3 PROFIT

This will vary according to risks and competition involved in different types of contracting, say, from bridge construction to simple warehouse construction. In general, 10 percent is acceptable as fair and reasonable for profit. When competition is keen, this amount may have to be reduced.

5.4 SEPARATE JOB EXPENSES FOR SALARIED PROFESSIONALS

The following list is presented as a guide and should be augmented to meet specific jobs. It is important for the contractor to have as many guide lists as possible so that he may refresh his mind on matters that may otherwise be overlooked.

Salaries

Project manager

Layout engineer

Surveyor

Rodman

Superintendent

Job sponsor (assistant superintendent)

Timekeeper

Foreman

There will be occasions when one or more of these men will be responsible for two or three small jobs at one time, and their earnings will have to be divided pro rata between jobs. Thus an employee who has worked on two projects in one given pay period may receive two separate checks aggregating his total earnings for such

periods of time. This method enables the contractor to keep track of the cost of each individual job.

Assume that a superintendent is looking after three small jobs at one time with contract prices as follows: Job No. 1, $100,000.00; Job No. 2, $86,000.00; and Job No. 3, $64,000.00. If the jobs are running concurrently and are to be completed in three months time, the total cost of all the jobs would be $250,000.00 and the salary earned on each job would be as follows:

Job No. 1   salary will be $\dfrac{100,000}{250,000}$ of 3 months total salary

Job No. 2   salary will be $\dfrac{86,000}{250,000}$ of 3 months total salary

Job No. 3   salary will be $\dfrac{64,000}{250,000}$ of 3 months total salary

Where jobs start and terminate at different times (as they usually do) a further adjustment will have to be made. *It is important that these salaries be apportioned correctly so that the estimated cost of each project can be reconciled with the actual cost.*

Other specific job costs listed in Sec. 5.5 below.

5.5 GUIDE LIST OF OVERHEAD EXPENSES CHARGEABLE TO EACH INDIVIDUAL JOB

The following items should appear on the first of the general estimate sheets as "Preliminaries":

Salaries

Superintendent

Job sponsor

Foreman, carpenter, mason, and others

Timekeeper and storeman

Watchman

Cleanup man

Temporary Buildings

Office

Toolshed

Store sheds

Workshops

Toilet facilities

Fencing

Job signs

Roads and trackage

Ramps

Platforms

Temporary doors and stairs

Barricades

Protecting new and adjoining properties

Job billboards for advertising

Legal

Public liability insurance

Workmen's compensation

Fire insurance

Bonds: performance, and so on

Attorneys and lawyers (barristers and solicitors)

Social Security benefits

Accountants and auditors

Real estate fees

Building Permits

Town

Out-of-town

Special permits for roads, and so on

Sewer connections

Excavating sidewalks and roads

Utilities

Temporary light and power

Poles and meters

Temporary heat: electric, gas, or salamanders

Temporary water meter and taxes

Digging of well, haulage by tanker

Temporary telephone

Testing

Soil compaction

Concrete

Decibel readings

Progress Reports

Written

Photographs

Progress diary

Out-of-Town Jobs

Traveling

Premium rates for staff and workmen

Room and board

Freight and trucking of own or hired materials and equipment

Professional Services

Survey

Engineers

Accountant

Attorney

Engineers/Architects

Protection of Work

Curing, covering, water-spraying, or keeping concrete ice free

Protecting finished work such as wood floors and staircases

Attending On Other Trades

Cut away and make good

Breaking and repairing sidewalks and streets

Plant

Own equipment

Concrete mixers, wheelbarrows, buggies, hoses, cutoff saws

Hired equipment

Pumps

Maintenance and haulage of equipment

Company transport

Completion Dates

Penalties or rewards

Final Cleanup

Replace and clean all damaged glass

Janitor service

Computer Services

Estimate annual expenses

Landscape and Lawn

Shrubs

5.6 SALARIES

On very large projects the contractor's temporary office is, in reality, a branch office of the company. All kinds of people work from this branch, from the office manager down to a junior bookkeeper and the janitor.

Salaries under such conditions may be made up in the field office.

When a superintendent is looking after two or three separate jobs, his salary should be estimated pro rata for each job. Where men are moved from one job to another, many firms pay them by separate checks, one for each job. This is to enable the different jobs to be kept separate in the costing department.

5.7 TEMPORARY BUILDINGS

Some of these buildings may be mobile and are moved from one job to the next with a saving on original cost; however, the cost of haulage must be allowed for. Even on small jobs, offices may be loaded on flats and moved from job to job. This is organization.

5.8 EQUIPMENT USED ON THE JOB

Rental equipment is a direct charge to the job on which it is used. The contractor's own equipment is charged out to each separate job from the main yard at so much per day. As a very simple example, assume that a hand electric saw which cost $100 is charged to a job at the rate of $1.00 per day. This $1.00 is entered in a *sinking fund*; after the saw has been in use for about 100 days, it is probably worn out, but there is another $100 available for a new one. As the saw is wearing out, the sinking fund is growing.

5.9 CONCRETE TESTING

Samples of concrete are taken (in cylinders 6 in. in diameter by 12 in. in length) during the actual placing of concrete on the job. These samples are put to test and the expense of such tests is sometimes a charge against the contractor. The estimator must establish whether or not he has to include these charges in his estimate.

5.10 PROGRESS REPORTS

Many owners (governments in particular) require written reports and accompanying photographs to be submitted at stated periods. These items are a charge against the contractor and the estimator must be aware that it should be allowed for on his estimates.

5.11 ATTENDING ON OTHER TRADES

It is always stated on specifications that the general contractor shall attend on all other trades and cut away and make good as necessary. Assume that a plumber wants to run a pipe where there are obstructions. This could involve a mason, a carpenter, an electrician, and a sheet-metal man. Someone must be held responsible for making good. This is the responsibility of the general contractor and an allowance on all estimates must be made for such attendance.

5.12 COMPLETION DATE

The owner expects his new building to be delivered on a given date. From that date, he anticipates a start to making a profit. Unless the building is handed over by the contracted time, the general contractor may have to pay a penalty for each day of delay. This is the reason for the progress schedule (a most important document dealt with in Sec. 12.6). These are liquidated damages.

5.13 GUIDE LIST OF MAJOR OVERHEAD EXPENSES NOT CHARGEABLE TO ANY ONE JOB

Office

Rent or interest on invested capital

Depreciation of office building and contents

Fuel

Light

Telephone

Stationery

Postal services

Business machines

Furnishings

Fire and public liability insurance

Property tax

Heat

Telex

Office supplies

Business machines

Office furnishings

Staff

Salaries:

Executives

Accountant

Estimators

Draftsmen

Stenographers

Clerks

Janitors

Staff traveling expenses

Advertising

Radio

Television

Magazines and journals

Daily newspapers

Club association dues

Billboards

Literature

Trade magazines

Trade journals

Trade reports

Company library (new publications)

Association dues

Tools and Equipment

Purchases

Maintenance

Depreciation. *Note: There is a fixed table of allowances.*

Legal Retention Fees

Attorneys, barristers, and lawyers

Professional Services

Architects

Engineers

Surveyors

Certified public accountant

Auditors

5.14 GENERAL OVERHEAD EXPENSES

At the end of the fiscal year of a company, the total general overhead expenses for that year should be computed, and then the percentage of general overhead that is chargeable to every job for the ensuing year should be calculated:

Step 1: Find the total overhead expenses for last year.

Step 2: Find the sum total of all successful bids for last year.

Step 3: Find the percentage of overhead expenses for last year against the sum total of all successful bids for last year.

Example

Assume that the total overhead for a business last year was \$15,000 and the sum total of successful bids was \$250,000.

$$\text{Then } \frac{\text{overhead}}{\text{successful bids}} \times 100 = \text{percentage of overhead.}$$

$$\text{Thus } \frac{15,000}{250,000} \times 100 = 6 \text{ percent.}$$

It must be assumed that there will be as big a volume of work (or more) in the new year as in the old year. A fixed 6 percent charge for general overhead expenses for every bid during the new year must be made. For small or large jobs, the 6 percent general overhead charge must be made.

Example

A bid is to be placed for the erection of a storehouse. The cost of job overhead expenses, materials cost, and labor cost is \$85,000. To this figure is added 6 percent overhead. Thus:

$85,000

$\underline{\quad 5,100}$ general overhead expenses at 6 percent

$90,100$ bid before the profit is added

5.15 INDIVIDUAL JOB OVERHEAD EXPENSES (CHECK)

Check the guide list for individual job overhead expenses. Every item should appear on the first sheet or sheets of the general estimate sheet. All these items belong solely to one job.

5.16 FINAL PROFIT

In the building industry, profit is subject to change according to competition in the field at any one time. Risk would also be an important consideration. Earthwork is the greatest risk. In fact, some earthwork on large undertakings is so risky that the successful bidder for all the work above ground is given the earthwork at cost plus 10 percent profit.

In general, throughout the industry a profit of 10 percent is considered fair and reasonable. It must be remembered by all estimators, however, that the incidence of bankruptcy is far greater among builders and contractors than most other lines of business.

Profit is the last thing to be added to the estimate. Don't forget it! In the foregoing example, the figures for the final bid would appear thus:

$85,000	net bid
5,100	general overhead expenses at 6 percent
90,100	
9,010	profit at 10 percent
$99,110	Final Bid

Remember that a contracting company only wants work at its own price. Most contracts are let to the lowest bid.

QUESTIONS

1. List ten general overhead expenses.

2. List ten job overhead expenses.

3. State where on the estimate both general and job overhead expenses are listed.

4. Why may the margin of profit be different for different types of projects? Give two examples.

5. State how the salary of a superintendent is charged if he is simultaneously looking after the construction of a small school, a small warehouse, and a garage.

6. Explain the use of a sinking fund for the charging of a 3½ cu ft capacity concrete mixer for each day's use on any one job.

7. How would the use of a contractor's temporary (mobile) buildings be charged to any one job?

8. At whose expense are concrete specimen testings usually made?

9. Define a progress report and its accompanying photographs.

10. What is meant when it is specified that the general contractor shall cut away, make good, and attend on other trades?

11. What is the significance to the contractor, the architect, and owner of a stipulated completion date for a newly contracted building?

12. Give an example of how you would assess the general overhead expenses for your business for the next year if, for last year, the successful bids totaled $900,000 and the overhead expenses totaled $39,000.

13. Give an example of how you would arrive at your final bid for a project estimated at $40,000 *net* cost, general overhead expenses at 4½ percent and profit at 7 percent.

6

ACCURACY TESTS AND QUICK CALCULATIONS

This chapter has been designed to help you get the feel of estimating numbers approximately, using arithmetic symbols.

An estimator's first approach to any part of an estimate should be, "What is the approximate estimate?" A brick wall 25'-0" long and 8'-0" high should feel like 200 sq ft of wall, with about 625 bricks per 100 sq ft (for a 4-inch wall), which is about 1250 bricks. The estimator should also consider that bricks will require additives for the mortar in freezing weather, brick ties, scaffold, a bricklayer and his helper, their fringe benefits, and the overhead expenses of the company.

The first few pages of this chapter deal with mental calculations; then a series of accuracy tests follows with allotted marks for each test. These are easy. They are so easy that many persons using them make errors because they treat them with contempt. I have presented these tests to hundreds of students and have yet to have one class where every student got 100 percent on any one test. Try them!

The estimator is never called on to solve complicated mathematical problems, but he must handle numbers with facility and confidence. Most estimating offices will not permit their staffs to use the slide rule because it is too easy to misplace the decimal point. Excellent calculating machines are provided, but remember that machines will only produce the right answers if fed the right information. Many persons who have become used to slide rule and machine calculating build up a fear of handling number with paper and pencil. When tape of a machine calculation is completed, it should be checked over immediately for correctness of the information fed into it.

6.1 RAPID MENTAL CALCULATIONS

Example

The following steps show the mental process used to multiply *two numbers of equal value* together, such as 16×16.

Step 1: Change one number to a number ending in zero; thus $16 + 4 = 20$.

Step 2: Deduct from the other number the amount that was added to the first number; thus $16 - 4 = 12$.

Step 3: Multiply the altered numbers together (mentally): $12 \times 20 = 240$.

Step 4: Add the square of the figure that was added and subtracted; thus $4 \times 4 = 16$. Add 16 to 240 (mentally) $= 256$.

$$13 \times 13 = 10 \times 16 = 160 + (3 \times 3) = 169$$

14×14	25×25	37×37
15×15	26×26	38×38
16×16	27×27	39×39
17×17	28×28	41×41
18×18	29×29	42×42
19×19	31×31	54×54
21×21	32×32	63×63
22×22	33×33	78×78
23×23	34×34	81×81
24×24	36×36	95×95

Now check your work by the long method. Check your time. Do some every day. Allow yourself ten points for each correct answer and see how many points you can make out of each column.

A man spends five minutes shaving every day for 40 years. Estimate mentally how many 8-hour working days this represents.

Try This: Multiply 17×16.

Step 1: Find 16×16 mentally.

Step 2: Add 1×16 to the result of Step 1.

Quick Mental Calculations

To multiply a number ending in a zero with another number as 120×27:

Step 1: The answer must end in a zero, so write a zero thus: 0.

Step 2: Discard the zero from 120, leaving 12.

Step 3: Multiply $12 \times 27 = 324$, which must now be placed before the zero: 3240.

Multiply by mental process the following:

110×34	441×30	773×50
90×28	60×431	20×5562
80×56	756×90	9784×70
40×115	544×60	543×80
30×179	80×987	60×5436

Check your answers freehand.

To multiply two numbers where one or a combination of them ends in two zeros as 120×270:

Step 1: The answer must end in two zeros, so write them thus: 00.

Step 2: Discard the zeros from the original numbers, thus $12 \times 27 = 324$, which must now be placed before the two zeros thus: 32,400.

Multiply by mental process the following:

240×120	2300×90	690×700
80×980	80×4500	8723×600
466×700	700×4500	$54,700 \times 400$
401×600	5000×400	90×6574
850×110	600×7890	120×9672
692×500	4678×800	2769×1100

Check your answers freehand.

6.2 PERCENTAGES

A man purchases a house with an $18,000 mortgage at 6 percent per annum. What are the interest charges for one year?

Step 1: An interest charge of 6 percent per annum is equal to a rate of $60 per $1000 per annum.

Step 2: An $18,000 mortgage at 6 percent per annum is equal to $18 \times 60 = \$1080$ per annum.

Complete the following table by mental process:

Loan	Rate per cent per annum	Charges per $1000 per annum	Total Charges for Total Loan per annum
$14,000	$4\frac{1}{2}\%$	$45	$630
$16,000	5		
$18,000	$5\frac{1}{2}$		
$20,000	6		
$22,000	$6\frac{1}{2}$		
$25,000	7		

Mark 5 points per correct answer: possible 50 points.

6.3 TO FIND THE AREA OF PLANE FIGURES

(a) Hexagon

Square the short diameter and multiply by 0.866.

Your thinking should be: 12 × 12 = 144 sq in. less what?

(b) Octagon

Square the short diameter and multiply by 0.707.

Your thinking should be: 12 × 12 = 144 sq in. less what?

(c) Parallelogram

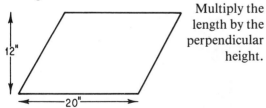

Multiply the length by the perpendicular height.

Your thinking should be: an oblong 20 × 12 = 240 sq in.

(d) Trapezoid

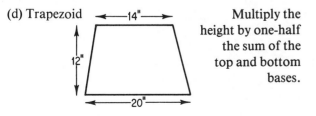

Multiply the height by one-half the sum of the top and bottom bases.

Your thinking should be: $\frac{14 + 20}{2} \times 12 = 204$ *sq in.*

Note: You will be using this formula a great deal in your estimating for land leveling in Chap. 10.

6.4 TO FIND THE APPROXIMATE AREA OF A CIRCLE

Step 1: Examine Fig. 6.1, where the area of the 12″ × 12″ *square is 144 sq in.*

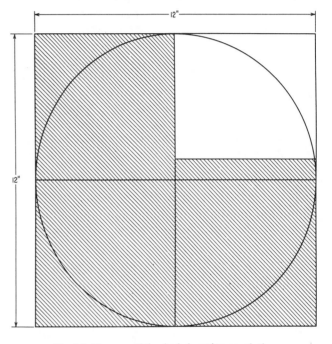

Fig. 6.1 The area of the shaded portion equals the area of the circle.

Step 2: For all practical estimating purposes, the area of a circle is πR^2, where π is $3\frac{1}{7}$ (approximately) and R equals the radius.

Step 3: The area of the circle shown is $3\frac{1}{7} \times 6 \times 6 = 113$ sq in.

Step 4: The approximate area of the circle by mental process is:

(a) the area of three squares, each 6″ × 6″, = 108 sq in.

(b) add $\frac{1}{7}$ of one square 6″ × 6″ = $\frac{36}{7}$ (= approx. 5) + 108 = 113 sq in.

For a rough estimate to guide your thinking, the area of any circle is a little more than three times the square of the radius. Thus the area of a circle 30′-0″ × 30′-0″ is a little more than 3 × 15 × 15—say, 700 sq ft. The area just could not be in the seventies nor seven thousands of sq ft. If you do use a slide rule, remember this. You must make this type of estimator's thinking your own.

Problems

By mental process estimate the approximate areas of each of the following circles:

(a) dia. 16′-0″; (b) dia. 20′-0″; (c) dia. 24′-0″; (d) dia. 28′-0″

6.5 TO FIND THE CIRCUMFERENCE OF A CIRCLE

Multiply D (diameter) by $3\frac{1}{7}$. The dimension C (circumference) of the circle shown in Fig. 6.1 is $12 \times 3\frac{1}{7} \simeq 38$. (The symbol $\simeq$ means "approximately equals.") *Your thinking should be: 3 × 12 plus what?*

6.6 TO FIND THE AREA OF AN ELLIPSE

Step 1: Examine Fig. 6.2, where the area of the rectangle containing an ellipse is $12 \times 16 = 192$ sq in.

Step 2: The area of an allipse is $ab\pi$, where π is $3\frac{1}{7}$. Note that a is half the major axis and b is half the minor axis.

Step 3: Then $3\frac{1}{7} \times 8 \times 6 \simeq 151$ sq in. (see the shaded portion of Fig. 6.2). *Your thinking should be: 3 × 8 × 6 = 144 plus what?*

6.7 TO FIND THE CIRCUMFERENCE OF AN ELLIPSE

Multiply one-half the sum of the major and minor axes by $3\frac{1}{7}$. The dimension C of the ellipse shown in Fig. 6.2 is

$$\frac{12 + 16}{2} \times 3\frac{1}{7} \simeq 44 \text{ in.}$$

Your thinking should be: 3 × 14 = plus what?

6.8 TO FIND THE VOLUME OF A SPHERE

Step 1: The approximate volume V of a sphere is equal to $\frac{2}{3}$ V of a cylinder with equal diameter D and height H.

Step 2: The approximate volume V of a 12″ sphere is equal to the base area of a cylinder with 12″ diameter D and H $\frac{2}{3}$ that of its diameter D.

Step 3: The area of the base is $3\frac{1}{7} \times 6 \times 6$ and the volume V of the sphere is $3\frac{1}{7} \times 6 \times 6 \times 8 \simeq 905$ cu in. *Your first thinking should be: 3 × 6 × 6 × 8 = 18 × 48 = 864 cu in. plus what?*

6.9 TO FIND THE SURFACE AREA OF A SPHERE

The surface area of a sphere is approximately equal to the lateral surface of a cylinder of the same diameter D and height H as the sphere. The surface area of a 12″-diameter sphere is $12 \times 3\frac{1}{7} \times 12 \simeq 453$ sq in. *Your first thinking should be: 12 × 3 × 12 = 432 plus what?*

Fig. 6.2 The area of the shaded portion equals the area of the ellipse.

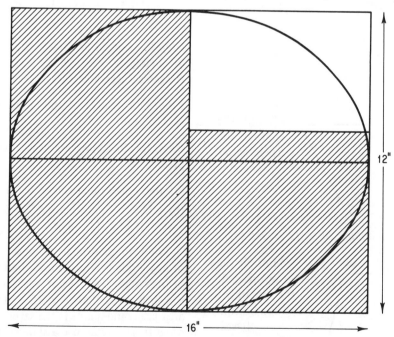

6.10 TO CHECK MULTIPLICATION

Example

Multiply 1379 by 3741 and check the answer by casting out nines. *It is important to read this paper completely and then study it.*

Step 1: Multiply:

		Check (sum of digits) in each row	Product	Digital remainder
(a)	1379	2	2 × 6 = 12	3
(b)	3741	6		
	1379			
	5516.			
	9653..			
	4137...			
(c)	5158839			

The answer line adds up to 39, which divided by 9 leaves 3   3

Step 2: Add together crosswise the figures in line (a) of Step 1: 1 + 3 + 7 + 9 = 20, which when divided by nine (cast out nines) yields a remainder of 2. This figure 2 is shown under the *Check* column in Step 1.

Step 3: Repeat the operation and write the remainder in the *Check* column as in Step 2 for line (b).

Step 4: Multiply the digital remainders of lines (a) and (b): 2 × 6 = 12 (see Step 1, *Product* column). Now cast out nines from the product of 12, leaving a final digital remainder of 3 in the *Digital remainder* column as in Step 1.

Step 5: Add all the figures (crosswise) in the answer line (c), Step 1. Cast out the nines, which leaves an *answer-line remainder of 3.*

Step 6: If the last *digital remainder of lines (a) and (b) and the answer-line remainder* (see Step 1) are equal, the multiplication is correct. In this case both the remainders are 3.

Problems

Multiply and check:

3584 × 753   2478 × 4853   78,901 × 234

To multiply by ¾: Calculate one-half the number and add one-half of *that* number; thus: ¾ of 4672 = (½ × 4672) + ½ (½ × 4672) = 2336 + 1168 = 3504. Try finding ¾ of the following:

56	49	473.08	1777.68
64	53	654.032	2413.44
76	97	998.72	8194.54

6.11 TO CHECK ADDITION

Step 1: Commencing with the units, add from the bottom upward. Write the carrying figure below the last line:

```
123456
210987
345678
432109
567890
1680120
12311  Carrying figures
```

Step 2: Check by commencing with the units, adding from the top downward and striking off the correct figure if correct.

If the work checks off correctly, there is no error unless it be a compensating error.

6.12 TO ADD COLUMNS OF DIMENSIONED FIGURES

Neatly list the dimensions, then add them in pairs and pairs again to resolutions as follows:

ft	in.
7	9¼
3	6½
5	11⅝
9	10¼
5	5½
7	7⅛
8	9⅝
9	11¼

11-3¾
15-9⅞
13-0⅝
18-8⅞

27-1⅝
31-9½

58-11⅛

6.13 THE TWENTY-SEVEN TIMES TABLE

As an estimator you will frequently be converting cu ft to cu yd. You should learn the 27 times table, which is as follows:

1 × 27 =	27	7 × 27 =	189
2 × 27 =	54	8 × 27 =	216
3 × 27 =	81	9 × 27 =	243
4 × 27 =	108	10 × 27 =	270
5 × 27 =	135	11 × 27 =	297
6 × 27 =	162	12 × 27 =	324

6.14 TO CHECK SUBTRACTION

Step 1: Subtract:

987654	Named minuend
456789	Named subtrahend
530865	Named remainder

Step 2: Add the subtrahend to the remainder; thus:

$$456789 = \text{subtrahend, Step 1}$$
$$530865 = \text{remainder, Step 1}$$
$$987654 = \text{minuend, Step 1}$$

If minuends are equal, unless a compensating error has been made, the work is correct.

6.15 TO CHECK DIVISION

Step 1: Divide 1234475 by 67; thus:

divisor) dividend (quotient

```
67)1234475(18425
   67
   564
   536
    284
    268
     167
     134
      335
      335
```

Step 2: Check by multiplying the quotient by the divisor; thus:

$$18425 \times 67 = 1234475$$

If the dividends are equal, unless a compensating error has been made, the work is correct.

An estimator's time is never wasted by checking and double checking all calculations. It is better that all work be checked by someone other than the originator of the work. One mistake may mean either not getting a contract or, even worse, losing money on an underpriced proposal bid.

6.16 MISCELLANEOUS CAPACITIES, MEASURES, AND WEIGHTS

(a) *Water*
  1. An American standard gal (gallon) weighs 8.337 lb and contains 231 cu in., and there are 7.48 gal in 1 cu ft—say 7½ gal per cu ft.
  2. An Imperial gal weighs 10 lb and contains 227.418 cu in., and there are 6.2321 gal in 1 cu ft, say 6¼ gal per cu ft.

(b) *Cement* weighs 94 lb per sack, which contains 1 cu ft.

(c) *Concrete* (cement, sand, and stone) weighs 144 ± lb per cu ft.

(d) *Gravel* or screened crushed stone weighs 2,500 lb per cu yd.

(e) *Earth* after excavation swells 5 percent to 50 percent according to character. Study soil-analysis reports in specifications (see Chap. 10).

6.17 ACCURACY TESTS

The following accuracy tests have been devised to help you to get used to the handling of numbers. An estimator will always have mental arithmetic to do, and you should develop a hunch for approximate estimates and then prove your hunch. I have given the following tests to hundreds of students, but I have never had one class where everybody got 100 percent correct.

It is recommended that you do one exercise per day, record your time, and then in a few weeks' time redo them, again recording your time. Then, and only then, do them by machine calculation and record your accuracy and time.

ACCURACY TEST NO. 1

Values	Question
17 (4×2) + (3×3)	1
3 (6×½)	2
10 (5×2)	3
10 (2×5)	4
Total	**40**

Add:

85	85	48	91	429	159	626
76	49	55	74	629	726	193
56	72	61	49	576	494	313
78	59	38	93	916	736	154
65	85	76	87	837	573	487
56	44	92	74	359	828	931
28	17	54	63	181	418	497
99	98	31	84	666	752	916
36	65	73	95	284	616	432
61	43	28	69	935	594	271
98	87	41	72	483	328	567
75	54	66	19	757	464	383
64	39	93	85	529	727	298

Subtract:

159	365	512	703	518	748
87	193	284	396	230	599

Multiply:

9546	1758	8345	2957	1684
32	61	75	49	72

Divide correct to three decimal places:

$$79)\overline{8734651} \qquad\qquad 63)\overline{2087198}$$

ACCURACY TEST NO. 2

Values	Question
20 (10×2)	1
6 (3×2)	2
8 (2×3) (1×2)	3
6 (6×1)	4
Total 40	

From mental calculation write in the answers:

$1\frac{1}{4} + 3\frac{3}{4} + 2\frac{1}{4} + 3 =$ $\qquad$ $10\frac{3}{11} + 7\frac{6}{11} - 8\frac{5}{11} =$

$2\frac{1}{3} + 3\frac{1}{3} + 4\frac{2}{3} + \frac{2}{3} =$ $\qquad$ $8\frac{1}{5} + 5\frac{3}{5} - 10\frac{2}{5} =$

$3\frac{5}{6} + 2\frac{1}{6} + 1\frac{3}{6} + 5\frac{4}{6} =$ $\qquad$ $7\frac{7}{8} + 4\frac{3}{8} - 9\frac{1}{8} =$

$8\frac{1}{5} + 4\frac{3}{5} + 5\frac{2}{5} + 7\frac{4}{5} =$ $\qquad$ $4\frac{2}{3} + 3\frac{2}{3} - 1\frac{1}{3} =$

$5\frac{4}{9} + 2\frac{2}{9} + 1\frac{6}{9} + 8\frac{7}{9} =$ $\qquad$ $5\frac{5}{9} + 6\frac{7}{9} - 8\frac{8}{9} =$

Divide:

$8\overline{)4160384}$ $\qquad$ $7\overline{)6311529}$ $\qquad$ $9\overline{)1335960}$

Multiply:

$$83754 \times 45 \qquad 37583 \times 63 \qquad 74185 \times 22$$

Find the total of the following:

6 prs hinges at $.52 per pair
43 lbs of nails at $.27 per lb
450 bd ft of lumber at $120.00 per M
78 lin ft of door trim at $.17 per lin ft
12 rolls of tar paper at $3.35 per roll

Total _____

ACCURACY TEST NO. 3

Values	Question	
12	1	**Add mentally and write in answers:**

$2\frac{1}{6}+3\frac{2}{3} =$ $10\frac{1}{5}+3\frac{3}{10} =$

$4\frac{2}{3}+1\frac{4}{9} =$ $8\frac{5}{6}+3\frac{2}{3} =$

$5\frac{1}{2}+4\frac{1}{4} =$ $4\frac{1}{4}+5\frac{5}{8} =$

$7\frac{3}{5}+5\frac{1}{10} =$ $3\frac{4}{9}+2\frac{2}{3} =$

$8\frac{3}{4}+3\frac{5}{8} =$ $7\frac{1}{3}+3\frac{11}{12} =$

$10\frac{1}{2}+6\frac{9}{10} =$ $6\frac{5}{8}+4\frac{1}{2} =$

Values	Question	
12 (6 × 1)	2	**Find the total value of each of the following:**

7 Pints of LINSEED OIL at $.48 per pint =
7 „   „   „   „   at $.65 per pint =
7 „   „   „   „   at $.73 per pint =
7 „   „   „   „   at $.88 per pint =
7 „   „   „   „   at $.90 per pint =
                                    Total _____

(6 × 1)

8 Pints of TURPENTINE at $.75 per pint =
5 „   „   „   at $.86 per pint =
4 „   „   „   at $.95 per pint =
3 „   „   „   at $.68 per pint =
12 „   „   „   at $.82 per pint =
                                    Total _____

Values	Question	
6 (2 × 3)	3	**Divide correct to two decimal places:**

$847)\overline{3086491} =$ $592)\overline{8316408} =$

Values	Question	
10	4	**Find the value of:**
(a) 5		
(b) 5		

(a) $(\frac{3}{5}$ of $8\frac{3}{4}) - (\frac{1}{4}$ of $8\frac{2}{5}) =$

(b) $(2\frac{1}{3} \times 3\frac{2}{3}) + (4\frac{1}{6} \times 6\frac{3}{7}) =$

Total   40

ACCURACY TEST NO. 4

Values	Question	
12 (3×4)	1	**Add:**

$$
\begin{array}{lll}
\$339.65 & \$827.43 & \$205.11 \\
573.68 & 169.39 & 464.38 \\
637.05 & 493.06 & 217.68 \\
326.54 & 575.61 & 592.39 \\
529.73 & 749.48 & 888.64 \\
609.74 & 393.74 & 516.81 \\
147.65 & 629.41 & 793.88 \\
821.49 & 287.79 & 475.61 \\
\hline \\
\hline
\end{array}
$$

Values	Question	
4 (4×1)	2	**Subtract:**

$$
\begin{array}{llll}
\$136.94 & \$173.72 & \$483.27 & \$261.09 \\
45.74 & 79.16 & 219.88 & 175.83 \\
\hline \\
\hline
\end{array}
$$

Values	Question	
10 (5×2)	3	**Find the total value of:**

15 tons slag at $8.35 per ton =
$24\frac{1}{2}$ tons slag at $7.40 per ton =
$33\frac{2}{3}$ tons slag at $9.00 per ton =
45 tons slag at $7.70 per ton =   _____

Total value =   _____

Values	Question	
14 ($21 \times \frac{2}{3}$)	4	**Find by mental calculation answers for the following:**

$$
\begin{array}{lll}
\frac{3}{4}+\frac{1}{3} = & \frac{5}{6}+\frac{4}{9} = & \frac{1}{3}+\frac{1}{2} = \\[4pt]
\frac{2}{3}+\frac{1}{2} = & \frac{3}{4}+\frac{5}{8} = & \frac{2}{5}+\frac{1}{4} = \\[4pt]
\frac{4}{5}+\frac{3}{4} = & \frac{5}{8}+\frac{1}{4} = & \frac{5}{8}+\frac{3}{4} = \\[4pt]
\frac{5}{8}+\frac{2}{3} = & \frac{11}{12}+\frac{3}{8} = & \frac{11}{12}+\frac{7}{8} = \\[4pt]
\frac{4}{7}+\frac{3}{5} = & \frac{7}{10}+\frac{4}{5} = & \frac{7}{8}+\frac{4}{5} = \\[4pt]
\frac{7}{8}+\frac{1}{12} = & \frac{5}{9}+\frac{5}{12} = & \frac{2}{5}+1\frac{1}{3} = \\[4pt]
\frac{11}{12}+\frac{3}{4} = & \frac{5}{8}+\frac{3}{4} = & \frac{6}{7}+\frac{3}{8} =
\end{array}
$$

Note: In questions such as at No. 3 above, an estimator's first thinking should be an approximate cost.

E.g., 15 tons of slag at $8.35 per ton is obviously something a little more than $15 \times 8 = \$120.00$, plus . . .

| Total | 40 | |

ACCURACY TEST NO. 5

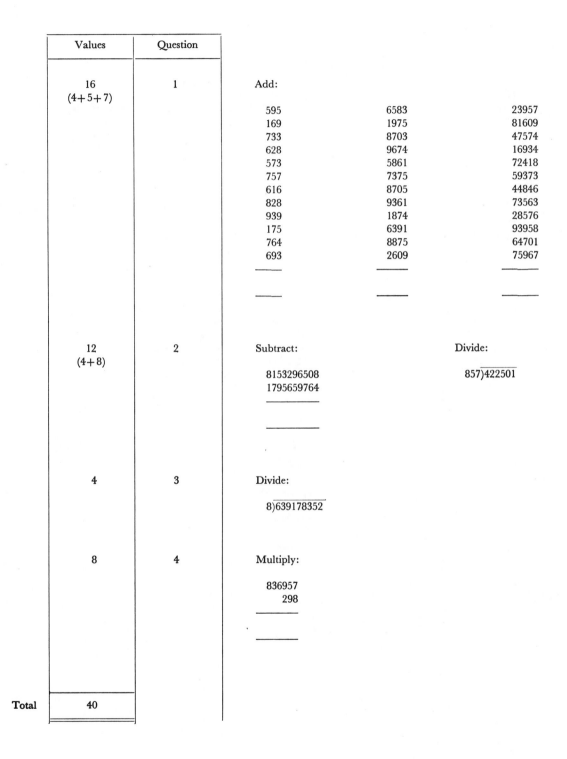

Values	Question	
16 (4+5+7)	1	**Add:**

595   6583   23957
169   1975   81609
733   8703   47574
628   9674   16934
573   5861   72418
757   7375   59373
616   8705   44846
828   9361   73563
939   1874   28576
175   6391   93958
764   8875   64701
693   2609   75967
———   ———   ———

———   ———   ———

Values	Question	
12 (4+8)	2	**Subtract:** **Divide:**

Subtract: Divide:

8153296508 $857 \overline{)422501}$
1795659764
—————

—————

| 4 | 3 | **Divide:** |

$8 \overline{)639178352}$

| 8 | 4 | **Multiply:** |

836957
  298
————

————

| **Total** | **40** | |

ACCURACY TEST NO. 6

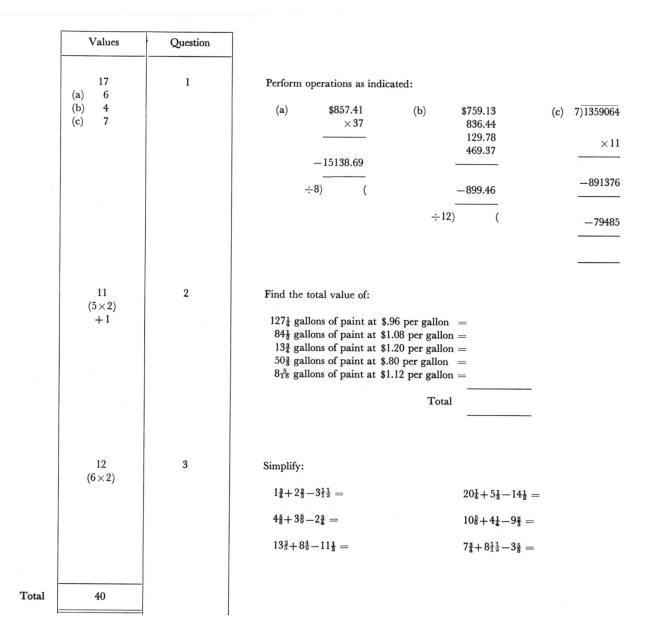

	Values	Question
	17	1
(a)	6	
(b)	4	
(c)	7	
	11	2
	(5×2)	
	+1	
	12	3
	(6×2)	
Total	**40**	

Perform operations as indicated:

(a) $857.41
   ×37
   ――――
   −15138.69
   ――――
   ÷8)   (

(b) $759.13
   836.44
   129.78
   469.37
   ――――
   −899.46
   ――――
   ÷12)   (

(c)   7)1359064
   ×11
   ――――
   −891376
   ――――
   −79485
   ――――
   ――――

Find the total value of:

$127\frac{1}{4}$ gallons of paint at $.96 per gallon =

$84\frac{1}{2}$ gallons of paint at $1.08 per gallon =

$13\frac{3}{4}$ gallons of paint at $1.20 per gallon =

$50\frac{3}{8}$ gallons of paint at $.80 per gallon =

$8\frac{5}{16}$ gallons of paint at $1.12 per gallon =

                    Total   ――――――
                            ――――――

Simplify:

$1\frac{3}{4}+2\frac{2}{3}-3\frac{11}{12} =$ $20\frac{1}{4}+5\frac{1}{3}-14\frac{1}{2} =$

$4\frac{5}{8}+3\frac{5}{9}-2\frac{3}{4} =$ $10\frac{5}{8}+4\frac{1}{4}-9\frac{2}{3} =$

$13\frac{2}{5}+8\frac{4}{9}-11\frac{1}{3} =$ $7\frac{3}{4}+8\frac{11}{12}-3\frac{5}{8} =$

ACCURACY TEST NO. 7

Values	Question
8 (4×2)	1
12 $(8 \times 1\frac{1}{2})$	2
8 (2×4)	3
12 (3×4)	4
Total 40	

1. Find the total value:

1563 cu yd of excavation at \$.96 per cu yd =
778 cu yd of excavation at \$1.13 per cu yd =
852 cu yd of excavation at \$1.36 per cu yd =
_____

Total _____

2. Multiply:

$216 \times 4\frac{1}{4} =$ $\qquad$ $516 \times 11\frac{2}{3} =$

$640 \times 2\frac{5}{8} =$ $\qquad$ $488 \times 6\frac{7}{8} =$

$366 \times 7\frac{1}{2} =$ $\qquad$ $705 \times 4\frac{3}{5} =$

$444 \times 6\frac{3}{4} =$ $\qquad$ $1640 \times 7\frac{7}{8} =$

3. Multiply: Divide (correct to two decimal places):

5937169
354
_____

_____

$749)\overline{49481570}$

4. Perform the operations as indicated:

(a) 759165
     $\times 7$
     _____

     -86591
     _____

8) _____

(b) \$658.46
     $\times 8$
     _____

     $+ \$1512.31$
     _____

11) _____

(c) 759.064
     $\times 11$
     _____

     -2418.907
     _____

7) _____

ACCURACY TEST NO. 8

	Values	Question
	10	1
(a)	4	
(b)	6	
	10	2
(a)	3	
(b)	4	
(c)	3	
	20	3
	(4 × 5)	
Total	40	

Find:

(a) $5\frac{3}{5} + 6\frac{2}{3} + 7\frac{4}{9} =$

(b) $1\frac{1}{2} \times 2\frac{7}{9} \times 7\frac{1}{5} =$

Find the remainder in each of the following:

(a) \$25.00 − (\$1.58 + \$3.09 + \$.93 + \$4.15 + \$2.64)

$$= \$25.00 - \qquad\qquad =$$

(b) \$51.80 − $\big[(15 \times \$1.95) + (37 \times \$.46)\big]$

$$= \$51.80 - \qquad\qquad =$$

(c) \$120.00 − $\big[(10\% \text{ of } \$3.25) + (\$16.20 + \$8.89 + \$19.95)\big]$

$$= \$120.00 - \qquad\qquad =$$

Add and divide:

$408.27	$864.05	73615	138.015
517.35	159.26	87492	69.24
633.48	347.10	19648	17.013
194.74	815.69	75764	436.7
626.59	293.83	83859	83.641
173.81	717.46	62605	7.46
616.74	593.63	89963	85.7406
359.43	148.59	57649	9.631
764.95	464.87	61675	174.3809
587.49	627.53	93968	58.75

5)   9)   6)   7)

Ans   Ans   Ans   Ans

ACCURACY TEST NO. 9

Values	Question
10 (5×2)	1
5 (5×1)	2
20 (a) (5×2) (b) (5×2)	3
5 $(2 \times 2\frac{1}{2})$	4
Total	40

Find the daily totals:

Mon	Tues	Wed	Thur	Fri.
$3.64	$5.73	$3.13	$7.46	$38.38
5.83	6.36	4.63	1.19	17.42
7.27	1.94	1.95	5.35	56.16
4.64	3.63	3.74	9.26	49.37
2.87	8.26	6.59	1.74	94.63
7.46	5.43	7.27	6.39	35.18
9.09	7.62	3.93	2.85	67.35
16.58	5.64	1.84	9.64	16.26
7.46	1.82	6.65	1.72	51.72
———	———	———	———	———
———	———	———	———	———

Subtract:

$17.63	$465.19	$837.41	$1000.00	$91.05
9.25	139.62	694.72	349.16	18.72
———	———	———	———	———
———	———	———	———	———

Multiply and find totals:

(a) $\quad 34 \times \$1.72 =$
$\qquad 19 \times \ 8.13 =$
$\qquad 63 \times \ 4.75 =$
$\qquad 22 \times \ 2.80 =$

   The Total

(b) $\quad 27 \times \$3.38 =$
$\qquad 14 \times \ 9.07 =$
$\qquad 50 \times \ 6.60 =$
$\qquad 73 \times \ 4.85 =$

   The Total

153,728 in. = ...yd ....ft ....in.

314,094 oz = ...cwt ....lb ....oz

ACCURACY TEST NO. 10

	Values	Question
	15	1
	$(10 \times 1\frac{1}{2})$	
	15	2
(a)	(6×1)	
(b)	$(6 \times 1\frac{1}{2})$	
	7	3
(a)	3	
(b)	4	
	3	4
Total	40	

Add and Subtract:

(a) $35\frac{2}{3} + 16\frac{4}{7} =$ (b) $100 - 11\frac{3}{7} =$

     $41\frac{5}{11} + 13\frac{3}{7} =$ $29\frac{1}{8} - 5\frac{1}{8} =$

     $15\frac{5}{6} + 10\frac{4}{5} =$ $73\frac{1}{3} - 16\frac{4}{11} =$

     $17\frac{1}{3} + 9\frac{3}{5} =$ $51\frac{5}{16} - 8\frac{1}{4} =$

     $14\frac{5}{8} + 11\frac{1}{4} =$ $46\frac{11}{12} - 13\frac{7}{8} =$

Multiply as indicated and add:

(a) $16 \times \$.65 =$ (b) $13 \times \$4.15 =$

     $37 \times .49 =$ $24 \times 2.05 =$

     $10 \times .33 =$ $31 \times 1.93 =$

     $55 \times .76 =$ $78 \times .65 =$

     $48 \times .98 =$ $56 \times 2.49 =$

       Total $=$ Total $=$

Multiply:

(a)   185936   (b)   6157938

       647   293

Cancel:

$$\frac{18 \times 45 \times 64 \times 108}{32 \times 24 \times 90} =$$

ACCURACY TEST NO. 11

Values	Question	
15 (5×3)	1	Add:

Add:

3687	4641	3218	5261	6214
1536	3090	1086	3937	1945
2749	5838	5947	1759	6367
4697	2647	6128	6484	5858
1758	6375	3386	9346	2749
9334	4596	4275	7275	3067
6461	2753	3659	3858	8378
1583	8361	9213	4345	9226
9264	1919	5484	9296	7553
3456	5758	6367	7564	3338
7277	6267	7576	4293	4845
9385	5934	1598	8656	8157

Question 2 — Values 7 (2×3) +1

Find the total value of:

1 carload of potatoes, 43,260 lb. net, at $1.42 per bus. and
1 carload of potatoes, 51,480 lb. net, at $1.67 pre bus.

There are 60 lb. in one bushel

= .... + ... =

Question 3 — Values 18 (18×1)

Find the values of:

$\frac{5}{8} + \frac{2}{3} =$ $\qquad$ $\frac{2}{3} + \frac{4}{5} =$ $\qquad$ $\frac{5}{6} - \frac{1}{3} =$

$\frac{1}{5} + \frac{4}{9} =$ $\qquad$ $\frac{4}{5} - \frac{3}{8} =$ $\qquad$ $\frac{4}{7} \times \frac{3}{8} =$

$\frac{2}{3} - \frac{1}{8} =$ $\qquad$ $\frac{11}{16} - \frac{5}{12} =$ $\qquad$ $\frac{10}{13} - \frac{5}{11} =$

$\frac{1}{2} + \frac{11}{16} =$ $\qquad$ $\frac{7}{8} + \frac{1}{12} =$ $\qquad$ $1\frac{1}{4} - \frac{11}{16} =$

$\frac{5}{9} + \frac{2}{5} =$ $\qquad$ $\frac{10}{11} - \frac{3}{7} =$ $\qquad$ $3\frac{1}{3} \div 1\frac{1}{6} =$

$\frac{3}{4} - \frac{5}{8} =$ $\qquad$ $\frac{7}{8} - \frac{4}{9} =$ $\qquad$ $\frac{2}{5}$ of $3\frac{3}{4} =$

| Total | 40 | |

ACCURACY TEST NO. 12

Values	Question
7 (3×2) +1	1
10 (3×3) +1	
12 (4 × 3)	2
11 (4×2½) +1	3
Total 40	

Find the total value of:

(a) 37 door lock sets at $3.90 each =
    51 door lock sets at $3.25 each =
    48 door lock sets at $3.40 each = _____

                    The total value = _____

(b) 7400 lineal feet of spruce at $99 per M =
    1575 lineal feet of spruce at $74 per M =
    3150 lineal feet of lath at $30 per M = _____

                    The total value = _____

Divide (correct to two decimal places as required):

(a) $37\overline{)362748}$ (b) $62\overline{)518363}$

(c) $86\overline{)713094}$ (d) $95\overline{)328410}$

How many inches in each of the following:

*10 rods, 3 yds, 1 ft, 9 in. = ...in.

48 rods, 1 yd, 2 ft, 10 in. = ...in.

32 rods, 4 yd, 1 ft, 6 in.   = ...in.

75 rods, 2 yd, 0 ft. 8 in.   = ...in.

                    The total   = _____ in.

*5½ yd = 1 rod, pole or perch.

ACCURACY TEST NO. 13

Values	Question
18 (8×2) +2	1
12 (a) 4 (b) 4 (c) 4	2
10 (3×3) +1	3

| Total | 40 |

Make out the following Time Sheet:

Name	Mon	Tues	Wed	Thur	Fri	Sat	Hours at $3.80	Amt. $ cts
F. Martin	8	$6\frac{1}{2}$	8	$7\frac{1}{2}$	6	4		
G. Scott	$7\frac{1}{4}$	$5\frac{1}{2}$	7	$6\frac{1}{2}$	8	4		
E. Wylie	8	7	7	8	7	5		
K. Grant	8	$5\frac{3}{4}$	$6\frac{1}{2}$	7	$6\frac{1}{2}$	8		
G. Fair	7	$6\frac{3}{4}$	8	$7\frac{1}{2}$	$7\frac{1}{4}$	6		
L. Boyle	$6\frac{1}{2}$	$8\frac{1}{4}$	7	$3\frac{3}{4}$	8	5		
M. Eady	$5\frac{3}{4}$	$6\frac{1}{2}$	8	$4\frac{1}{2}$	7	$6\frac{1}{4}$		
B. Cody	8	6	5	7	8	5		
Total amount of hrs at $3.80 =								

Simplify:

(a) $\frac{3}{4} + \frac{2}{3} + \frac{1}{8} =$

(b) $2\frac{1}{3} + 4\frac{5}{8} + 10\frac{3}{4} =$

(c) $\frac{5}{8} + 3\frac{1}{2} + 3\frac{1}{4} + 4\frac{1}{2} =$

Find the total of:

31,500 lb slag at $6.40 per ton =
44,700 lb slag at $8.00 per ton =
58,400 lb slag at $5.90 per ton =
_____

The total value   =
_____

Note: Question 3.
First make an approximate of cost as a guide, e.g., 31,500 lb of slag at $6.40 per ton is approximately 16 tons at $6.00 = $96.00. Use this method in estimating as a guide. The actual cost is $100.80

ACCURACY TEST NO. 14

Values	Question	

12
(2 × 6)
	1	**Compound Addition:**
		
		(a) yd ft in.
		    8 1 7
		  13 2 6
		   7 2 1
		  27 1 11
		   8 1 7
		
		(b) gal qt pt
		   5 3 1
		   3 2 0
		  13 1 1
		   6 3 1
		   8 2 0

8
(2 × 4)
	2	**Compound Subtraction:**
		
		(a) 135 yd 2 ft 5 in.
		    93 yd 2 ft 11 in.
		
		(b) 351 yd 0 ft 10 in.
		   196 yd 2 ft 7 in.

15 (3 × 5)	3	Find (1) by cancelling, the number of cords in each of the following piles, and (2) the value of each pile: (1 cord = 128 cu ft) (a) A pile of wood 32 ft long, 12 ft wide and 8 ft high at $6.40 per cord. The number of cords = The value of .... cords = (b) A pile of wood 80 ft long, 16 ft wide and 6 ft high at $8.00 per cord. The number of cords = The value of .... cords = (c) A pile of wood 48 ft long, 8 ft wide and 8 ft high at $12.50 per cord. The number of cords = The value of .... cords =
5	4	**Simplify:** (a) $5\frac{3}{4} - 1\frac{2}{3} =$ (b) $2\frac{5}{12} \times 2\frac{2}{3} =$
Total 40		

ACCURACY TEST NO. 15

Values	Question
12 (3 × 4)	1
12 (3 × 4)	2
12 (3 × 4)	3
4	4
Total 40	

Add:

$ 316.08	$ 751.09	$ 53.136
1584.61	1063.47	8.041
758.02	394.68	163.57
93.74	87.43	84.637
869.37	145.48	9.053
351.85	274.33	2510.8
109.43	58.47	63.04
7314.38	1543.85	139.57
852.45	738.84	400.095
196.14	69.69	8.1
3427.58	451.75	1040.538

Multiply:

$8715.40	$4539.46	$1564.90
158	98	271

Find by mental calculation and give your answer in feet and inches:

(a) $\frac{1}{2}$ of 3 ft 6$\frac{1}{4}$ in. =

(b) $\frac{1}{2}$ of 9 ft 3$\frac{3}{4}$ in. =

(c) $\frac{1}{4}$ of 27 ft 2$\frac{3}{4}$ in. =

Find the value of 1 barrel (36 gal) of vinegar at $.16 a pint.

The value = ... pints × $.16 =

ACCURACY TEST No. 16

	Values		Question
		15	1
(a)		9	
(b)		6	
		5	2
		8	3
	(4×2)		
		5	4
		7	5
(a)		4	
(b)		3	
Total		40	

The following two questions are compound addition:

(a)

	cu yds	cu ft	cu in.
	53	26	1014
	49	18	936
	37	27	109
	148	9	95
	70	14	1416
	83	23	809
	368	17	95

(b)

	cu yds	cu ft
	16	93
	33	26
	51	37
	19	92
	74	83
	26	109
	53	73

The following question is compound subtraction:

8 miles, 1503 yds, 1 ft, 10 in.
3 miles, 1695 yds, 2 ft, 8 in.

Change to cords and cubic feet as indicated:
(One cord = 128 cu ft)

6321 cu ft = ... cds .... cu ft

1465 cu ft = ... cds .... cu ft

7352 cu ft = ... cds .... cu ft

4930 cu ft = ... cds .... cu ft

Multiply the sum of 186.9 and 73.15 by their difference.

Answer = ... × .... =

Simplify:

(a) $2\frac{1}{5} + 3\frac{2}{3} + 6\frac{3}{8} =$

(b) $\frac{5}{8}$ of $\frac{2}{3} - \frac{3}{4}$ of $\frac{7}{18} =$

ACCURACY TEST NO. 17

Values	Question
12 (12 × 1)	1
6 (6 × 1)	2
16 (4 × 4)	3
6 (2 × 3)	
	4
Total 40	

1. Find mentally:

$\frac{3}{8}$ of 152 = $\frac{4}{5}$ of 225 = $\frac{1}{3}$ of 1011 =

$\frac{4}{9}$ of 162 = $\frac{3}{4}$ of 180 = $\frac{2}{7}$ of 518 =

$\frac{3}{5}$ of 310 = $\frac{2}{3}$ of 369 = $\frac{4}{11}$ of 693 =

$\frac{4}{7}$ of 448 = $1\frac{1}{2}$ of 888 = $\frac{8}{13}$ of 689 =

2. Subtract mentally each number from 5000 and put the remainder under each number

3373	4271	1394	2706	1927	4298
___	___	___	___	___	___
___	___	___	___	___	___

3. Add the following:

ft	in	ft	in	ft	in	ft	in
4	$9\frac{1}{4}$	16	$10\frac{7}{8}$	19	$9\frac{3}{8}$	21	$7\frac{1}{2}$
6	$7\frac{1}{2}$	17	$4\frac{1}{4}$	17	$8\frac{1}{8}$	7	$9\frac{1}{4}$
5	$3\frac{3}{8}$	5	$9\frac{1}{2}$	7	$9\frac{15}{16}$	12	$11\frac{3}{4}$
7	$11\frac{5}{8}$	7	$0\frac{3}{8}$	8	$10\frac{3}{8}$	17	$10\frac{1}{8}$
4	$9\frac{1}{2}$	16	$9\frac{7}{8}$	17	$7\frac{7}{8}$	36	$6\frac{5}{8}$
16	$3\frac{7}{8}$	13	$3\frac{3}{4}$	14	$4\frac{5}{16}$	9	$11\frac{1}{4}$

4. Change as indicated:

(a) 83,192 cu in. = ... cu yd, ... cu ft, ... cu in.

(b) 136 sections = 136 × .... acres = ... acres

ACCURACY TEST NO. 18

Values	Question	
10 (10×1)	1	**Find by mental calculation:**

Find by mental calculation:

10% of $\frac{1}{2}$ of $840 =   .05 of $\frac{3}{4}$ of 168 =

25% of $\frac{2}{5}$ of $638 =   .33$\frac{1}{3}$ of $\frac{3}{5}$ of 320 =

40% of $\frac{1}{3}$ of $720 =   .75 of $\frac{1}{3}$ of 204 =

16$\frac{2}{3}$% of $\frac{3}{4}$ of $1120 = $\frac{2}{3}$ of $\frac{5}{8}$ of 84 =

66$\frac{2}{3}$% of $\frac{5}{8}$ of $600 =   1$\frac{1}{4}$ of .08 of 175 =

4
(4×1) **2**

Subtract:

$195.00	$291.35	$639.21	$735.44
36.72	107.27	136.94	359.71
———	———	———	———
———	———	———	———

12
(3×4) **3**

Find the L.C.M. of

(a) 8, 5, 12, 16 and 20 =

(b) 4, 5, 6, 10 and 24 =

(c) 3, 8, 9, 12 and 16 =

6
$(4 \times 1\frac{1}{2})$ **4**

Simplify:

$\dfrac{3\frac{3}{8}}{1\frac{4}{5}} =$ $\dfrac{6\frac{1}{7}}{1\frac{10}{13}} =$ $\dfrac{7\frac{2}{7}}{1\frac{3}{14}} =$ $\dfrac{28\frac{1}{4}}{7\frac{2}{3}} =$

8
(2×4) **5**

Find:

(a) $13\frac{2}{3} + 5\frac{1}{5} + 10\frac{8}{15} \div 16\frac{1}{3} =$

(b) $8\frac{1}{9} + 6\frac{4}{5} - 11\frac{7}{15} \div \frac{8}{27} =$

Total   40

ACCURACY TEST NO. 19

Values	Question
20 (4×5)	1
20 $(8 \times 2\frac{1}{2})$	2
Total 40	

Add:

(a)	(b)	(c)	(d)
$100.94	$818.03	$ 69.16	$501.07
36.61	175.64	151.85	179.85
18.95	96.58	76.49	95.76
43.49	29.36	41.85	214.14
136.15	253.81	259.64	196.84
87.43	97.46	108.47	606.16
36.65	35.71	63.91	59.38
94.71	164.69	345.63	414.57
215.48	253.07	187.49	139.06
75.83	300.94	64.72	75.64
193.64	162.18	481.59	218.69
28.57	95.60	209.10	75.82
381.49	525.72	185.73	494.07

Simplify: (Note: In the following questions, complete the operations necessary above and below the lines first. This is important.)

$$\frac{\frac{3}{4}+\frac{2}{5}}{1\frac{2}{3}} = \qquad \frac{1\frac{5}{8}-\frac{5}{6}}{\frac{5}{12}} = \qquad \frac{16\frac{11}{13}}{5\frac{3}{14}} =$$

$$\frac{.15+.075}{.15} = \qquad \frac{.64-.316}{.18} = \qquad \frac{.1444 \div 4}{.19} =$$

$$(\tfrac{1}{5} \text{ of } 3\tfrac{3}{4}-\tfrac{5}{8}) \div \tfrac{1}{8} =$$

$$(\tfrac{5}{9} \text{ of } 3\tfrac{3}{7}+1\tfrac{2}{3}) \div \tfrac{2}{3} =$$

ACCURACY TEST NO. 20

Values	Question
(a) $\begin{array}{r}16\\8\\(4\times 1\frac{1}{2})\\+2\end{array}$ (b) do	1
$\begin{array}{c}16\\(4\times 4)\end{array}$	2
8	3
Total 40	

Find the total:

(a)
$$.13 \times 4.6 =$$
$$.05 \times 8.4 =$$
$$.012 \times .71 =$$
$$.71 \times 7.31 =$$

      Total

(b)
$$.075 \times 140.0 =$$
$$1.6 \times 15.8 =$$
$$.04 \times 16.4 =$$
$$8.5 \times 1.24 =$$

Simplify:

(a) $8\frac{2}{3} + \frac{1}{5} \div 1\frac{1}{15} =$

(b) $23 + 17\frac{4}{5} - 11\frac{2}{3} =$

(c) $3\frac{2}{5} \times 3\frac{8}{9} - 6\frac{5}{6} =$

(d) $(.016 + 8.43 + 1.004 - 1.125) \div .037 =$

By multiplying, find the yearly interest on the following sums at the rates shown:

$560	$475	$928	$63.75
8%	6%	7%	4%

$956	$83.60	$1650	$735
$5\frac{1}{2}$%	$6\frac{1}{4}$%	11%	8%

QUESTIONS

1. Using the following sets of figures give an example in each instance of the error of making a transposition of figures:

$367.69 $967.63 $396.76

2. Explain the use of the work-up sheet for estimating purposes.

3. Explain the following terms in estimating wood products:
   (a) lineal feet
   (b) board feet
   (c) square feet

4. Mentally calculate the approximate areas of circles having the following diameters:

5. By mental calculation, multiply the following numbers:

13×13 23×23 27×27
29×29 31×31 45×45

6. Multiply 7435×2346 and show proof of your answer by casting out of nines.

7. By mental calculation, multiply:

$125,000 \times 400$ 3640×70 $45,600 \times 300$

8. By mental calculation, divide:

375 by 25   3750 by 125

Answers to Accuracy Tests

Accuracy Test No. 1

1. 877   797   756   955   7581   7415   6068

2. 72   172   228   307   288   149

3. 305472   107238   625875   144893   121248

4. 110565.203   33130.127

Accuracy Test No. 2

1. $10\frac{1}{4}$ $9\frac{4}{11}$ 11 $3\frac{2}{5}$ $13\frac{1}{6}$ $3\frac{1}{8}$ 26   7 $18\frac{1}{9}$ $3\frac{4}{9}$

2. 520048   901647   148440

3. 3768930   2367729   1632070

4.
```
    3.12
   11.61
   54.00
   13.26
   40.20
  _____
  122.19
```

Accuracy Test No. 3

1. $5\frac{5}{6}$ $13\frac{1}{2}$ $6\frac{1}{9}$ $12\frac{1}{2}$ $9\frac{3}{4}$ $9\frac{7}{8}$ $12\frac{7}{10}$ $6\frac{1}{9}$ $12\frac{3}{8}$
   $11\frac{1}{4}$ $17\frac{2}{5}$ $11\frac{1}{8}$

2.
3.36	6.00
4.55	4.30
5.11	3.80
6.16	2.04
6.30	9.84
25.48	25.98

3. 3644.027   14047.986

4. $2\frac{9}{20}$ $35\frac{5}{9}$

Accuracy Test No. 4

1. 3985.53   4125.91   4154.50

2. 91.20   94.56   263.39   85.26

3.
```
   125.25
   181.30
   303.00
   346.50
  _____
   956.05
```

4. Read down:

$1\frac{1}{12}$	$1\frac{5}{18}$	$\frac{5}{6}$
$\frac{9}{10}$	$1\frac{7}{12}$	$\frac{13}{20}$
$1\frac{7}{36}$	$\frac{29}{36}$	$1\frac{7}{12}$
$1\frac{7}{24}$	$1\frac{7}{24}$	$1\frac{5}{36}$
$1\frac{6}{35}$	$1\frac{1}{2}$	$1\frac{27}{40}$
$\frac{23}{24}$	$1\frac{7}{36}$	$1\frac{11}{15}$
$1\frac{2}{3}$	$1\frac{3}{8}$	$1\frac{9}{35}$

Accuracy Test No. 5

1. 7470   77986   683476

2. 6357636744   493

3. 79897294

4. 249413186

Accuracy Test No. 6

1. Read down:

31724.17	2194.72	194152
16585.48	1295.26	2135672
2073.19	107.935	12442.96
		11648.11

2.
```
   122.16
    91.26
    16.50
    40.30
     9.31
   _____
   279.53
```

3.
$\frac{1}{2}$	$11\frac{1}{12}$
$5\frac{55}{72}$	$5\frac{5}{12}$
$10\frac{23}{45}$	$13\frac{1}{24}$

Accuracy Test No. 7

1. Read down:

1500.48
879.14
1158.72
3538.34

2.

918	6020
1680	3355
2745	3243
2997	12915

3. 2101757826 66063.51

4.

5314155	5267.68	8349.704
5227564	6779.99	5930.797
$653445\frac{1}{2}$	616.36	847.256

Accuracy Test No. 8

1. $19\frac{32}{45}$ 30

2. \$ 25.00 − 12.39 = 12.61
   \$ 51.80 − 46.27 = 5.53
   \$120.00 − 45.37 = 74.63

3. Read down:

4882.85	5032.01	706238	1080.5715
976.57	$559.11\frac{2}{9}$	$117706\frac{1}{3}$	$154.3673\frac{4}{7}$

Accuracy Test No. 9

1. 64.84 46.43 39.73 45.60 426.47

2. 8.38 325.57 142.69 650.84 72.33

3.

58.48	91.26
154.47	126.98
299.25	330.00
61.60	354.05
573.80	902.29

4.

4270 yd	0 ft	8 in.
196 cwt	30 lb	14 oz

Accuracy Test No. 10

1. Read down:
   $52\frac{5}{21}$ $54\frac{68}{77}$ $26\frac{16}{45}$ $25\frac{7}{8}$ $88\frac{4}{7}$
   $24\frac{1}{24}$ $56\frac{32}{33}$ $43\frac{1}{16}$ $33\frac{1}{24}$

Read down:

10.40	53.95
18.13	49.20
3.30	59.83
41.80	50.70
47.04	139.44
\$120.67	\$353.12

3. 120300592 1804275834

4. 81

Accuracy Test No. 11

1. 61187 58179 57937 72074 67697

2. 1023.82 + 1432.86 = \$2456.68

3. Read across:
   $1\frac{7}{24}$ $1\frac{7}{15}$ $\frac{1}{2}$ $\frac{29}{45}$ $\frac{17}{40}$ $\frac{3}{14}$ $\frac{13}{24}$
   $\frac{13}{48}$ $\frac{45}{143}$ $1\frac{3}{16}$ $\frac{23}{24}$ $\frac{9}{16}$ $\frac{43}{45}$ $\frac{37}{77}$
   $2\frac{6}{7}$ $\frac{1}{8}$ $\frac{31}{72}$ $1\frac{1}{2}$

Accuracy Test No. 12

1.

(a)	(b)
144.30	732.60
165.75	116.50
163.20	94.50
473.25	943.60

2. (a) 9804 (b) 8360.69 (c) 8291.791 (d) 3456.947

3. 2109 9574 6498 14930 Total 33111

Accuracy Test No. 13

1. Read across:
   40 hr; \$152.00 $38\frac{1}{4}$ hr; \$145.35 42 hr; \$159.60
   $41\frac{3}{4}$ hr; \$158.65 $42\frac{1}{2}$ hr; \$161.50 $38\frac{1}{2}$ hr; \$146.30
   38 hr; \$144.40 39 hr; \$148.20
   Total: 320 hr; \$1216.00

2. $1\frac{13}{24}$ $17\frac{17}{24}$ $11\frac{7}{8}$

3. 100.80 178.80 172.28 Total 451.88

Accuracy Test No. 14

1. 66 yd 0 ft 8 in. 38 gal 0 qt 1 pt

2. 41 yd 2 ft 6 in. 154 yd 1 ft 3 in.

3. 24 cords = \$153.60 60 cords = \$480.00
   24 cords = \$300.00

4. $4\frac{1}{12}$ $6\frac{4}{9}$

Accuracy Test No. 15

1. \$15873.65 \$5579.08 \$4480.58

2. \$1377033.20 \$444867.08 \$424087.90

3. $1'\text{-}9\frac{1}{8}''$ $4'\text{-}7\frac{7}{8}''$ $6'\text{-}9\frac{11}{16}''$

4. 288 pints × \$0.16 = \$46.08

Accuracy Test No. 16

1. (a) 813 cu yd 1 cu ft 1018 cu in. (b) 291 cu yd

2. 4 miles 1567 yd 2 ft 2 in.

3.

49 cords	49 cu ft	11 cords	57 cu ft
57 cords	56 cu ft	38 cords	66 cu ft

260.05 × 113.75 = 29580.6875

5. $12\frac{29}{120}$ $\frac{1}{8}$

Accuracy Test No. 17

1. Read across:
   57 180 337 72 135 148 186
   246 252 256 814 424

2. 1627 729 3606 2294 3073 702

3. Read across:

$44'\text{-}9\frac{1}{8}''$ $77'\text{-}2\frac{5}{8}''$ $86'\text{-}2''$ $106'\text{-}8\frac{1}{2}''$

4. 1 cu yd   21 cu ft   248 cu in.

$136 \times 640 = 87040$ acres

Accuracy Test No. 18

1. Read down:

42.00	63.80	96.00	140.00	249.975
6.30	64.00	51.00	35.00	17.50

2. 158.28   184.08   502.27   375.73

3. 240   120   144

4. $1\frac{7}{8}$ $3\frac{5}{7}$ 6 $3\frac{3}{4}$

5. $1\frac{4}{5}$ $11\frac{5}{8}$

Accuracy Test No. 19

1. 1449.94   3008.79   2245.63   3271.05

2. Read across:

69.100 $1\frac{9}{10}$ $3\frac{3}{13}$ 1.5   1.8   0.19   1 $5\frac{5}{14}$

Accuracy Test No. 20

1. (a) 0.598   (b) 10.5
      0.420   25.28
      0.00852   0.656
      5.1901   10.54
      ‾‾‾‾‾‾   ‾‾‾‾‾
      6.21662   46.976

2. (a) $12\frac{1}{11}$ (b) $(23 + 17\frac{4}{5}) - 11\frac{2}{3} = 29\frac{2}{15}$

(c) $6\frac{7}{18}$ (d) 225

3. Read across:

44.80   28.50   64.96   2.55   52.58
5.23   181.50   58.80

7

BOARD AND LINEAL MEASURE; STOCK MOLDINGS; PLYWOOD, FIBERBOARDS, AND NAILS

This chapter is designed to give you a complete understanding of the method of shopping for and computing lumber.

One type of the estimator's personal work-up sheet (which he retains) and the general estimate sheet as finally presented to the chief estimator are shown with a lumber bill work-up completed. Several lumber problems are presented for you to estimate, followed by a table of allowance for waste in wood.

Lastly, the chapter explains and shows sections of stock moldings and concludes with information about nails.

Your work from the outset must be set out methodically and neatly so that anyone can refer to it at any future time and check the method and resultant computations.

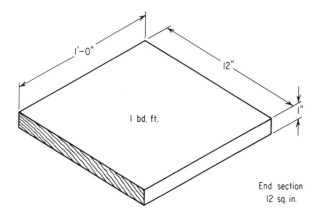

Fig. 7.1 One foot of board measure.

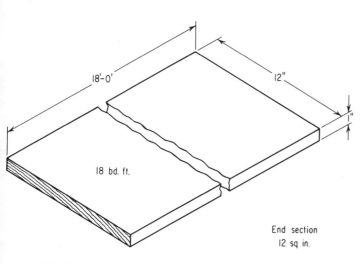

Fig. 7.2 Eighteen feet of board measure.

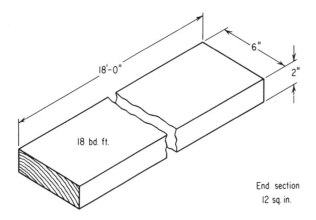

Fig. 7.3 Eighteen feet of board measure.

7.1 DIMENSION LUMBER

Dimension lumber in North America is marketed by the thousand-feet-board measure, abbreviated fbm or bm; sometimes it may be quoted as M bd ft. One fbm as cut on the saw is shown in Fig. 7.1. Note that the end section (area of the end) is 12 sq in. One fbm of lumber is equivalent to a piece of board 1-in. thick, 12-in. wide, and 1′-0″ long.

Note very carefully that when lumber is cut on the saw, the sizes as shown on the drawings are the actual sizes. When lumber has been dressed on the planing machine, it is reduced in size, but marketed in its original sawn sizes.

As an example, the trade method of writing certain lumber is 10/1 × 12–18′-0″. It is written in this order: number of pieces, thickness, width, and lastly the length. *This is very important and must be remembered.*

The thickness and width of dimension lumber is written in plain figures without the inch (″) sign, it being understood that the figures refer to its original sawn size. Lumber 2″ thick and 4″ wide off the saw is 1⅝″ thick and 3⅝″ wide off the planning machine; but is is called 2 × 4.

Example

Estimate the fbm in one piece of lumber 1 × 12 and 18′-0″ long, written 1/1 × 12–18′-0″ (see Fig. 7.2).

The end section is reckoned as 12 sq in.; therefore, for every foot of length there is 1 fbm. In 18′-0″ there are 18 fbm.

To convert an order into fbm, multiply the number of pieces by the thickness and width, multiply this product by the length in feet, and then divide by twelve. Thus

$$1/1 \times 12\text{--}18'\text{-}0'' = 1 \times \frac{1 \times 12}{12} \times 18 = 18 \text{ fbm}$$

Example

Estimate the fbm in 10/2 × 6–18′-0″. The end section 2 × 6 equals that of a board with an end section $\frac{1 \times 12}{12}$. Thus

$$10/2 \times 6\text{---}18'\text{-}0'' = 10 \times \frac{2 \times 6}{12} \times 18 = 180 \text{ fbm}$$

(See Fig. 7.3.)

Example

Estimate the fbm in 10/1 × 6–18′-0″.

$$10 \times \frac{1 \times 6}{12} \times 18 = 10 \times \frac{1}{2} \times 18 = 90 \text{ fbm}$$

Note that $\dfrac{1 \times 6}{12}$ is half the end section of a $\dfrac{1 \times 12}{12}$ board. (See Fig. 7.4.)

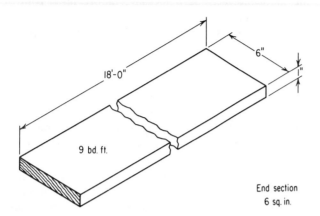

Fig. 7.4 Nine feet of board measure.

Example

Estimate the fbm in 10/1 × 5—18'-0".

$$10 \times \frac{1 \times 5}{12} \times 18 = 75 \text{ fbm}$$

(See Fig. 7.5.)

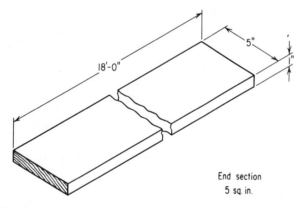

Fig. 7.5 Seven and one-half feet of board measure.

Example

$$10/2 \times 10\text{—}18'\text{-}0'' = 10 \times \frac{2 \times 10}{12} \times 18 = 300 \quad \text{fbm}$$

Note here that the end section is $\dfrac{2 \times 10}{12} = 1\frac{2}{3}$ that of a 1 × 12 board.

7.2 TRANSPOSITION OF FIGURES

Examine the following figures very carefully. By inspection and transposition of figures, the number of fbm may be reckoned mentally; thus:

$$10/2 \times 10\text{-}18'\text{-}0'' = 10 \times \frac{2 \times 10}{12} \times 18$$

which by transposition of figures is

$$10 \times \frac{2 \times 18}{12} \times 10 = 10 \times 3 \times 10 = 300 \text{ fbm}$$

Many such problems for small orders (but not all) may be resolved mentally by transposition of figures and setting them over the number 12.

Example

As Written	Mental Process	Answer
10/1 × 2 – 18'-0"	$\dfrac{10 \times 36}{12}$	30
12/1 × 5 – 16'-0"	$\dfrac{12}{12} \times 80$	80
10/1 × 4 – 18'-0"	$10 \times \dfrac{72}{12}$	60
18/2 × 8 – 14'-0"	$\dfrac{36}{12} \times 112$	336
6/1 × 10 – 14'-0"	$\dfrac{60}{12} \times 14$	70

Problem

By mental process complete the following:

$$4/2 \times 3 - 10'\text{-}0''$$
$$5/1 \times 4 - 18'\text{-}0''$$
$$2/1 \times 3 - 10'\text{-}0''$$
$$9/2 \times 4 - 8'\text{-}0''$$
$$2/3 \times 6 - 14'\text{-}0''$$

Board Measure

Find by mental calculation the number of fbm in each of the following lumber orders.

Values: (4 × 5)(1) 24/2 × 6 – 10'-0"
plus 5 for the   8/3 × 5 – 14'-0"
total   13/2 × 4 – 18'-0"
Total points per   10/1 × 4 – 18'-0"
set of five: 25   18/2 × 2 – 20'-0"

                            Total   _____

                  _____

(2) 108/2 × 4 – 10'-0"
    72/1 × 3 – 20'-0"
    21/4 × 6 – 8'-0"
    57/6 × 6 – 10'-0"
    19/5 × 8 – 12'-0"

                       Total   _____

(3)   14/1 × 6 – 16'-0"
      9/2 × 8 – 14'-0"
      20/1 × 3 – 18'-0"
      3/3 × 12 – 20'-0"
      17/2 × 6 – 16'-0"   _____
                    Total

                          _____

(4)   6/2 × 4 – 20'-0"
      6/2 × 4 – 14'-0"
      36/2 × 4 – 8'-0"
      5/2 × 6 – 14'-0"
      20/2 × 4 – 18'-0"   _____
                    Total

                          _____

Possible points: 100   Actual points: _____

7.3 WORK-UP SHEETS

A work-up sheet is what the name implies. It is used by the estimator to make all his calculations freehand, or if using the calculating machine, the typescript calculations should be boss-stitched to this paper so that it may be checked, preferably by another person. See p. 136, where the work-up for a priced lumber bill is shown; on p. 138 is shown the completed general estimate sheet.

The work-up sheets must be so neat and self-explanatory that at any future time any other person could follow the computations step by step. After the estimators are in agreement, the quantities are transferred to the general estimate sheet (see p. 138).

When using the work-up sheets, adhere rigidly to the following points:

(a) Date every sheet.

(b) Number every sheet. If there are forty sheets, the first one would be numbered "1 of 40" and so on up to "40 of 40."

(c) Every sheet in the series must have the same information under the heading "Building and Location."

(d) When the estimate is complete, check the series of sheets. (I know a firm that submitted an accepted bid only to find that they had forgotten to include one sheet with $14,000 worth of labor and materials.)

(e) *Remember that it is a cardinal rule among estimators that no computations are ever altered or destroyed until conclusively proved to be incorrect.*

7.4 HOW TO USE THE WORK-UP SHEETS

(a) The left-hand margin should be used for main headings.

(b) If you are not a legible writer, you should print your work.

(c) The center of the paper should be used for calculations only.

(d) The right-hand column should be used for answers only.

(e) There should be a space of one or two lines left between the calculations of each main heading.

Remember that the work-up sheets are to show the method of calculating and that the general estimate sheets are made up from the work-up sheets.

7.5 CHECKING ESTIMATES

It is common practice for two estimators to estimate (take off) the quantities for a project independently of each other. *This checking is not only against errors in computation, it is also against any item being missed on the estimate.* If an item is missed, it is a three-way loss: first, the item has to be purchased; second, the labor costs for its installation may exceed the purchase cost; and third, if the item had been included, there would have been a profit on it.

The work-up sheets should only leave the estimator's desk to be used in defense of the method and correctness of the calculations. See the general estimate sheet on p. 000 where (without any computations) all the necessary information from the work-up sheet is shown. *Remember that the work-up sheets are yours, but the general estimate sheets go from you to the chief estimator.* When checking estimates, it requires men of fair and open minds to reach complete agreement on all phases of the estimate.

7.6 NEW LUMBER STANDARDS

The following pages have been excerpted from a special report on new lumber standards which were developed, simultaneously, in the United States and Canada. The material is here presented through the courtesy of the Canadian Wood Council.

Lumber Standards and Design Data

Lumber Standards

"Sawn lumber" is the product of a sawmill manufactured by sawing, chipping or any combination thereof, edging, resawing or cross cutting to length. It may also be processed through a planing mill for surfacing one or more sides and, possibly, precision end trimming. This includes various lumber products such as dimension lumber, timber, boards and finish lumber.

Standards for softwood lumber provide a common base for classification, grading, measurement and assignment of stresses for these lumber products. Lumber standards, in effect, regulate the product for the benefit of producers, designers, builders, building officials and consumers. Although this publication applies to Canadian lumber standards, some information is provided about standards in the U.S. Current lumber standards in Canada and in the U.S. are similar; they were developed simultaneously, primarily because Canada exports more than half of its lumber to the U.S. Documents and authorities governing softwood lumber standards in Canada and in the U.S. are shown in Table 1.

SPECIES AND SPECIES GROUPS

Many softwood lumber species in Canada are harvested, manufactured and marketed together. Some of these have similar performance properties; they can be easily used together or interchangeably. In fact, some species cannot be distinguished from others by visual inspection after manufacture. For convenience, certain commercial species are combined into a single species designation and marketed under a group name. The commercial species groups and the characteristics of the species in each are shown in Table 2.

Although the northern aspen and black cottonwood species groups are technically hardwoods, they are graded and marketed under softwood standards. Some other hardwoods, such as maple, beech, birch and oak, can also be graded under softwood standards so that allowable unit stresses can be assigned.

Allowable unit stresses are assigned to each of the commercial species groups. The assigned stresses for the group are based on the value for the weakest species for each applicable property such as bending, compression, tension and shear. Sometimes, certain species of lumber are marketed individually. Allowable unit stresses for individual species, properly identified by grade stamps, are obtained by applying factors to the stresses for the group in which it is listed. Allowable stresses are not provided for the black cottonwood species group because it is not usually marketed in Canada; generally, it is weaker than the northern aspen group.

LUMBER CATEGORIES AND GRADES

Lumber grading is an orderly system of classification to provide material for a specific end use. For convenience in distribution and merchandising, lumber grades are limited in number, yet satisfy consumer needs.

There are five categories for dimension lumber (2 to 4 inches in thickness), two categories for timber (5 inches or more in least nominal dimension), one category for decking and four categories for boards; all except the board categories are shown in Table 3. Boards are not "stress-graded" and can be graded under any of five different grading rules, all of which are contained in the **NLGA** *Standard Grading Rules for Canadian Lumber,* published by the **National Lumber Grades Authority.** If allowable stresses are required, the boards must be graded as dimension lumber.

All categories of lumber are divided into grades, to additionally separate material according to quality for a particular end use. The principal uses of the grades within each category are given in Table 3. Use of material need not be restricted to the principal uses of the category that it is graded in; for example, a timber graded as "beams and stringers" might be used as a column, and a piece of dimension lumber graded as "light framing" might be used in a truss where "structural light framing" is more often used. Usually, if there is a choice, the appropriate category is structurally most efficient, i.e., more bending strength is obtained out of a piece graded for "beams and stringers" than one graded for "posts and timbers." The allowable stresses used must be those assigned to the grade and category that the material was graded in.

This piled lumber clearly illustrates good storage practices — covered piles for quicker air drying and proper support from ground and between layers to maintain shape and straightness.

TABLE 1

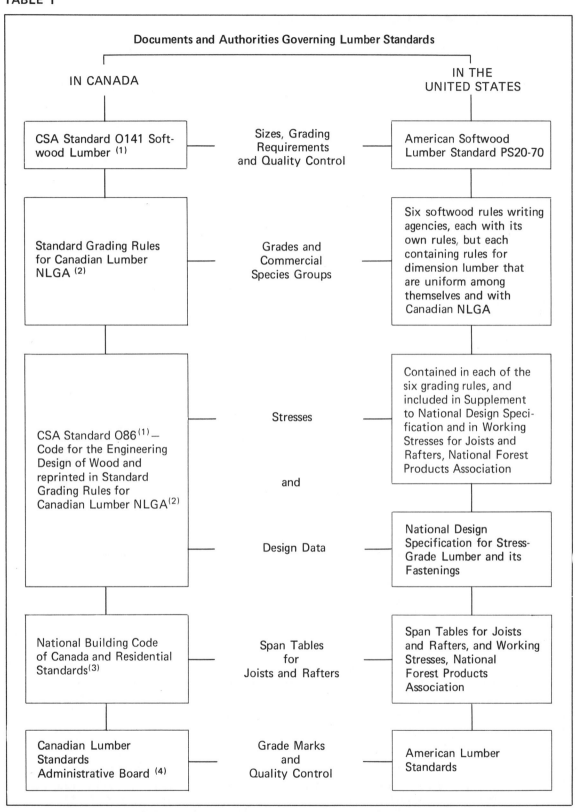

Documents and Authorities Governing Lumber Standards

IN CANADA		IN THE UNITED STATES
CSA Standard O141 Soft-wood Lumber (1)	Sizes, Grading Requirements and Quality Control	American Softwood Lumber Standard PS20-70
Standard Grading Rules for Canadian Lumber NLGA (2)	Grades and Commercial Species Groups	Six softwood rules writing agencies, each with its own rules, but each containing rules for dimension lumber that are uniform among themselves and with Canadian NLGA
CSA Standard O86 (1) — Code for the Engineering Design of Wood and reprinted in Standard Grading Rules for Canadian Lumber NLGA (2)	Stresses	Contained in each of the six grading rules, and included in Supplement to National Design Speci-fication and in Working Stresses for Joists and Rafters, National Forest Products Association
	and	
	Design Data	National Design Specification for Stress-Grade Lumber and its Fastenings
National Building Code of Canada and Residential Standards(3)	Span Tables for Joists and Rafters	Span Tables for Joists and Rafters, and Working Stresses, National Forest Products Association
Canadian Lumber Standards Administrative Board (4)	Grade Marks and Quality Control	American Lumber Standards

Notes: (1) Canadian Standards Association (CSA), 178 Rexdale Blvd., Rexdale, Ontario M9W 1R3.
(2) National Lumber Grades Authority (NLGA), 1500-1055 West Hastings St., Vancouver, British Columbia V6E 2H1.
(3) Division of Building Research, National Research Council (DBR/NRC), Montreal Road, Ottawa, Ontario K1A 0R7.
(4) Canadian Lumber Standards Administrative Board (CLS AB) 1475-1055 West Hastings St., Vancouver, British Columbia V6E 2E9.

TABLE 3 Lumber Categories, Grades and Uses in Canada

Lumber Product	Grade Category	Nominal Sizes (Inches)	Grades	Principal Uses
Dimension Lumber	Light Framing	2″ to 4″ Thick, 2″ to 4″ Wide	Construction	Widely used for general framing purposes. Pieces are of good appearance but graded primarily for strength and serviceability.
			Standard	
			Utility	Used in non-load bearing walls where economical construction is desired for such purposes as studding, blocking, plates and bracing.
			Economy*	Temporary or low cost construction where strength and appearance are not important.
	Structural Light Framing	2″ to 4″ Thick, 2″ to 4″ Wide	Select Structural	Intended primarily for use where high strength, stiffness and good appearance are desired, such as trusses.
			No. 1	
			No. 2	For most general construction uses.
			No. 3	Appropriate for use in general construction where appearance is not a factor, such as studs in non-load bearing walls.
	Stud	2″ to 4″ Thick, 2″ to 6″ Wide	Stud	Special purpose grade intended for all stud uses.
			Economy Stud*	Temporary or low cost construction where strength and appearance are not important.
	Structural Joists and Planks	2″ to 4″ Thick, 5″ and Wider	Select Structural	Intended primarily for use where high strength, stiffness and good appearance are desired.
			No. 1	
			No. 2	For most general construction uses.
			No. 3	Appropriate for use in general construction where appearance is not a factor.
			Economy*	Temporary or low cost construction where strength and appearance are not important.
	Appearance	2″ to 4″ Thick, 2″ and Wider	Appearance	Intended for use in general housing and light construction where lumber permitting knots but of high strength and fine appearance is desired.
Decking	Decking	2″ to 4″ Thick, 6″ and Wider	Select	For roof and floor decking where strength and fine appearance are required.
			Commercial	For roof and floor decking where strength is required but appearance is not so important.
Timber	Beams and Stringers	5″ and Thicker, Width More Than 2″ Greater Than Thickness	Select Structural	For use as heavy beams in buildings, bridges, docks, warehouses and heavy construction where superior strength is required.
			No. 1	
			Standard*	For use in rough, general construction.
			Utility*	
	Posts and Timbers	5″ x 5″ and Larger, Width Not More Than 2″ Greater Than Thickness	Select Structural	For use as columns, posts and struts in heavy construction such as warehouses, docks and other large structures where superior strength is required.
			No. 1	
			Standard*	For use in rough, general construction.
			Utility*	

Note: All grades are "stress graded" meaning that working stresses have been assigned (and span tables calculated for dimension lumber) except those marked *.

LUMBER SIZES AND PROPERTIES OF SECTIONS

Lumber thickness and width are referred to as "nominal size" in inches of the traditional rough piece of wood before being dressed or planed to actual size (Figure 1). Green or unseasoned lumber has a moisture content of more than 19%; dry lumber, a maximum of 19% which is dried to average 15%. Green lumber is dressed to a slightly larger size than dry lumber so that when used under similar moisture conditions, green lumber will shrink to equivalent size and load capacity of lumber surfaced dry. Size differentials between green and dry dressed lumber are listed in Table 4; lumber sizes — nominal, green and dry — for common sizes of lumber are given in Table 5. Timbers are always dressed green, $1/2$ inch under nominal size.

In design, dry sizes are used for section properties and allowable stresses, even if the material is surfaced green, unless the average moisture content of the wood over a year exceeds 19%. Generally, material dressed green will dry to less than 19% MC before construction of a building is complete. Section properties for dry lumber are given in Table 5.

QUALITY CONTROL

The quality of lumber products is controlled to ensure that the end product meets minimum specified standards. Overall control is the responsibility of the Canadian Lumber Standards Administrative Board which reports to the Canadian Standards Association Certification Policy Board. Eleven lumber manufacturing associations and three independent grading agencies are licensed by CLS to grade mark lumber (Figure 2). The CLS Administrative Board ensures that graders are qualified and that grading associations or agencies supervise their graders. A check-grading service is also provided to periodically verify work of graders.

Dimension lumber, decking and boards for wall sheathing, subflooring and roof sheathing are grade marked about 2 feet from one end of each piece so that the mark will be clearly visible during construction of the building, before sheathing and cladding are in place. Only grade marked lumber is accepted by Central Mortgage and Housing Corporation for use in housing in Canada financed under the National Housing Act; it is also specified in Part 9 of NBCC for houses and small buildings. In the U.S., the Federal Housing Administration has a similar requirement.

Although the grade mark design varies between agencies (Figure 2) each grade mark shows the name (or symbol or both) of the grading agency; the number (or name or both) of the mill; the species or species group designation; the grade; and if on lumber thinner than 4-inch nominal, either S-GRN (for surfaced green), S-DRY (for surfaced dry) or MC15 (for lumber specially dried to 15% or less moisture content).

Timber, which is 5 inches or more in least nominal dimension, may be grade marked or a certificate may

FIGURE 1

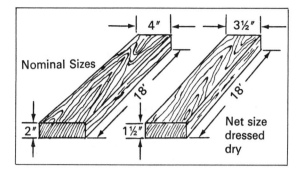

Nominal Sizes

4"   3½"   18'   18'

2"   1½"   Net size dressed dry

TABLE 4 Size Differentials for Dressed Sizes

Lumber Products	Nominal Width or Thickness (Inches)	Size Differential Between Green and Dry Dressed Sizes (Inches)
Boards and Dimension Lumber	1½ or less	$1/32$
	2 to 4½	$1/16$
	5 to 7	$1/8$
	8 or more	$1/4$
Timbers	5 or more	Always dressed green, ½" under nominal size

be obtained from an inspection agency for a complete shipment. Identification requirements should be included in the specifications, especially if structurally graded material (Select Structural or No. 1 grades) is required such as for columns and beams.

Both visual grading and mechanical stress grading are provided for in lumber standards. At present, most lumber is visually graded. The grader examines each piece on all four faces and assigns a grade based on the natural characteristics of the wood such as slope of grain, size and location of knots, checks and splits. Sometimes, lumber is mechanically stress-graded; lumber is passed through a machine and transverse deformation of the piece determines stiffness and strength of the piece. All grading is done in accordance with the **NLGA** *Standard Grading Rules for Canadian Lumber.*

Lumber that is resawn lengthwise or remanufactured must be regraded because changes in location of the characteristics will affect the performance of the material. End-jointed dimension lumber is acceptable if produced by plants certified under **CSA O268,** *Qualification Code for Manufacturers of Glued End Jointed Structural Lumber,* and properly identified.

7.7 EXAMPLE OF THE WORK-UP SHEET

Using the work-up sheet, take off the cost of the following lumber bill. All the work must be shown clearly so that it can be checked by another estimator. See p. 139 for an example.

$$82/1 \times 4 \; - \; 16\,'\text{-}0\,'' \; @ \; \$90 \text{ per M fbm}$$
$$76/2 \times 3 \; - \; 12\,'\text{-}0\,'' \; @ \; \$82 \text{ per M fbm}$$
$$56/1 \times 10 - 14\,'\text{-}0\,'' \; @ \; \$94 \text{ per M fbm}$$
$$44/2 \times 10 - 16\,'\text{-}0\,'' \; @ \; \$86 \text{ per M fbm}$$
$$160/2 \times 6 \; - \; 8\,'\text{-}0\,'' \; @ \; \$84 \text{ per M fbm}$$

Examine the computation of the first item.

Step 1:

$$82/1 \times 4 - 16\,'\text{-}0\,'' = \frac{82 \times 1 \times 4 \times 16}{12} = 437\tfrac{1}{3}$$
$$= 437 \text{ fbm}$$

Step 2:

$$437 \; @ \; \$90 \text{ per M fbm} = \frac{437 \times 90}{1000} \text{ or } 437 \times 0.090$$
$$= \$39.33.$$

Step 3: Use "horse sense." By inspection one may see that 437 fbm is something less than half a thousand. The price is $90 per M, so something less than $45 is the answer. The answer is $39.33, which is something less than half of $90. *This type of thinking is a must for estimators.*

Problem

Using the work-up sheets, work up the following unpriced lumber orders. Allow 1 point for each separate computation plus 1 point for the correct total of each bill, making 6 points per problem. The total number of possible points 24 for four correct bills.

Carefully note that it is trade practice to discard thirds of fbm; for example, 133⅓ fbm becomes 133 fbm and 266⅔ fbm becomes 267 fbm.

Lumber Order No. 1:
$$24/2 \times 6 — 18\,'\text{-}0\,''$$
$$8/3 \times 5 — 14\,'\text{-}0\,''$$
$$13/2 \times 4 — 18\,'\text{-}0\,''$$
$$10/1 \times 4 — 18\,'\text{-}0\,''$$
$$18/2 \times 2 — 20\,'\text{-}0\,''$$

Lumber Order No. 2:
$$108/2 \times 4 — 10\,'\text{-}0\,''$$
$$72/1 \times 3 — 20\,'\text{-}0\,''$$
$$21/4 \times 6 — 8\,'\text{-}0\,''$$
$$57/6 \times 6 — 10\,'\text{-}0\,''$$
$$19/5 \times 8 — 12\,'\text{-}0\,''$$

Lumber Order No. 3:
$$14/1 \times 6 — 16\,'\text{-}0\,''$$
$$9/2 \times 6 — 14\,'\text{-}0\,''$$
$$20/1 \times 3 — 18\,'\text{-}0\,''$$
$$3/3 \times 12 — 20\,'\text{-}0\,''$$
$$17/2 \times 6 — 16\,'\text{-}0\,''$$

Lumber Order No. 3:
$$6/2 \times 4 — 20\,'\text{-}0\,''$$
$$6/2 \times 4 — 14\,'\text{-}0\,''$$
$$36/2 \times 4 — 8\,'\text{-}0\,''$$
$$5/2 \times 4 — 14\,'\text{-}0\,''$$
$$20/2 \times 4 — 10\,'\text{-}0\,''$$

7.8 PRICING LUMBER

Discard thirds of fbm: 257⅓ becomes 257 fbm. Take up to the next higher number two-thirds fbm: 266⅔ becomes 267.

In computing cash, discard half-cents: $10.555 becomes $10.55. Take up to the next higher cent items ending in over a half-cent: $10.556 becomes $10.56. Most estimators now totally discard all cents.

7.9 TRANSFERRING NUMBERS

On transferring numbers be aware of the very common error of making a reversion of figures. As an example, it is very easy to read 428/2 × 4 – 16′-0″ and it is just as easy to miswrite this as 482/2 × 4 – 16′-0″.

An estimator should check and recheck all his work or, better still, have someone else check it also.

7.10 THE GENERAL ESTIMATE SHEET

The general estimate shown on p. 161 is extensively used in North America, although some companies prefer to devise and use their own types. Irrespective of the type used, it must be complete with all the information necessary for anyone to be able to read it at any time and, if necessary, to check the calculated quantities. No sign or symbol of any description should be used on the general estimate sheet that can possibly be left off. It should contain only relevant and pertinent information, so that others can use the sheets with no difficulty. Extraneous matter must be excluded.

The general estimate sheet must be:

(a) completely headed for building, location, architects, subject (*Note:* the subject is useful for future filing—say, schools with schools, churches with churches, and so on—this is also useful for checking prices with previous estimates of comparable jobs);

(b) given an estimate number—some companies start a new series each year;

(c) given a sheet number (*Note:* the sheets should be numbered from one to the number of the final sheet and each sheet should be numbered thus: ''Sheet 1 of (last number) 39,'' ''Sheet 2 of 39,'' to ''Sheet 39 of 39'');

(d) dated, and bear the estimator's name and the initials of the checker.

Under column 1 each main heading should be underlined and terminate with a colon (:), and subheadings should be indented but not underlined, thus:

<u>Excavations:</u>

        Boiler Room:
        East Wing:

and so on. See the general estimate for brickwork in Chapter 16. *The estimator must be neat, thorough, and methodical.*

Problem

Transfer the following priced lumber order to a work-up sheet and then to a general estimate sheet. Use the score sheet, 25 points for each order and 100 points for the complete paper.

	Values		*Score*
(1)	20 plus 5 for Total	16/2×6 — 16'-0" @ \$96 per M fbm 4/3×6 — 16'-0" @ \$104 per M fbm 72/2×10 — 16'-0" @ \$106 per M fbm 104/2×12 — 18'-0" @ \$108 per M fbm 86/3×10 — 18'-0" @ \$115 per M fbm	_____
(2)	20 plus 5 for Total	26/2×4 — 18'-0" @ \$98 per M fbm 26/2×4 — 10'-0" @ \$94 per M fbm 32/2×8 — 18'-0" @ \$102 per M fbm 64/2×4 — 18'-0" @ \$99 per M fbm 106/2×6 — 16'-0" @ \$101 per M fbm	_____
(3)	20 plus 5 for Total	74/3×12 — 14'-0" @ \$80 per M fbm 74/4×12 — 14'-0" @ \$84 per M fbm 74/6×12 — 18'-0" @ \$86 per M fbm 74/8×12 — 20'-0" @ \$88 per M fbm 74/10×13 — 22'-0" @ \$92 per M fbm	_____

(4)	20 plus 5 for Total	500/2×4 — 18'-0" @ \$86 per M fbm 70/2×6 — 18'-0" @ \$80 per M fbm 20/2×4 — 18'-0" @ \$94 per M fbm 8/2×6 — 8'-0" @ \$88 per M fbm 52/2×10 — 16'-0" @ \$88 per M fbm	

Final score   _____

Many companies discard cents on the general estimate sheets. When the figure is less than a half-dollar, it is discarded; and when the figure is more than a half-dollar, it is taken up to the next higher dollar.

7.11 ALLOWANCE FOR WASTE IN WOOD

A 1×8 common board is understood to be $1'' \times 8''$ board (end section) as it leaves the saw. The finished product, after having been through the planing machine, will finish $\frac{3}{4}'' \times 7\frac{1}{2}''$. As a consequence, an allowance must always be made for waste both in production and in application on the job by the carpenters. The following table gives the allowance for some of the wood products after application.

7.12 TABLE OF ALLOWANCE FOR WASTE IN WOOD

	Add
Common 1 × 8 boards laid square	10%
Common 1 × 8 boards laid diagonally	17
Shiplap 1 × 6 laid square	20
Shiplap 1 × 6 laid diagonally	25
Shiplap 1 × 8 laid square	15
Shiplap 1 × 8 laid diagonally	20
Shiplap 1 × 10 laid square	13
Shiplap 1 × 10 laid diagonally	17
Drop siding 6″	20
Lap siding 6″ with 4″ to the weather	50
Lap siding 6″ with 4½″ to the weather	25
Flooring T & G (Tongue and Groove) 6″	20

7.13 ALLOWANCE FOR WASTE IN WOOD FOR HARDWOOD FLOORING ½″ OR ¹³⁄₁₆″

	Add
1½″ *face*	50%
1¾″ face	40
2″ face	38

7.13 (continued)

	Add
2¼″ face	33
2½″ face	30
1½″ face ⅜″ thickness	33
1¾″ face ⅜″ thickness	30
2″ face ¼″ thickness	25
Building paper (400-sq-ft roll) allow	8

(Note that building paper is purchased by weight per roll.)

7.14 STOCK MOLDINGS

The names, the spellings, and the uses of stock moldings should be memorized. Study the sections and be aware that many of these molds may be found in mediums other than wood. Many of them come from the Greeks and have been in use for over 2,000 years.

The following molding patterns were supplied by Western Wood Products Association, 700 Yeon Building, Portland, Oregon 97204 and are reproduced here by permission.

CROWNS

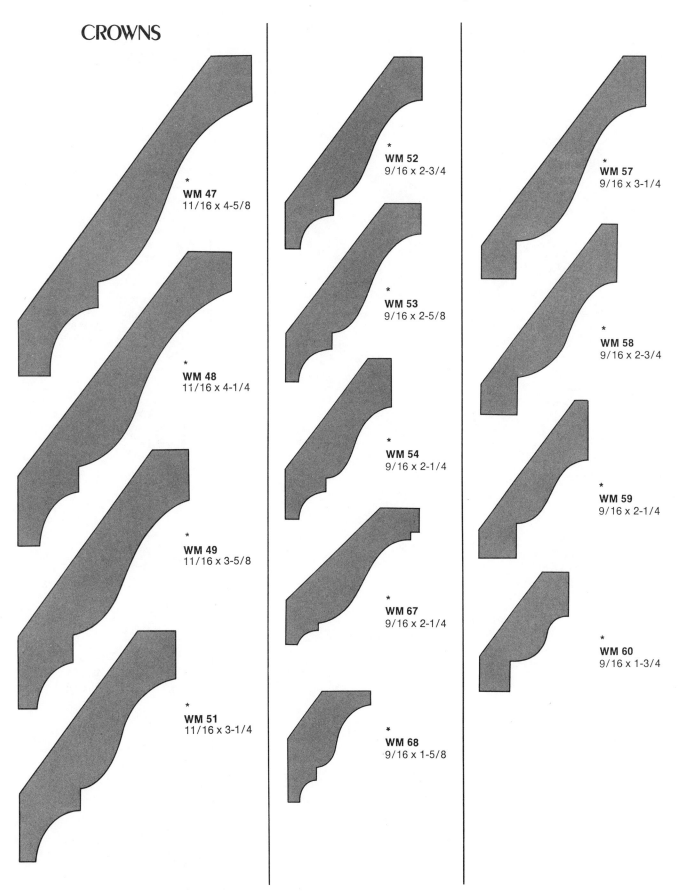

*
WM 47
11/16 x 4-5/8

*
WM 48
11/16 x 4-1/4

*
WM 49
11/16 x 3-5/8

*
WM 51
11/16 x 3-1/4

*
WM 52
9/16 x 2-3/4

*
WM 53
9/16 x 2-5/8

*
WM 54
9/16 x 2-1/4

*
WM 67
9/16 x 2-1/4

*
WM 68
9/16 x 1-5/8

*
WM 57
9/16 x 3-1/4

*
WM 58
9/16 x 2-3/4

*
WM 59
9/16 x 2-1/4

*
WM 60
9/16 x 1-3/4

BEDS

COVES

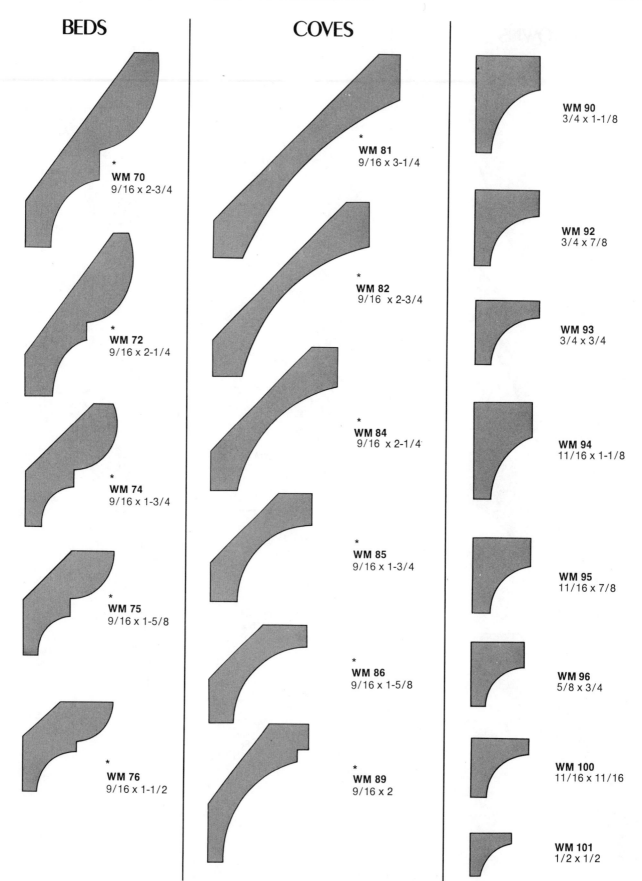

WM 70
9/16 x 2-3/4

WM 72
9/16 x 2-1/4

WM 74
9/16 x 1-3/4

WM 75
9/16 x 1-5/8

WM 76
9/16 x 1-1/2

WM 81
9/16 x 3-1/4

WM 82
9/16 x 2-3/4

WM 84
9/16 x 2-1/4

WM 85
9/16 x 1-3/4

WM 86
9/16 x 1-5/8

WM 89
9/16 x 2

WM 90
3/4 x 1-1/8

WM 92
3/4 x 7/8

WM 93
3/4 x 3/4

WM 94
11/16 x 1-1/8

WM 95
11/16 x 7/8

WM 96
5/8 x 3/4

WM 100
11/16 x 11/16

WM 101
1/2 x 1/2

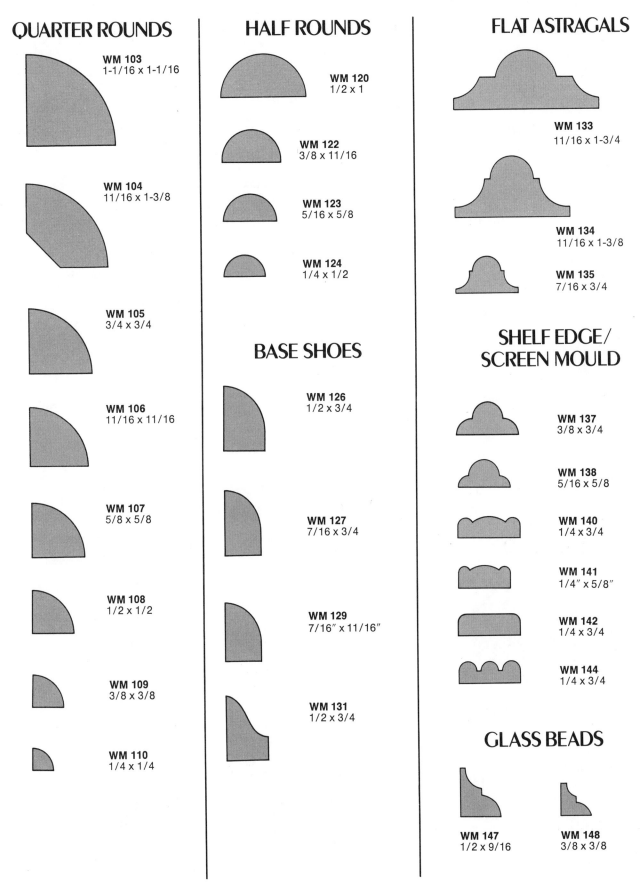

QUARTER ROUNDS

WM 103
1-1/16 x 1-1/16

WM 104
11/16 x 1-3/8

WM 105
3/4 x 3/4

WM 106
11/16 x 11/16

WM 107
5/8 x 5/8

WM 108
1/2 x 1/2

WM 109
3/8 x 3/8

WM 110
1/4 x 1/4

HALF ROUNDS

WM 120
1/2 x 1

WM 122
3/8 x 11/16

WM 123
5/16 x 5/8

WM 124
1/4 x 1/2

BASE SHOES

WM 126
1/2 x 3/4

WM 127
7/16 x 3/4

WM 129
7/16" x 11/16"

WM 131
1/2 x 3/4

FLAT ASTRAGALS

WM 133
11/16 x 1-3/4

WM 134
11/16 x 1-3/8

WM 135
7/16 x 3/4

SHELF EDGE/ SCREEN MOULD

WM 137
3/8 x 3/4

WM 138
5/16 x 5/8

WM 140
1/4 x 3/4

WM 141
1/4" x 5/8"

WM 142
1/4 x 3/4

WM 144
1/4 x 3/4

GLASS BEADS

WM 147
1/2 x 9/16

WM 148
3/8 x 3/8

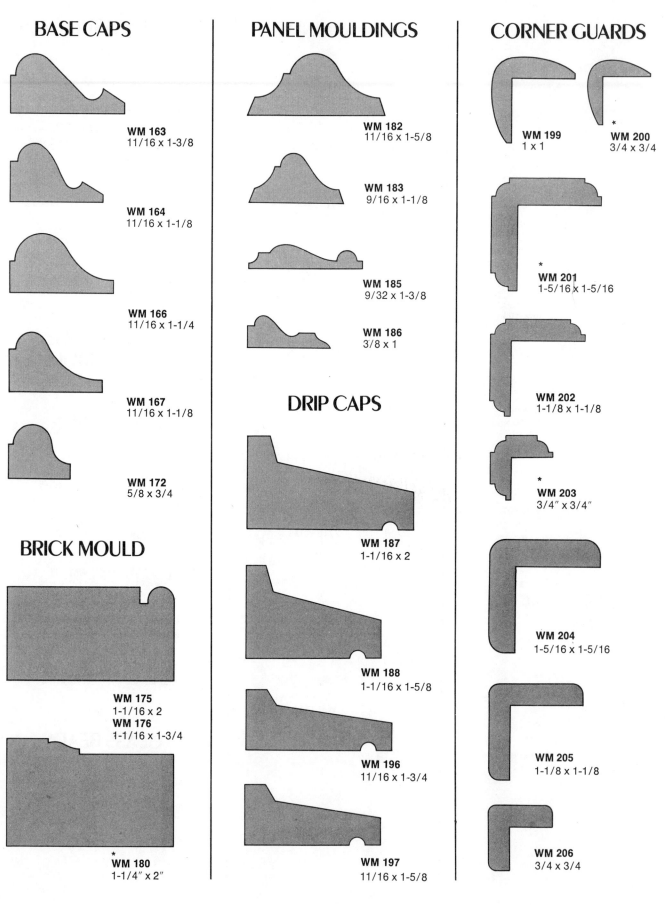

BASE CAPS

WM 163
11/16 x 1-3/8

WM 164
11/16 x 1-1/8

WM 166
11/16 x 1-1/4

WM 167
11/16 x 1-1/8

WM 172
5/8 x 3/4

BRICK MOULD

WM 175
1-1/16 x 2
WM 176
1-1/16 x 1-3/4

*
WM 180
1-1/4" x 2"

PANEL MOULDINGS

WM 182
11/16 x 1-5/8

WM 183
9/16 x 1-1/8

WM 185
9/32 x 1-3/8

WM 186
3/8 x 1

DRIP CAPS

WM 187
1-1/16 x 2

WM 188
1-1/16 x 1-5/8

WM 196
11/16 x 1-3/4

WM 197
11/16 x 1-5/8

CORNER GUARDS

WM 199
1 x 1

*
WM 200
3/4 x 3/4

*
WM 201
1-5/16 x 1-5/16

WM 202
1-1/8 x 1-1/8

*
WM 203
3/4" x 3/4"

WM 204
1-5/16 x 1-5/16

WM 205
1-1/8 x 1-1/8

WM 206
3/4 x 3/4

SHINGLE/
PANEL MOULDINGS

WM 207
11/16 x 2-1/2

WM 209
11/16 x 2

WM 210
11/16 x 1-5/8

WM 212
11/16 x 2-1/2

WM 213
9/16 x 2

WM 217
11/16 x 1-3/4

WM 218
11/16 x 1-1/2

BATTENS

WM 224
9/16 x 2-1/4

WM 229
11/16 x 1-5/8

HAND RAIL

*
WM 230
1-1/2 x 1-11/16

*
WM 231
1-1/2 x 1-11/16

*
WM 240
1-1/4 x 2-1/4

ROUNDS

WM 232 1-5/8
WM 233 1-5/16
WM 234 1-1/16

SQUARES

WM 236 1-5/8 x 1-5/8
WM 237 1-5/16 x 1-5/16
WM 238 1-1/16 x 1-1/16
WM 239 3/4 x 3/4

SCREEN/
S4S STOCK

WM 241
1-1/16 x 2-3/4
WM 243
1-1/16 x 1-3/4
WM 246
3/4 x 2-3/4
WM 247
3/4 x 2
WM 248
3/4 x 1-3/4
WM 249
3/4 x 1-5/8
WM 250
3/4 x 1-1/2
WM 251
3/4 x 1-3/8
WM 252
3/4 x 1-1/4
WM 254
1/2 x 3/4

S4S stock also available in 7/16", 1/2",
9/16", 5/8" and 11/16" standard thickness.

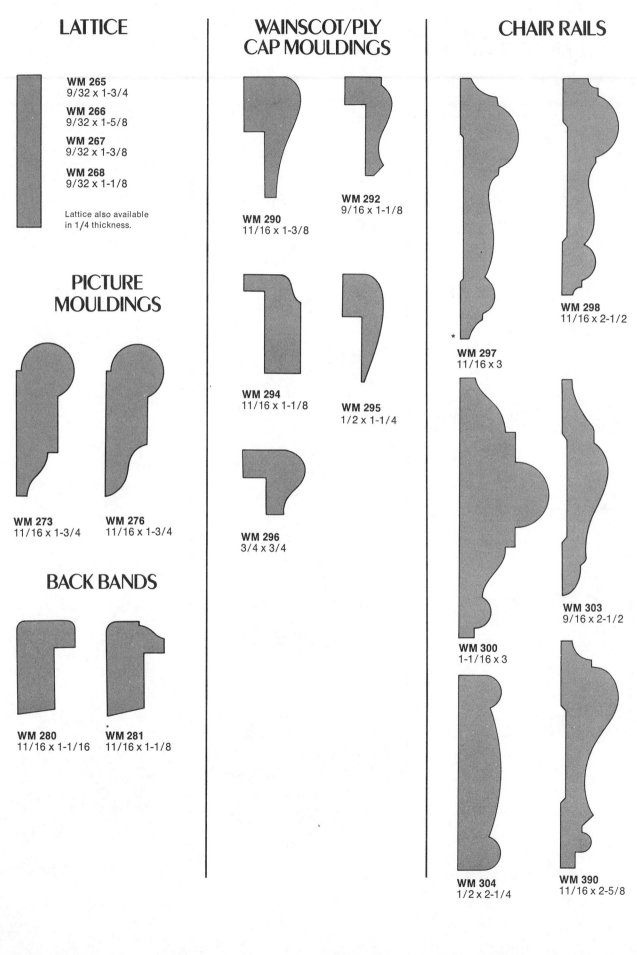

LATTICE

WM 265
9/32 x 1-3/4

WM 266
9/32 x 1-5/8

WM 267
9/32 x 1-3/8

WM 268
9/32 x 1-1/8

Lattice also available in 1/4 thickness.

PICTURE MOULDINGS

WM 273
11/16 x 1-3/4

WM 276
11/16 x 1-3/4

BACK BANDS

WM 280
11/16 x 1-1/16

*
WM 281
11/16 x 1-1/8

WAINSCOT/PLY CAP MOULDINGS

WM 290
11/16 x 1-3/8

WM 292
9/16 x 1-1/8

WM 294
11/16 x 1-1/8

WM 295
1/2 x 1-1/4

WM 296
3/4 x 3/4

CHAIR RAILS

*
WM 297
11/16 x 3

WM 298
11/16 x 2-1/2

WM 300
1-1/16 x 3

WM 303
9/16 x 2-1/2

WM 304
1/2 x 2-1/4

WM 390
11/16 x 2-5/8

CASING

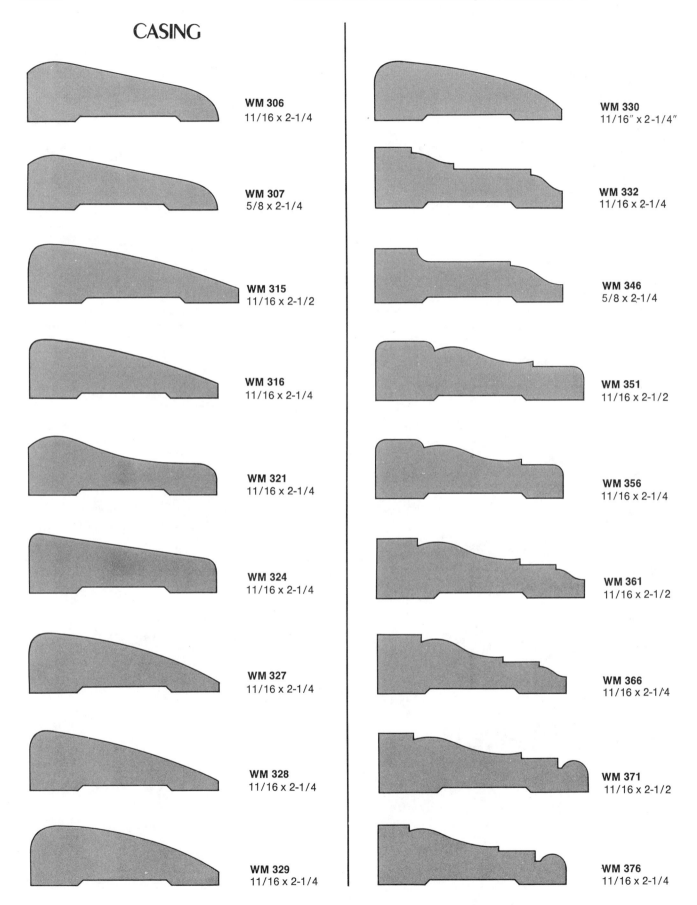

WM 306
11/16 x 2-1/4

WM 307
5/8 x 2-1/4

WM 315
11/16 x 2-1/2

WM 316
11/16 x 2-1/4

WM 321
11/16 x 2-1/4

WM 324
11/16 x 2-1/4

WM 327
11/16 x 2-1/4

WM 328
11/16 x 2-1/4

WM 329
11/16 x 2-1/4

WM 330
11/16" x 2-1/4"

WM 332
11/16 x 2-1/4

WM 346
5/8 x 2-1/4

WM 351
11/16 x 2-1/2

WM 356
11/16 x 2-1/4

WM 361
11/16 x 2-1/2

WM 366
11/16 x 2-1/4

WM 371
11/16 x 2-1/2

WM 376
11/16 x 2-1/4

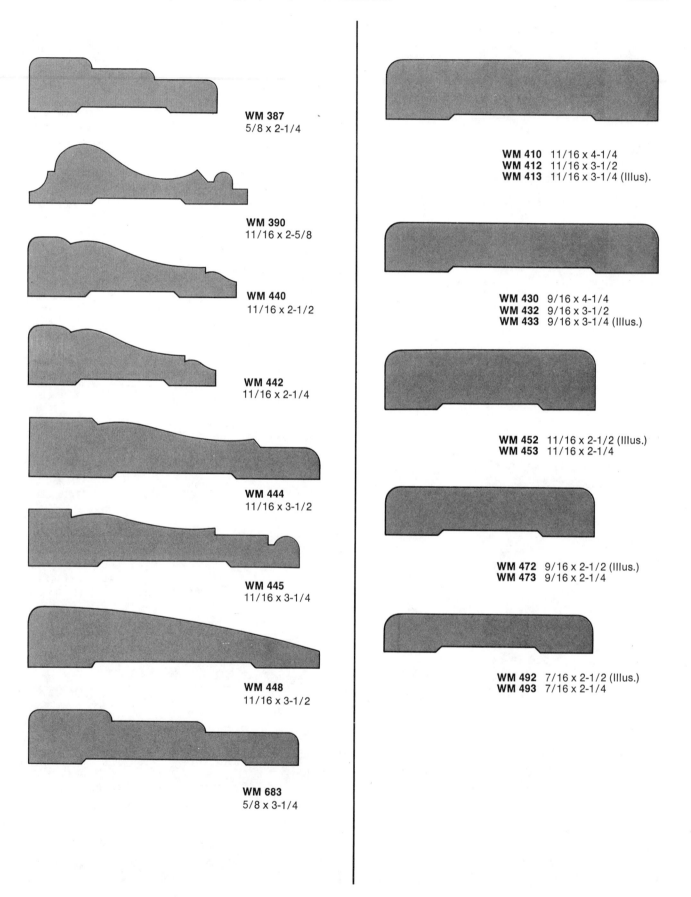

WM 387
5/8 x 2-1/4

WM 390
11/16 x 2-5/8

WM 440
11/16 x 2-1/2

WM 442
11/16 x 2-1/4

WM 444
11/16 x 3-1/2

WM 445
11/16 x 3-1/4

WM 448
11/16 x 3-1/2

WM 683
5/8 x 3-1/4

WM 410   11/16 x 4-1/4
WM 412   11/16 x 3-1/2
WM 413   11/16 x 3-1/4 (Illus).

WM 430   9/16 x 4-1/4
WM 432   9/16 x 3-1/2
WM 433   9/16 x 3-1/4 (Illus.)

WM 452   11/16 x 2-1/2 (Illus.)
WM 453   11/16 x 2-1/4

WM 472   9/16 x 2-1/2 (Illus.)
WM 473   9/16 x 2-1/4

WM 492   7/16 x 2-1/2 (Illus.)
WM 493   7/16 x 2-1/4

BASE MOULDINGS

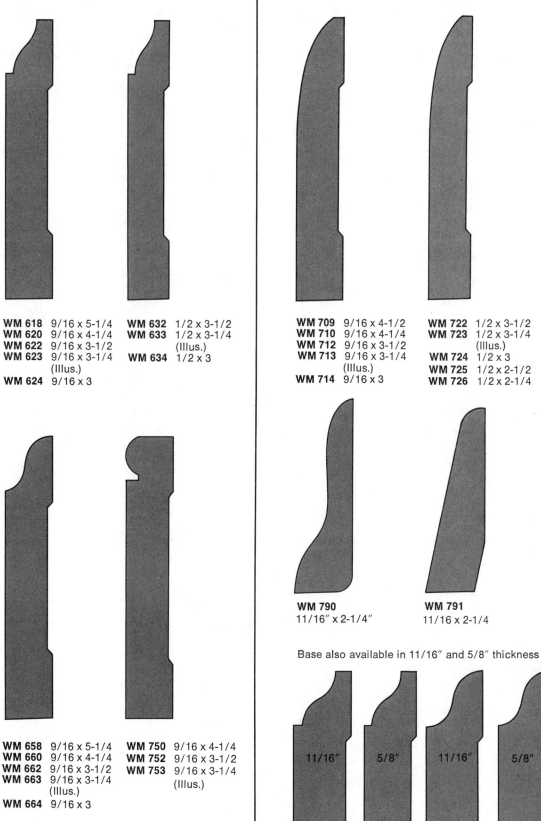

WM 618	9/16 x 5-1/4	**WM 632**	1/2 x 3-1/2
WM 620	9/16 x 4-1/4	**WM 633**	1/2 x 3-1/4
WM 622	9/16 x 3-1/2		(Illus.)
WM 623	9/16 x 3-1/4	**WM 634**	1/2 x 3
	(Illus.)		
WM 624	9/16 x 3		

WM 709	9/16 x 4-1/2	**WM 722**	1/2 x 3-1/2
WM 710	9/16 x 4-1/4	**WM 723**	1/2 x 3-1/4
WM 712	9/16 x 3-1/2		(Illus.)
WM 713	9/16 x 3-1/4	**WM 724**	1/2 x 3
	(Illus.)	**WM 725**	1/2 x 2-1/2
WM 714	9/16 x 3	**WM 726**	1/2 x 2-1/4

WM 790
11/16″ x 2-1/4″

WM 791
11/16 x 2-1/4

WM 795
11/16 x 2-1/4

Base also available in 11/16″ and 5/8″ thickness

WM 658	9/16 x 5-1/4	**WM 750**	9/16 x 4-1/4
WM 660	9/16 x 4-1/4	**WM 752**	9/16 x 3-1/2
WM 662	9/16 x 3-1/2	**WM 753**	9/16 x 3-1/4
WM 663	9/16 x 3-1/4		(Illus.)
	(Illus.)		
WM 664	9/16 x 3		

STOPS

WM 813   7/16 x 2-1/4
WM 814   7/16 x 1-3/4
WM 815   7/16 x 1-5/8
WM 816   7/16 x 1-3/8 (Illus.)
WM 817   7/16 x 1-1/4
WM 818   7/16 x 1-1/8
WM 820   7/16 x 7/8

WM 823   3/8 x 2-1/4
WM 824   3/8 x 1-3/4
WM 825   3/8 x 1-5/8
WM 826   3/8 x 1-3/8 (Illus.)
WM 827   3/8 x 1-1/4
WM 828   3/8 x 1-1/8
WM 830   3/8 x 7/8
WM 831   3/8 x 3/4

WM 843   7/16 x 2-1/4
WM 844   7/16 x 1-3/4
WM 845   7/16 x 1-5/8
WM 846   7/16 x 1-3/8 (Illus.)
WM 847   7/16 x 1-1/4
WM 848   7/16 x 1-1/8
WM 850   7/16 x 7/8
WM 851   7/16 x 3/4

WM 853   3/8 x 2-1/4
WM 854   3/8 x 1-3/4
WM 855   3/8 x 1-5/8
WM 856   3/8 x 1-3/8 (Illus.)
WM 857   3/8 x 1-1/4
WM 858   3/8 x 1-1/8
WM 860   3/8 x 7/8
WM 861   3/8 x 3/4

WM 873   7/16 x 2-1/4
WM 874   7/16 x 1-3/4
WM 875   7/16 x 1-5/8
WM 876   7/16 x 1-3/8 (Illus.)
WM 877   7/16 x 1-1/4
WM 878   7/16 x 1-1/8
WM 880   7/16 x 7/8
WM 881   7/16 x 3/4

WM 883   3/8 x 2-1/4
WM 884   3/8 x 1-3/4
WM 885   3/8 x 1-5/8
WM 886   3/8 x 1-3/8 (Illus.)
WM 887   3/8 x 1-1/4
WM 888   3/8 x 1-1/8
WM 890   3/8 x 7/8
WM 891   3/8 x 3/4

WM 903   7/16 x 2-1/4
WM 904   7/16 x 1-3/4
WM 905   7/16 x 1-5/8
WM 906   7/16 x 1-3/8 (Illus.)
WM 907   7/16 x 1-1/4
WM 908   7/16 x 1-1/8
WM 910   7/16 x 7/8
WM 911   7/16 x 3/4

WM 913   3/8 x 2-1/4
WM 914   3/8 x 1-3/4
WM 915   3/8 x 1-5/8
WM 916   3/8 x 1-3/8 (Illus.)
WM 917   3/8 x 1-1/4
WM 918   3/8 x 1-1/8
WM 920   3/8 x 7/8
WM 921   3/8 x 3/4

WM 933   7/16 x 2-1/4
WM 934   7/16 x 1-3/4
WM 935   7/16 x 1-5/8
WM 936   7/16 x 1-3/8 (Illus.)
WM 937   7/16 x 1-1/4
WM 938   7/16 x 1-1/8
WM 940   7/16 x 7/8
WM 941   7/16 x 3/4

WM 943   3/8 x 2-1/4
WM 944   3/8 x 1-3/4
WM 945   3/8 x 1-5/8
WM 946   3/8 x 1-3/8 (Illus.)
WM 947   3/8 x 1-1/4
WM 948   3/8 x 1-1/8
WM 950   3/8 x 7/8
WM 951   3/8 x 3/4

WM 953   7/16 x Width
         Specified

WM 954   3/8 x Width
         Specified

PANEL STRIPS/MULLION CASINGS

SHELF CLEAT

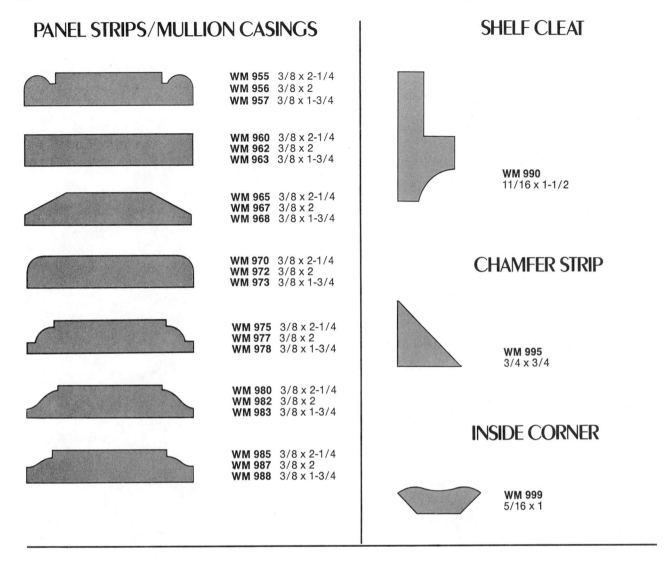

WM 955	3/8 x 2-1/4
WM 956	3/8 x 2
WM 957	3/8 x 1-3/4

WM 960	3/8 x 2-1/4
WM 962	3/8 x 2
WM 963	3/8 x 1-3/4

WM 965	3/8 x 2-1/4
WM 967	3/8 x 2
WM 968	3/8 x 1-3/4

WM 970	3/8 x 2-1/4
WM 972	3/8 x 2
WM 973	3/8 x 1-3/4

WM 975	3/8 x 2-1/4
WM 977	3/8 x 2
WM 978	3/8 x 1-3/4

WM 980	3/8 x 2-1/4
WM 982	3/8 x 2
WM 983	3/8 x 1-3/4

WM 985	3/8 x 2-1/4
WM 987	3/8 x 2
WM 988	3/8 x 1-3/4

WM 990
11/16 x 1-1/2

CHAMFER STRIP

WM 995
3/4 x 3/4

INSIDE CORNER

WM 999
5/16 x 1

FLAT STOOLS

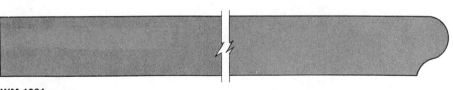

WM 1021
11/16" x width specified

RABBETED STOOLS

T-ASTRAGALS

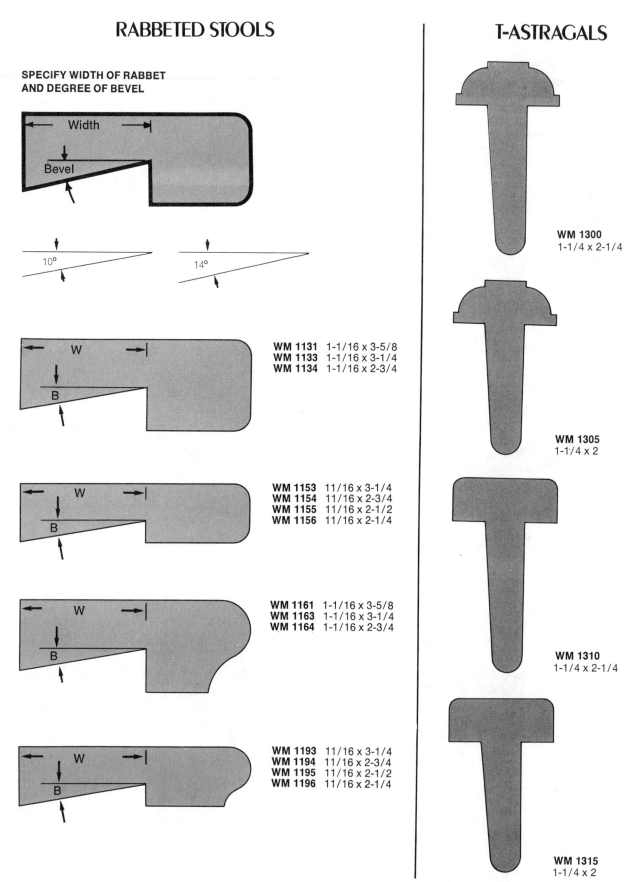

**SPECIFY WIDTH OF RABBET
AND DEGREE OF BEVEL**

Width

Bevel

10°

14°

WM 1131 1-1/16 x 3-5/8
WM 1133 1-1/16 x 3-1/4
WM 1134 1-1/16 x 2-3/4

W

B

WM 1153 11/16 x 3-1/4
WM 1154 11/16 x 2-3/4
WM 1155 11/16 x 2-1/2
WM 1156 11/16 x 2-1/4

W

B

WM 1161 1-1/16 x 3-5/8
WM 1163 1-1/16 x 3-1/4
WM 1164 1-1/16 x 2-3/4

W

B

WM 1193 11/16 x 3-1/4
WM 1194 11/16 x 2-3/4
WM 1195 11/16 x 2-1/2
WM 1196 11/16 x 2-1/4

W

B

WM 1300
1-1/4 x 2-1/4

WM 1305
1-1/4 x 2

WM 1310
1-1/4 x 2-1/4

WM 1315
1-1/4 x 2

MOULDING WEIGHTS

ESTIMATED WEIGHT PER 1000 LINEAL FEET FOR LINEAL SASH, S4S, ROUND EDGE PATTERNS, ETC

WIDTH	THICKNESS											
	1/4	9/32	5/16	3/8	7/16	1/2	9/16	5/8	11/16	3/4	1-1/8	1-3/8
1/4	11	—	—	—	—	—	—	—	—	—	—	—
5/16	14	—	—	—	—	—	—	—	—	—	—	—
3/8	17	19	21	25	—	—	—	—	—	—	—	—
1/2	22	25	28	34	39	45	—	—	—	—	—	—
5/8	28	31	35	42	49	56	63	70	70	—	—	—
3/4	34	35	42	50	59	67	75	84	84	101	—	—
7/8	39	44	49	59	69	78	88	98	98	118	157	—
1	45	50	56	67	78	90	101	112	112	134	180	—
1-1/8	50	57	63	76	88	101	113	126	126	151	227	—
1-1/4	56	63	70	84	98	112	126	140	140	168	252	—
1-3/8	62	69	78	92	108	123	138	154	154	185	277	338
1-1/2	67	75	84	101	118	134	151	168	168	202	302	370
1-5/8	73	81	91	109	127	146	164	182	182	218	328	400
1-3/4	78	83	98	118	137	157	177	196	196	235	353	431
2	—	—	—	134	159	179	201	224	224	269	403	493
2-1/4	—	—	—	151	176	202	227	252	252	302	454	554
2-1/2	—	—	—	—	196	224	252	280	280	336	504	616
2-5/8	—	—	—	—	206	236	265	294	294	353	529	641
2-3/4	—	—	—	—	216	247	278	308	308	370	554	678
3	—	—	—	—	—	269	302	336	336	403	605	739
3-1/4	—	—	—	—	—	297	328	364	364	437	655	800
3-1/2	—	—	—	—	—	—	353	392	392	470	706	862
3-5/8	—	—	—	—	—	—	—	406	406	487	731	893
3-3/4	—	—	—	—	—	—	—	420	420	504	756	924
4-1/4	—	—	—	—	—	—	—	476	476	571	857	1047
4-3/4	—	—	—	—	—	—	—	532	532	638	958	1170
5-1/2	—	—	—	—	—	—	—	616	616	739	1008	1356
5-3/4	—	—	—	—	—	—	—	644	644	773	1159	1417

ESTIMATED WEIGHT PER 1000 LINEAL FEET FOR MOULDED PATTERNS.

WIDTH	THICKNESS											
	1/4	9/32	5/16	3/8	7/16	1/2	9/16	5/8	11/16	3/4	1-1/8	1-3/8
1/4	9	—	—	—	—	—	—	—	—	—	—	—
5/16	12	—	—	—	—	—	—	—	—	—	—	—
3/8	14	16	17	21	—	—	—	—	—	—	—	—
1/2	19	21	23	28	33	37	—	—	—	—	—	—
5/8	23	26	29	35	41	47	53	59	59	—	—	—
3/4	28	32	35	42	49	56	63	70	70	84	—	—
7/8	33	37	41	49	57	66	74	82	82	98	—	—
1	37	42	47	56	66	75	83	94	94	112	—	—
1-1/8	42	47	53	63	74	84	95	105	105	126	190	—
1-1/4	47	53	59	70	82	94	106	117	117	140	211	—
1-3/8	52	58	64	77	90	103	116	129	129	154	232	283
1-1/2	56	63	70	84	98	112	126	140	140	169	253	309
1-5/8	61	68	76	91	104	122	137	152	152	183	274	335
1-3/4	66	74	82	98	117	131	147	164	164	197	295	360
2	—	—	—	112	131	150	169	187	187	225	337	412
2-1/4	—	—	—	126	148	169	211	211	211	253	379	463
2-1/2	—	—	—	—	164	187	222	234	234	281	421	515
2-5/8	—	—	—	—	170	197	231	246	246	295	442	541
2-3/4	—	—	—	—	180	206	252	257	257	309	463	566
3	—	—	—	—	—	—	274	281	281	337	505	618
3-1/4	—	—	—	—	—	—	—	304	304	365	548	669
3-1/2	—	—	—	—	—	—	—	328	328	393	590	721
3-5/8	—	—	—	—	—	—	—	339	339	407	611	750
3-3/4	—	—	—	—	—	—	—	351	351	421	632	772
4-1/4	—	—	—	—	—	—	—	374	374	477	716	875
4-3/4	—	—	—	—	—	—	—	445	445	534	800	978
5-1/2	—	—	—	—	—	—	—	515	515	618	927	1356
5-3/4	—	—	—	—	—	—	—	538	538	646	969	1420

THE ESTIMATED WEIGHTS ARE FOR ALL WESTERN SOFTWOODS EXCEPT LARCH AND DOUGLAS FIR. FOR THESE TWO SPECIES AND SOUTHERN PINE ADD 16% TO THE ABOVE WEIGHTS.

BUNDLING SCHEDULE

PIECES PER BUNDLE

WIDTH	THICKNESS					
	5/16″ AND UNDER	3/8″ AND 7/16″	1/2″ AND 9/16″	5/8″ TO 3/4″	1-1/16″	1-5/16″
1/2″ AND UNDER	75**	75**	50*	50*		
5/8″	75**	50*	50*	40*		
3/4″	75**	50*	50*	40*		
7/8″ AND 1″	50*		40*	40*	20	
1-1/8″	40*	40*	40	30	20	
1-1/4″	40*	30*	30	30	12	12
1-3/8″	40*	30	30	25	12	12
1-1/2″	40*	30	25	25	12	9
1-5/8″	30*	30	25	20	12	9
1-3/4″	30*	30	25	20	12	9
1-7/8″ AND 2″	30	20	20	12	12	8
2-1/8″ AND 2-1/4″	30	20	20	12	8	8
2-3/8″	30	20	20	12	8	8
2-1/2″	30	20	20	12	8	8
2-5/8″	20	16	16	12	6	6
2-3/4″	20	16	16	12	6	6
3″	16	16	10	10	6	6
3-1/4″	16	10	10	10	6	6
3-1/2″ TO 3-3/4″	12	10	10	8	6	6
4″ AND WIDER	10	10	8	8	6	4

*SOME SHIPPERS BUNDLE 500 LINEAL FEET.

**SOME SHIPPERS BUNDLE 1000 LINEAL FEET.

EXCEPTIONS TO BUNDLING SCHEDULE			PIECES
	3/4 x 3/4″	S4S	30
	3/4 x 1-1/4 AND 1-3/8″	S4S	20
	3/4 x 1-1/2 TO 1-3/4″	S4S	15
	3/4 x 2 TO 2-3/4″	S4S	10
	1-1/16″	S4S OR FULL ROUND	12
	1-5/16 AND 1-5/8″	S4S OR FULL ROUND	9
	1-1/16 x 1-5/8 TO 2″	DRIP CAP	16
	1-1/8 AND 1-3/8″	CORNER GUARD	20

LINEAL BUNDLING	LENGTH IN FEET	"500 FT." BUNDLE PIECES LIN. FT.*		"1000 FT." BUNDLE PIECES LIN. FT.*		
	3	84 =	252	168 =	504	*Actual Measure.
	4	63 =	252	125 =	500	
	5	50 =	250	100 =	500	
	6	42 =	252	84 =	504	
	7	36 =	252	72 =	504	
	8	32 =	256	63 =	504	
	9	56 =	504	112 =	1008	
	10	50 =	500	100 =	1000	
	11	46 =	506	91 =	1001	
	12	42 =	504	84 =	1008	
	13	39 =	507	77 =	1001	
	14	36 =	504	72 =	1008	
	15	34 =	510	67 =	1005	
	16	32 =	512	63 =	1008	

THIS BUNDLING SCHEDULE IS GENERALLY USED BY MOST MOULDING MANUFACTURERS AND MAY ALSO BE USED ON CUT-TO-LENGTH ITEMS.

CONVERSION TABLE
ENGLISH TO METRIC UNIT MEASUREMENT

ENGLISH MEASUREMENT	ACTUAL METRIC CONVERSION	SUGGESTED METRIC SIZE*
1/8″	3.18 mm	3.0 mm
1/4″	6.35	6.0
9/32″	7.14	7.0
5/16″	7.94	8.0
11/32″	8.73	9.0
3/8″	9.53	10.0
7/16″	11.11	11.0
15/32″	11.91	12.0
1/2″	12.70	13.0
9/16″	14.29	14.0
19/32″	15.08	15.0
5/8″	15.88	16.0
21/32″	16.67	17.0
11/16″	17.46	17.0
23/32″	18.26	18.0
3/4″	19.05	19.0
25/32″	19.84	20.0
13/16″	20.64	21.0
7/8″	22.23	22.0
1″	25.40	25.0
1-5/32″	29.39	29.0
1-1/4″	31.75	32.0
1-5/16″	33.34	33.0
1-13/32″	35.72	36.0
1-1/2″	38.10	38.0
1-9/16″	39.69	40.0
1-3/4″	44.45	44.0
2″	50.80	51.0
2-1/8″	53.98	54.0
2-1/4″	57.25	57.0
2-1/2″	63.50	64.0
3″	76.20	76.0
3-1/4″	82.55	83.0
3-1/2″	88.90	89.0
3-9/16″	90.49	90.0
3-3/4″	95.25	95.0
4-1/4″	107.95	108.0
4-1/2″	114.30	114.0
4-9/16″	115.89	116.0
4-3/4″	120.65	121.0
5-1/2″	139.70	140.0
5-3/4″	146.05	145.0

*To manufacture and market mouldings in metric sizes with minimum confusion, it is recommended to round actual metric conversion sizes to whole millimeters.

7.15 LINEAL FEET

Moldings are marketed by the lin ft. As an example, "quarter round" may be marketed at $1.50 per 100 lin ft. Assume an order for 300 lin ft of $1\frac{13}{16}'' \times 1\frac{13}{16}''$ quarter round, which would be entered on the lumber yard estimate sheet in the feet column as lin ft. It will be recalled that dimension lumber is shown as fbm.

Sometimes 2×2 dimension lumber is sold by the lin ft—for example, 250 lin ft of 2×2 @ $3.75 per 100 lin ft, which costs

$$\frac{250 \times 3.75}{100} = \$9.38$$

To change 250 lin ft of 2×2 into fbm, imagine 250 lin ft of 2×2 as if it were in one long piece; thus:

$$1/2 \times 2 - 250'\text{-}0'' = \frac{1 \times 2 \times 2 \times 250}{12} = 83 \text{ fbm}$$

7.16 PLYWOOD AND FIBERBOARDS

Most plywood and fiberboards are marketed by the M sq ft. Thus 1,000 sq ft of plywood at $350 per M is $350.

The common size for marketing plywood and fiberboards is in sheets $4'\text{-}0'' \times 8'\text{-}0''$. Some manufacturers may quote the size in inches, thus: $48'' \times 96''$. The cost of one sheet of $4'\text{-}0'' \times 8'\text{-}0''$ @ $350 per M would be $\frac{32 \times 350}{1,000}$ or $32 \times 0.35 = \$11.20$.

Special sizes of plywood and fiberboards may be obtained by special order. Read the architect's specifications carefully for unusual sizes. They may require a special order, and a prompt order will ensure that the material is on hand when required. Be sure to get a firm quotation for all fiberboards of special size. Read addendums to specifications.

7.17 NAILS

An estimator must have an adequate knowledge of nails, screws, and all types of fasteners. A course on materials would be a very great advantage for any estimator. He must read and keep up to date with the latest methods for securing members of structures together will all kinds of metallic devices.

Some large contractors employ one man whose sole job is to estimate the cost of hardware. As an example, consider the different types of hardware required for an interior house door or for the doors to a hospital, hotel, church, bank, or public building.

The estimator must have as many catalogues as competitive pricings may dictate.

NUMBER OF NAILS TO THE POUND

Size	Length	Common	Finishing	Casing
2d	$1''$	850		
3d	$1\frac{1}{4}''$	550	640	
4d	$1\frac{1}{2}''$	350	456	
5d	$1\frac{3}{4}''$	230	328	
6d	$2''$	180	273	228
7d	$2\frac{1}{4}''$	140	170	178
8d	$2\frac{1}{2}''$	100	151	133
9d	$2\frac{3}{4}''$	80	125	100
10d	$3''$	65	107	96
12d	$3\frac{1}{4}''$	50		60
16d	$3\frac{1}{2}''$	40		50
20d	$4''$	31		
30d	$4\frac{1}{2}''$	22		
40d	$5''$	18		
50d	$5\frac{1}{2}''$	14		
60d	$6''$	12		

7.18 SCREWS

Screws are another type of fastener commonly used in wood. They have greater holding power than nails, present a neater appearance, and are less apt to damage the material if removed or replaced. They are used extensively in furniture manufacturing and for fastening all kinds of trimming hardware.

Types of Screws

Screws are ordinarily made of steel but sometimes of brass or bronze. Silver, nickel, and gold-plated screws are also manufactured. Steel screws are also finished with blue, bronze, lacquered, galvanized, or tinned surface.

Three types of screw heads are commonly made: the slotted head, Robertson head, and Phillips head.

The slotted-head screw is the most commonly used. A slot across the head provides the means of driving with an ordinary screwdriver.

The Robertson-head screw has a square hole in the head, made in three different sizes, and must be driven with a special Robertson screwdriver.

The Phillips-head screw has an indented cross in the head and also requires a special screwdriver.

Slotted Screws

A number of different styles of slotted-head screws are made, depending on their use. The most common styles are flat, round, and oval heads. They are marketed by the gross.

(a) dowel screws with thread on both ends, used in place of wooden dowels;

TABLE OF NAIL SIZES WITH ALLOWANCES FOR DIMENSION LUMBER AND FINISHING CARPENTRY

Size and kind of material	Board measure	Type of nail	Length of nail	Lb of nails required 12″ o.c.	16″ o.c.	24″ o.c.
1×4 boards & shiplap	1000	common	2½″	60	48	30
1×6 boards & shiplap	1000	common	2½″	40	32	20
1×8 boards & shiplap	1000	common	2½″	31	27	16
1×10 boards and shiplap	1000	common	2½″	25	20	13
1×12 boards and shiplap	1000	common	2½″	31	24	16
1×4 T & G blind nailed	1000	common	2½″	30	24	15
1×6 T & G blind nailed	1000	common	2½″	20	16	10
1×8 T & G blind nailed and 1 face nail	1000	common	2½″	31	27	16
1×10 T & G blind nailed and 1 face nail	1000	common	2½″	25	20	13
1×12 T & G blind nailed and 1 face nail	1000	common	2½″	21	16	11
2×4 to 2×16 framing	1000	common	4″	20	16	10
			3½″	10	10	6
			3″	8	6	4
Built-up beams	100 lin	common	3½″		3	
2×6 T & G flooring	1000	common	4″	35	27	18
2×8 T & G flooring	1000	common	4″	27	20	14
6″ bevel siding	1000	siding	2″	15	13	
8″ bevel siding	1000	siding	2″	12	10	
10″ bevel siding	1000	siding	2¼″	45	35	
12″ bevel siding	1000	siding	2¼″	60	50	
1×3 softwood flooring	1000	Floor brads	2½″	42	32	21
1×4 softwood flooring	1000	Floor brads	2½″	32	26	16
1×6 softwood flooring	1000	Floor brads	2½″	22	18	11
1½″ hardwood flooring	1000	flooring	2″	13	10	
2″ hardwood flooring	1000	flooring	2″	11	8	
2¼″ hardwood flooring	1000	flooring	2¼″	20	14	
Base	100 lin	Finish	2½″		1	
Interior trim	100 lin	Finish	2½″		1	
Casing		Finish	2½″		1 lb per opening	
Outside trim	100 lin	Finish	2½″		2½	
Outside mouldings	100 lin	Finish				
Carpet strip	100 lin	Finish	1½″		½	
Metal lath	100 sq yd	Roofing	2″		17½	
Metal lath	100 sq yd	Staples	1″		12	
Gypsum lath	100 sq yd	Gyproc	1½″		10	
Gyproc wall board	100 sq yd	Gyproc	1½″		10	
Plywood ¼″	1000 sq ft	Finish or common	1½″		9	
Plywood ⅜″	1000 sq ft	Finish or common	2″		12	
Plywood ⅝″	1000 sq ft	Finish or common	2½″		20	
Shingles (see page 187)	per M	Shingle	1¼″		4½	

(b) drive or lag screws, with very coarse thread, and often a square head, to be driven or turned into predrilled holes;

(c) felloe screws, with thread all the way to the top, for special holding power.

Sizes of Screws

The size of screws is given in length in inches and the number of the gage, the gage denoting the diameter. Thus a 1-in. no. 12 screw is one inch long and 0.2158 inch in diameter. The gage numbers range from 0 to 30 and the lengths from ¼ in. to 6 in. The lengths vary by eighths of an inch up to one inch, by quarters of an inch up to three inches, and by halves up to five inches. Screws from ⅝-in. to 4½-in. long are made in about sixteen different gage numbers. See page 160.

WOOD SCREW TABLE

Length	Gage		Gage number	Actual size	Decimal	Approx. Fraction	Drill or Auger Bit size		
	Steel screws	Brass screws					A	B	C
$\frac{1}{4}$	0 to 4	0 to 4	0	O	.060	$\frac{1}{16}$	$\frac{1}{16}$		
$\frac{3}{8}$	0 to 8	0 to 6	1	O	.073	$\frac{5}{64}$ −	$\frac{3}{32}$		
$\frac{1}{2}$	1 to 10	1 to 8	2	O	.086	$\frac{5}{64}$ +	$\frac{3}{32}$	$\frac{1}{16}$	$\frac{3}{16}$
$\frac{5}{8}$	2 to 12	2 to 10	3	O	.099	$\frac{3}{32}$	$\frac{1}{8}$	$\frac{1}{16}$	$\frac{4}{16}$
$\frac{3}{4}$	2 to 14	2 to 12	4	O	.112	$\frac{7}{64}$	$\frac{1}{8}$	$\frac{1}{16}$	$\frac{4}{16}$
$\frac{7}{8}$	3 to 14	4 to 12	5	O	.125	$\frac{1}{8}$	$\frac{1}{8}$	$\frac{3}{32}$	$\frac{4}{16}$
1	3 to 16	4 to 14	6	O	.138	$\frac{9}{64}$	$\frac{5}{32}$	$\frac{3}{32}$	$\frac{5}{16}$
$1\frac{1}{4}$	4 to 18	6 to 14	7	O	.151	$\frac{5}{32}$ −	$\frac{5}{32}$	$\frac{1}{8}$	$\frac{5}{16}$
$1\frac{1}{2}$	4 to 20	6 to 14	8	O	.164	$\frac{5}{32}$ +	$\frac{3}{16}$	$\frac{1}{8}$	$\frac{6}{16}$
$1\frac{3}{4}$	6 to 20	8 to 14	9	O	.177	$\frac{11}{64}$	$\frac{3}{16}$	$\frac{1}{8}$	$\frac{6}{16}$
2	6 to 20	8 to 18	10	O	.190	$\frac{3}{16}$	$\frac{3}{16}$	$\frac{1}{8}$	$\frac{6}{16}$
$2\frac{1}{4}$	6 to 20	10 to 18	11	O	.203	$\frac{13}{64}$	$\frac{7}{32}$	$\frac{5}{32}$	$\frac{7}{16}$
$2\frac{1}{2}$	6 to 20	10 to 18	12	O	.216	$\frac{7}{32}$	$\frac{7}{32}$	$\frac{5}{32}$	$\frac{7}{16}$
$2\frac{3}{4}$	8 to 20	8 to 20	14	O	.242	$\frac{15}{64}$	$\frac{1}{4}$	$\frac{3}{16}$	$\frac{8}{16}$
3	8 to 24	12 to 18	16	O	.268	$\frac{17}{64}$	$\frac{9}{32}$	$\frac{7}{32}$	$\frac{9}{16}$
$3\frac{1}{2}$	10 to 24	12 to 18	18	O	.294	$\frac{19}{64}$	$\frac{5}{16}$	$\frac{1}{4}$	$\frac{10}{16}$
4	12 to 24	12 to 24	20	O	.320	$\frac{21}{64}$	$\frac{11}{32}$	$\frac{9}{32}$	$\frac{11}{16}$
$4\frac{1}{2}$	14 to 24	14 to 24							
5	14 to 24	14 to 24	24	O	.372	$\frac{3}{8}$	$\frac{3}{8}$	$\frac{5}{16}$	$\frac{12}{16}$

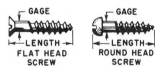

GAGE — LENGTH — FLAT HEAD SCREW · GAGE — LENGTH — ROUND HEAD SCREW · GAGE — LENGTH — OVAL HEAD SCREW

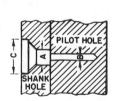

PILOT HOLE · SHANK HOLE · C · A · B

QUESTIONS

1. Dimension lumber is marketed in _____ units.

2. Baseboard and moldings are marketed in _____ units.

3. Plywood and fiberboards are marketed in _____ units.

4. Nails are marketed by the _____ units.

5. Screws are marketed by the _____ units.

6. Give an example of the transposition of the number $1689.98.

7. How many bd ft would be charged for the following orders for dimension lumber:
   (a) 100/2 × 4 – 14'-0"
   (b) 50/2 × 8 – 8'-0"

8. What are the actual sizes of dimension lumber known as:
   (a) two-by-four

(b) two-by-six

(c) two-by-eight

(d) two-by-twelve

9. Make a neat sectional view of the following moldings:

(a) quarter round

(b) carpet strip

(c) shiplap

(d) T & G board

(e) hardwood flooring

(f) baseboard

10. Define linear feet.

11. How many board feet are there in 300 lin ft of 2×4 dimension lumber?

12. Transfer the following to the blank work-up sheet and find the final cost:

(a) 57/2 × 4 – 16'-0" @ $116.00 per M

(b) 12 sheets of 4'-0" × 8'-0" ¾" plywood @ $238.00 per M

(c) 116 lin ft of carpet strip @ 2½ cents per lin ft

(d) 42 sacks of cement @ $1.22 per sack

(e) 75 lbs of nails @ 23 cents per lb

(f) 12 brass coat hooks @ 27 cents each

N.B.: Allow 10% discount for cash.

WORK-UP SHEET

Date_____

Sheet No._____of_____

Building	Location	Architect	Estimator

8 PERIMETERS, AREAS, AND EXCAVATIONS

This chapter is designed to give you an appreciation of making the most economical use of the basic data of perimeters and areas, and of applying that information so it can be readily translated into the requirements for estimating.

When a building is estimated, much of the basic data is used over and over again. This will become more evident throughout the text and especially in Chap. 9, concerning the typical wall section.

8.1 PERIMETERS

Examine the four plans in Figs. 8.1 through 8.4. The single-line drawings represent the building lines of houses. With the introduction of any inset to a basic plan, the floor area is reduced, but the number of corners is increased, and in some cases the perimeter will be longer. Count the corners in the plans shown in Figs. 8.1 through 8.4.

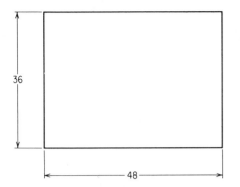

Fig. 8.1 Plan building lines.

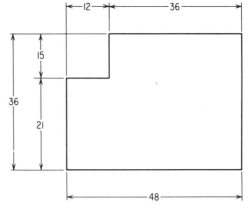

Fig. 8.2 Plan building lines.

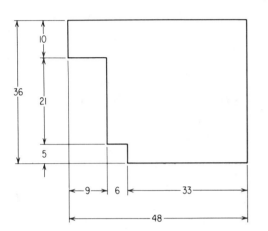

Fig. 8.3 Plan building lines.

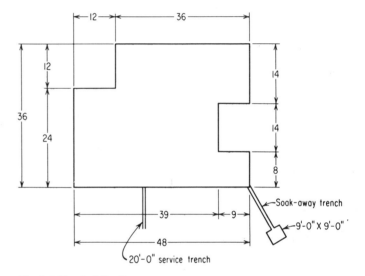

Fig. 8.4 Plan building lines.

Example

What is the perimeter in lin ft of the floor plan of Fig. 8.5?

Step 1: Basic perimeter is $60'\text{-}0'' + 40'\text{-}0'' \times 2 = 200$ lin ft.

Step 2: Allow for an extra wall at each end of the right-side recess. These are each $5'\text{-}0''$; $2 \times 5 = 10$ lin ft.

Step 3: Total basic perimeter   200
  Total of recess walls   10
        Total   210 lin-ft perimeter

Problem

Find the perimeters of the four plans of Figs. 8.1 through 8.4.

8.2 AREAS

Examine Fig. 8.5. Although this plan has a long perimeter, it also has a small area.

Example

Find the area of the plan of Fig. 8.5.

Step 1: Find the basic area of the plan: $60'\text{-}0'' \times 40'\text{-}0'' = 2,400$ sq ft.

Step 2: Deduct the areas of the offsets and recess shown shaded on plan:

Left-side offset	10×10	$= 100$
Bottom-left offset	25×8	$= 200$
Right-side recess	24×5	$= 120$
Total deductions		$= 420$ sq ft

Step 3: From the basic rectangular plan area, deduct

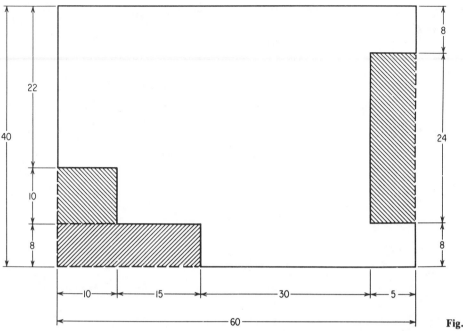

Fig. 8.5 Plan showing offsets and recess.

the areas of offsets and the recess: 2400 − 420 = 1980 sq ft.

Problem

Find the plan areas of Figs. 8.1 through 8.4.

8.3 EXCAVATIONS

There are more hazards in estimating excavations than in any other phase of building operations. Every month houses and buildings are being erected either on the incorrect lot or incorrectly placed on the proper lot. In all cases of doubt, a registered surveyor should be engaged.

Consider the situation in the center of our large cities, where it is often necessary to haul all the excavated material away in order to dump it legally, because there is no place to dump it on the building site. In this case, excavated material may not only have to be hauled away (through traffic) but some of it may have to be returned for backfill. In many cases there is also the risk of striking a water table, which may require constant pumping during the whole of the excavation process.

Remember safety precautions. Trenches more than 4'-0" in depth require shoring. Usually in cities you must provide fences, hoardings, and nighttime lighting, and in some cases a night watchman to protect the public and to guard the materials on the site.

Specifications

Many specifications state that all topsoil shall be removed clear of the building area and stockpiled on the site ready for landscaping.

Example

Estimate the number of cu yd of topsoil to be removed and stacked on site for the plan shown by Fig. 8.5.

The clearance is to be made 5'-0" clear of the basic plan lines and to a depth of 15". Assume the lot to be level and graded. *Note:* It is the usual practice on small jobs to remove all topsoil clear of the basic dimension plan. This includes the shaded portions shown on Fig. 8.5. *See a specimen general estimate sheet for excavations on p. 161* (Note that the sheet is not for this example.)

Step 1: The basic plan size is 40'-0" × 60'-0".

Step 2: Add to each end of the basic dimensions an extra 5'-0" to be cleared of topsoil. Thus:

and
$$5 + 40 + 5 = 50$$
$$5 + 60 + 5 = 70$$

$$50 \times 70 = 3500$$

Step 3: Multiply the area in sq ft by the depth of soil (15") to be removed:

$$3500 \times 1\frac{1}{4} = 4,375 \text{ cu ft}$$

Step 4: Reduce cu ft to the nearest higher cu yd:

$$\frac{4375}{27} = 162 \text{ cu yd (to the next higher cu yd)}$$

Problem

Estimate the number of cu yd of topsoil to be removed and stockpiled on site for the plan shown by Fig. 8.4. Assume the topsoil is 15″ deep. Use work-up sheets for your calculation and enter the information only on the general estimate sheets. Clear 5′-0″ from building lines.

Using local labor costs, give a priced subcontractor's estimate for the foregoing problem.

8.4 EXCAVATIONS MASS

Many jurisdictions state that all excavations for housing shall be made 2′-0″ clear of all building lines to allow the workmen clearance for erecting the walls.

Example

Allowing for excavations to be made 2′-0″ clear of the building lines and to a total depth of 5′-0″ below the grade, how many cu yd of earth would have to be removed from the plan shown in Fig. 8.6? Assume that the lot is level and to the correct grade.

Step 1: Find the basic area of the excavation required. Make a neat sketch on your work-up sheet and dimension it. *An estimator's time is never wasted by making sketches.*

Basic dimensions: 40′-0″ × 60′-0″; add 2′-0″ all around.

Area is 44 × 64 = 2816 sq ft.

Step 2: Deduct the area of the offsets and recess (shown shaded on plan):

Left-side offset	10 × 10 =	100
Bottom-left offset	25 × 8 =	200
Right-side recess	20 × 5 =	100
		400 sq ft

2816 − 400 = 2416 sq ft surface or excavation

Step 3: *Remember that 15″ of topsoil has already been removed.* The depth remaining is 5′-0″ − 1′-3″ = 3′-9″ or 3¾ ft.

Step 4: Multiply the surface area by the depth:

2416 × 3¾ = 9060 cu ft. Reduce to cu yd:
9060 ÷ 27 = 335$\frac{15}{27}$ cu yd

or 336 cu yd to the next higher cu yd.

Step 5: Remember that there is always some hand work to be done in excavating. This is for clearing the corners and bringing to final levels during the time that the carpenter's helper is assisting the carpenter. It will require one helper's time to that of two carpenters. Keep

very careful records of all this class of work. There are no quoted figures as good as your own. *Build up your own labor-cost data.*

Note very carefully: **It is of the utmost importance that excavations be carried down on the job to the correct depth. If they are carried too deep, it is understandable that architects and engineers insist that the difference in depth be made good with reinforced concrete. I know of several such cases. Such an unfortunate event may delay the completion date of the job, and it is disastrous for the morale of the crews and is a very serious direct loss of profit.**

Problem

Allowing for excavations to be made 2′-0″ clear of the building lines and to a total depth of 5′-0″ below grade (the topsoil has already been removed to a depth of 1′-3″), estimate the number of 8-cu-yd-truckloads of spoil to be hauled away after allowing for backfill. Assume the earth will swell 25 percent above its unexcavated volume when excavated (see Sec. 10.2, "Cut and Fill," for swell on cut earth). Complete this exercise for Figs. 8.1 through 8.4; using local subcontractor labor costs, estimate the cost for the above problem.

8.5 EXCAVATIONS: SERVICE TRENCHES

Sewer and water lines are usually laid in the same trench. Assuming that the service trench of Fig. 8.6 is 25″-0″ from the building line to the city property line, 2′-0″ wide, and by even grade from 7′-″ to 12′-0″, estimate the cu yd of excavation required.

Example

Step 1: Draw a dimensioned sketch (see Fig. 8.7). You will never waste your time as an estimator by making sketches.

The length of the trench to be excavated is 23′-0″, because 2′-0″ has already been removed for the basement excavation.

Step 2: Multiply the length by the width times the average depth: 23 × 2 × 9½ = 437 cu ft.

The estimated number of cu yd is 16.

Note very carefully that trench excavations are usually a different price from mass excavating and there is always some hand work to be done in clearing corners and trimming. Most jurisdictions state that where a trench is excavated to a depth of more than 4′-0″, shoring shall be provided. This material and labor must be estimated.

Be aware tht there are more hazards in estimating excavations than in any other phase of building

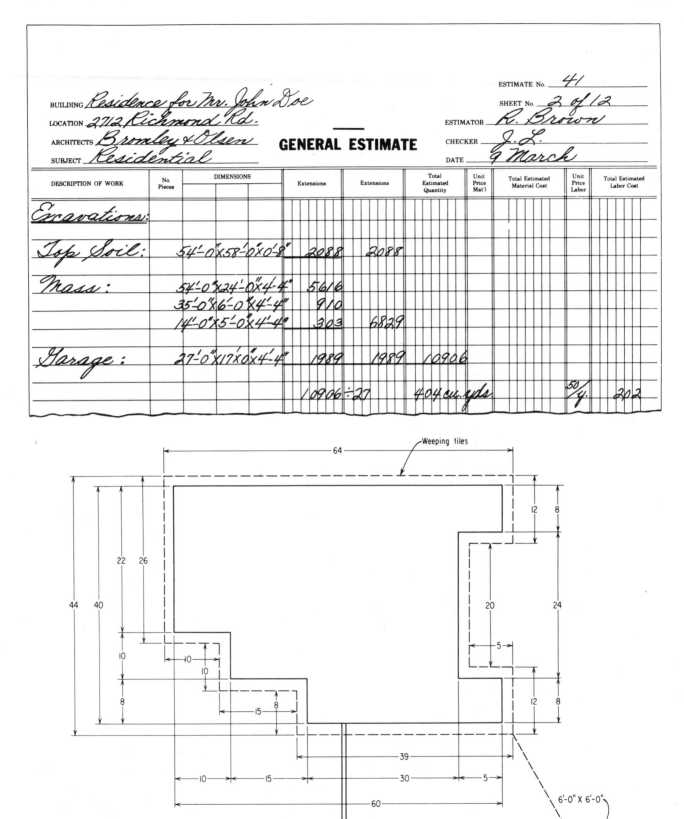

DESCRIPTION OF WORK	No. Pieces	DIMENSIONS			Extensions	Extensions	Total Estimated Quantity	Unit Price Mat'l	Total Estimated Material Cost	Unit Price Labor	Total Estimated Labor Cost
Excavations:											
Top Soil:		54'-0" x 58'-0" x 0'-8"			2088	2088					
Mass:		54'-0" x 24'-0" x 4'-4"			5616						
		35'-0" x 6'-0" x 4'-4"			910						
		14'-0" x 5'-0" x 4'-4"			303	6829					
Garage :		27'-0" x 17'-0" x 4'-4"			1989	1989	10906				
					10906 ÷ 27		404 cu. yds.			50/y.	202

ESTIMATE No. *41*

BUILDING *Residence for Mr. John Doe*

LOCATION *2712 Richmond Rd.*

ARCHITECTS *Bromley & Olsen*

SUBJECT *Residential*

GENERAL ESTIMATE

SHEET No. *2 of 12*

ESTIMATOR *R. Brown*

CHECKER *J. L.*

DATE *9 March*

Fig. 8.6

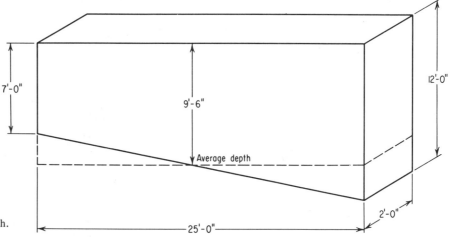

Fig. 8.7 Dimensioned sketch of a service trench.

construction. The cost of excavating increases enormously in price as the depth increases.

Problem

Estimate the number of cu yd of earth to be removed for a service trench for the plan shown in Fig. 8.6. The length from the building line to the city property is 20'-0". It is 2'-0" wide, and by even grade from 7'-0" at the buildings to 12'-0" at the city line. Throw the topsoil to one side, the spoil to the other side, and return the fill.

8.6 DRAINAGE TRENCH

A drainage trench is provided in areas where there is poor natural drainage, as in heavy clay. Its function is to carry water from the drain tiles placed outside the footings. It also usually carries storm water shed from the roof of the building. (See Fig. 9.1.)

Problem

Estimate the number of cu yd of earth to be excavated for a soak-away area 6'-0" × 6'-0" and 9'-0" deep. line, 2'-0" wide, and, by even grade, from 7'-0" at the building to 9'-0" at the soak-away area. Throw the topsoil to one side and the spoil to the other and return the fill. Use a work-up sheet and make a neat dimensioned sketch of the trench.

8.7 DRAINAGE AREA

A soak-away area is a pit excavated to take water from the weeping tiles and also the storm water shed from the roof. It is partly filled with very coarse gravel and covered over with fill and topsoil to grade. This is simply a dug pit. There are neither concrete floors nor walls.

Problem

Estimate the number of cu yd of excavation necessary for a soak-away area 6'-0" × 6'-0" and 9'-0" deep. Allow for backfilling to a depth of 3'-0" with coarse gravel and haul away the surplus spoil after backfilling and replacing the topsoil.

In your answer, show cu yd to be excavated, cu yd of gravel, and cu yd of spoil to be hauled.

Remember that in the above example you would have to purchase and haul the coarse gravel. When an estimator forgets to allow for an item, it is a three-way loss. The item has to be purchased, it must be installed, and, lastly, there is no allowance for profit.

Again you are reminded that there are more hazards in estimating the cost for excavations than in any other phase of building construction. The quantities are easy to estimate, but the problems below ground are often a gamble. On some large jobs, the successful bidder for above-ground work is sometimes given the below-ground work on a cost-plus-a-percentage contract because of the risks involved.

8.8 SWIMMING POOL

Problem

Get a plan and specifications for a middle-class home's swimming pool and estimate:

(a) excavation
(b) trenching for water and power
(c) concrete (or other)
(d) tiling
(e) built-in equipment such as diving boards
(f) handrail
(g) steps down
(h) plumbing installation
(i) filtering plant
(j) decorative lampposts
(k) safety fencing

QUESTIONS

1. Study the outline plan below and find:

   (a) the perimeter of the plan
   (b) the area of the plan
   (c) the number of cubic yards of topsoil to be cleared (and stockpiled on site) 5′-0″ clear of the building lines and to a depth of 0′-10″
   (d) the number of cubic yards of excavation for the basement. Excavate 2′-0″ clear of the building lines and to a finished depth of 5′-0″ from an original level grade.

2. Below what depth should the sides of trench excavations be shored?

3. State two precautions that should be taken in cities when leaving an open trench overnight.

4. What type of soil conditions would require the use of a soak-away area?

5. Explain how you would recommend that a soak-away area be constructed.

6. Define weeping tiles and show by a neat sectional drawing where they should be placed.

7. A service trench has to be dug (from a level lot) 36′-0″ long, 2′-6″ wide and, by even grade, from 8′-6″ to 12′-0″ deep. The depth of the topsoil which has to be removed and stockpiled on site is 0′-8″. Estimate:

   (a) the number of cubic yards of topsoil to be removed
   (b) the number of cubic yards of other earth to be removed.
   (c) the number of cubic yards of gravel that will be required for a depth of 1′-3″ in the bottom of the trench.

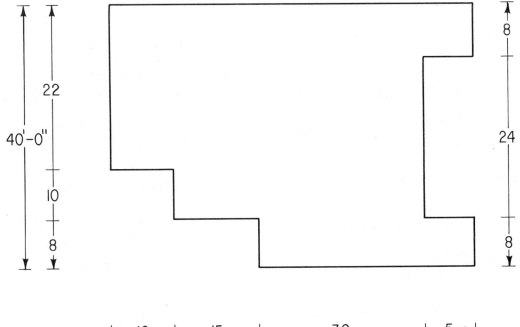

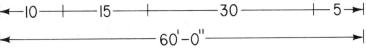

TYPICAL WALL SECTION AND GUIDE LIST FOR RESIDENTIAL CONSTRUCTION

The first part of this chapter is designed to enable a beginning estimator to read a typical wall section with complete understanding. By following the text very closely and at the same time examining the drawing, the estimator should get the feel of the method and order in which the building is to be erected.

The second part is a guide list for residential construction and should be used by the estimator to check that nothing has been forgotten on the estimate (take-off). It also covers, in order, the materials required for the erection of a house. The guide list should be constantly kept up to date with new materials and methods added as they become available. Most important, it must be remembered that if an item has been forgotten on the take-off, the labor, overhead, and profit on the item have also been forgotten.

Architectural drawings for proposed buildings show at least one typical wall section. You must learn the names and the correct spelling of all the parts. See Fig. 9.1 for a typical wall section.

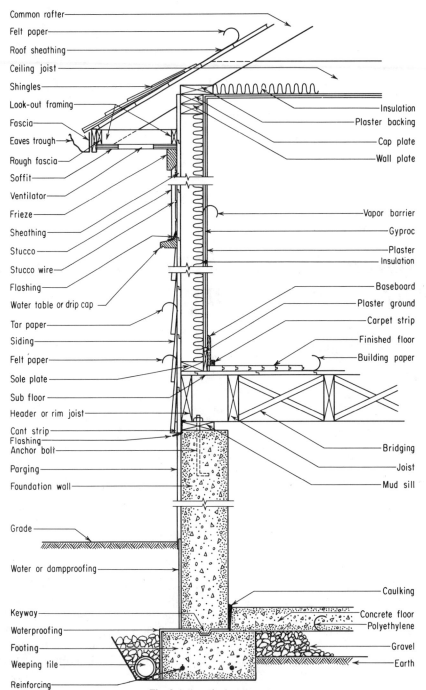

Common rafter
Felt paper
Roof sheathing
Ceiling joist
Shingles
Look-out framing
Fascia
Eaves trough
Rough fascia
Soffit
Ventilator
Frieze
Sheathing
Stucco
Stucco wire
Flashing
Water table or drip cap
Tar paper
Siding
Felt paper
Sole plate
Sub floor
Header or rim joist
Cant strip
Flashing
Anchor bolt
Parging
Foundation wall
Grade
Water or dampproofing
Keyway
Waterproofing
Footing
Weeping tile
Reinforcing

Insulation
Plaster backing
Cap plate
Wall plate

Vapor barrier
Gyproc
Plaster
Insulation

Baseboard
Plaster ground
Carpet strip
Finished floor
Building paper

Bridging
Joist
Mud sill

Caulking

Concrete floor
Polyethylene

Gravel
Earth

Fig. 9.1 A typical wall section.

9.1 BASIC DATA

When estimating, especially for small jobs, the basic data are:

(a) perimeter
(b) floor area
(c) wall area
(d) plan area of flat roofs
(e) roof pitched area

With this information and a reading of the typical wall section, a ready solution may be seen for many estimating problems. Consider the number of times that the perimeter may be used as a factor in estimating quantities. Read the wall section upward from the bottom. There are many books that cover house details in extensive depth. Consult your local library.

9.2 PERIMETER EXPRESSED IN LINEAL FEET—EXTERIOR WALLS

1. Weeping tiles equal the perimeter. Add for corners and drainage trench. Estimate by the lineal foot.

2. Rough gravel over the weeping tiles equals the area of the end section of the gravel in sq ft times the perimeter. Add for corners and the soak-away trench. The unit for estimating is the cu yd.

3. Concrete footing forms equals the perimeter times the depth of the form in feet, times two (two sides). This is square feet; convert to bd ft. Add for stakes, tie wires, nails, and so on.

4. Concrete footings (for small jobs) equals the perimeter times the end section of the concrete in sq ft. Reduce to the estimating unit of cu yds.

5. Reinforcing equals perimeter times the number of rods. Allow 16 times the diameter for lappings.

6. Damp-proofing equals the perimeter times the depth required in feet. This gives sq ft. See manufacturer's specifications for number of sq ft of coverage per 5-gal drum of emulsion.

7. Waterproofing equals perimeter times the depth in ft. This gives sq ft. There is a very great difference between damp-proofing and waterproofing. Damp-proofing is an emulsion applied with a brush; waterproofing is a mortar put on by the plasterers and is estimated by the sq yd.

8. The keyway is the same number of lineal feet as the perimeter. It may be wood, masonry, or metal. Read the specifications.

9. For concrete wall forms, take the perimeter times the height in ft, multiply by two (two forms, inside and outside). This gives the number of sq ft of forms touching wet concrete. See concrete wall forms for allowances for walers, studs, plates, ties, nails, and so on, Sec. 15.14.

10. Concrete wall-form ties equals the perimeter divided by the OCs of the ties, times the number in height. Estimate by the hundred or thousand. *Note that* OCs *throughout the text means on centers.*

11. Stripping concrete wall forms. See item (10) above for quantity. Estimate per M sq ft stripped and reconditioned. There are no estimates like your own for this kind of work. It depends on organization.

12. Parging equals perimeter times the depth in ft for sq ft. Reduce to the unit sq yd. This is the plasterer's job. Parging is often used where it is not necessary to waterproof.

13. The number of anchor bolts required equals the perimeter divided by the OCs. Add extra for corners.

14. Flashing at wall plates equals perimeter. Order in lin ft and add for laps.

15. Cant strip (under the lower course of siding) equals perimeter. Order in lin ft, usually 1 × 2 material.

16. Header or rim joist is equal to the perimeter in lin ft. Convert to bd ft and order in convenient lengths.

17. Outside wall plates, top and bottom, equals the perimeter times the number of plates for lin ft. Convert to lengths and order in bd ft. There is usually one bottom plate and two top plates. Plates for internal walls must be taken separately.

18. Felt or building paper equals perimeter times the height in ft. This gives sq ft. Add 8 percent for waste and deduct only for very large openings. Estimate per roll and check locally for number of sq ft per roll.

19. Water table or drip cap equals the perimeter. Order in lin ft and deduct for openings.

20. Siding equals perimeter times the height required in ft. This equals sq ft to be covered. See manufacturers' pamphlets for allowances for waste in milling and applying. Order by M sq ft.

21. Flashing over the water table equals perimeter. Deduct for large openings. Order in lin ft from the tinsmith.

22. Stucco wire equals perimeter times the height in ft. This gives sq ft. Deduct for openings. Order by the roll, and consult manufacturers' publications for quantity and quality per roll.

23. Stucco equals perimeter times the height in ft. Deduct for large openings. Reduce to the unit of sq yds for estimating.

24. Sheathing equals perimeter times the height of the wall including the floor assembly in ft. This equals sq ft. Deduct for large openings. Allow 3 percent waste for plywood sheathing. See Table of Allowances for dimension board sheathing, Chapter 7.

25. Frieze equals perimeter for a hip roof or front- and back-wall lengths for other roofs. Order by the lin ft.

26. Ventilators. If this is a continuous, fine-meshed ventilator under the eaves, take the perimeter for a hip roof, or front and back walls for other roofs, and multiply by the width of the mesh in ft. This gives sq ft. Order by the roll. Get your information for quality and quantity from the dealers.

27. Eaves assembly for a flat or regular hip roof is taken from the perimeter:
    (a) lookout framing,
    (b) soffit,
    (c) fascia board (rough),

(d) fascia board (finished),

(e) eaves trough.

In each case add for turning corners.

28. Paint in sq ft equals the perimeter times the height to be covered in ft, less large openings.

9.3 INTERIOR OF TYPICAL WALL SECTION

1. Caulking material equals the perimeter, and the unit is lin ft.

2. Vapor barrier equals the perimeter in ft times the height in ft. Add 8 percent for lap and order by the roll. Check for number of sq ft per roll locally.

3. Mud sill equals the perimeter. Add for interior walls as necessary. Order in length required by the bd ft.

9.4 PLAN AREAS

1. Gravel under the basement floor equals the plan area in sq ft times the depth of the gravel in ft, less the amount displaced by the outside and interior footings. Unit of estimating is cu yd.

2. Vapor barrier over the gravel and under the concrete floor equals the floor area. Add for lap according to the width of the rolls.

3. Basement concrete floor area equals the plan area less the displacement by the outside and interior footings or pier pads. The unit is usually priced per sq ft delivered, placed, and finished. Consult your local concrete finishers for unit price.

4. Subfloor equals the plan area. Remember it requires nails, and add for waste.

5. Building paper over the subfloor (checks dust from falling below). This equals the floor-plan area. Add 8 percent for lap per 400 sq ft of tar paper.

6. Ceiling insulation and finish equals the plan area. There are several types of insulation. Consult local merchants for coverage and prices.

7. A flat roof assembly equals the plan area plus an allowance for the overhang.

9.5 FLAT ROOF AREA INCLUDING THE OVERHANG

1. The decking sheathing equals the plan area of the roof. Order by the M sq ft.

2. Rigid insulation equals plan area. Order by the M sq ft. Add 2 percent for waste.

3. A bonded roof equals the plan area. Get a subcontractor's price.

9.6 PITCHED ROOF AREAS

1. Roof sheathing equals roof pitched area. Add 5 percent waste for plywood. Where a roof is cut up with a lot of hips and valleys, add 8 percent waste.

2. Building paper equals roof pitched area. Add 8 percent for lap if using 400 sq ft rolls. See manufacturers' lists.

3. Shingles and nails. These items equal the pitched area. Add for starter courses of shingles, ridge caps, and hip and valley allowances. Consult current publications from the A.I.A. (American Institute of Architects) file.

4. Shingle paint in sq ft equals the actual pitched roof area.

9.7 AN ESTIMATOR'S GUIDE LIST FOR HOUSING PROJECTS

The following guide list covers, in order of building, the operations and materials required. This should be used as a check that nothing on the plans and specifications has been omitted. One of the sources of error in estimating is in forgetting to take off some item. *Anything forgotten is a loss in cost of material and in its installation.*

Preliminaries

Examination of a building lot

Lot purchase:

  Survey

  Locating survey pegs

  Location of building on site

  Bench mark

  Batter boards

  Legal costs

  Real estate fees

  City taxes

  Plans and specifications

  Building permits

  Road closure permit

  City water permit

  Cost of alternate water supply

  Digging a well

Temporary Amenities

Buildings on site

Mobile office
Toilet facilities
Power pole and meter
Light
Heat

Builders' Loan Charges

Bank, real estate, insurance company
Credit union, other

Insurance

Bonds
Public liability
Workmen's compensation
Fire
Vehicle
Equipment
Bonds
Fringe benefits

Permanent Utilities

Heat
Light
Power
Telephone
Cable television
Sewer
Water

Advertising

Newspapers
Trade papers
Billboards
Radio and TV

Supervision

Superintendence
Job runner (junior estimator)
Foreman
Transportation costs

Equipment

Machinery (own or hired)
Ladders
Scaffold
Wheelbarrows
Hoses, picks, shovels, wrecking bars
Sawhorses

Topographical

Clearing the site
Demolitions
Soil tests
Establishing a bench mark
Grading
Lot layout
Batter boards

Excavations

Remove and stockpile topsoil
Cut, fill, and grade
Mass excavations:
    Basement
    Septic tank
    Soak-away area
    Swimming pool
Trenches:
    Sewer, water, power, telephone, cable television
Backfill and landscaping
Seeding and planting

Concrete and Formwork

Take off concrete and formwork together
Footings—house and garage
Reinforcing
Pier pads
Walls (internal and external)
Floors
Steps
Sidewalks and driveways
Septic tank
Concrete test specimens
Swimming pool

Driveways

Blacktop
Gravel
Stone flagging
Concrete
Other

Drain and Weeping Tile

Straight runs
Elbows, tees, inspection chambers
Fabric

Waterproofing and Damp-proofing

Parging
Emulsion

Progress Reports

Written
Photographs

Underpinning for First-Floor Joists

Wood posts
Wood partitions
Masonry walls
Lally posts
Masonry columns
Tele-posts
Steel columns

Beams

I beam
Wood, solid
Wood, laminated
Glu-lam
Reinforced concrete

Rough Floor Assembly

Joists
Joist hangers
Headers (rangers)

Tail joists
Bridging—wood or metal
Subfloor
Floor deadening
Building paper
Reinforced concrete

Walls

Plates
Studs—wood or metal
Expanding metal
Girths
Metal strapping for plumbing
Concrete block and reinforcing

Carport

Framing
Posts
Fire protection
    Footings ⎱
    Ceiling ⎰ (see local fire regulations)

Garage

Roof Eaves

Lookout framing
Soffit and soffit ventilators
Fascia board—rough and finished
Eaves: troughs and downpipe

Roofing Materials

Sheathing
Flashing
Building paper
Shingles:
    Wood or wood shakes
Concrete tile
Asbestos
Masonry
Asphalt tile
Asphalt roll roofing
Underlay for roofing

Corrugated metal

Plastic

Slate

Roofing Nails

Chimney

Concrete reinforced footing

Bricks—common, decorative, fireclay

Mortar, fireclay

Flue lining

Flat concrete arch

Metal for arch

Damper

Cleanout

Fuel chute

Ash dump

Parging (cement, lime, and sand)

Flue lining

Chimney pots

Chimney cap

Flashing (roof to chimney)

Insulation (chimney to wood)

Hearth

Cladding Exterior

Sheathing—board, plywood, other

Building paper

Siding

Stucco and stucco wire

Brick, veneer, and wood framing

Brick, solid

Brick and cinderblock

Stone

Masonry units

Stairs

Wood, metal, concrete

Stringers

Undercarriage

Treads

Nosings

Risers

Wedges

Newel posts

Handrail and brackets

Concrete forms and ties

Reinforcing

Cladding Interior

Insulation—floors, walls, and ceilings

Vapor barriers—floors, walls, and ceilings

Lath and plaster

Drywall and taping

Decorative finish

Expanding metal

Corner beads

Grounds for doors and windows

Outside Trim

Gable ends

Eaves

Frieze

Fascia and soffit

Outside trim

Drip cap

Drip-cap flashing

Ventilators

Window Units

Wood, metal, and storm

Basement, main floor, dormer

Flywire screens

Flashing to all openings

Headers—wood, metal, or concrete lintels

Doors

Frames—inside and outside

Doors:
    Front, back, storm, and screen
    Room
    Closet
    Linen cabinets
    Clothes and storage

Inside Trim

Baseboard

Carpet strip

Window and door trim

Window stool-apron-stop

Closet rods and shelves

Valance

Bathroom

Medicine cabinet

Vanity

Towel bars

Soap and grab

Toilet-paper holder

Shower-rail and curtain

Glass Block

Front entrance and patio

Exterior steps and handrails

Decorative Concrete Units

Bricks and metal ties

Stone

Plumbing wall

Cripples to doors and windows

Headers for doors and windows—exterior and interior

Milk chute

Fuel chute

Ceiling Joists

Blocking at wall plates

Backing

Bridging—wood or metal

Metal hangers

Hanging beams

Ceiling access door

Roof Framing

Gable studs

Barge boards

Barge-board moldings

Roof-bracing members

Purlins

Collar ties and bracing

Rafters—common, hip, jack, valley

Trusses

Vents

Dormer-window framing

Hanging beam or strong-back

Tiling

Living room

Kitchen

Bathroom

Den

Hall and other

Hardware

Form wire or ties

Screws and fastening devices

Door locks, hinges, checks, and stops

Towel rails

Cabinet tracks, hinges, catches, and pulls

Coat hangers

Handrail brackets

Weatherstripping

Letter-drop chute

Clothesline posts

Glue and sandpaper

Nails:
    Common and finishing
    Double-headed nails for formwork
    Roofing—galvanized, cooper, aluminum, or shingle
    Wall board

Attending On Other Trades

Cutting away and making good after plumbers, electricians, heating engineers, and so on—some allowances must be made.

Cabinets

Meters

Storage

Ironing boards

Den
Medicine
Books
Kitchen
Vanity

Special Fixtures

Dishwasher
Oven and range
Refrigerator
Deep freeze
Garbage-disposal unit
Range hood and fan
Planter boxes
Mirrors
Ceiling fans
Ceiling decorative light fixtures
Shower doors
Vacuum cleaning outlets
Cable television and telephone outlets
Others

Subtrades

Excavator
Concrete
Plumber
Chimney and fireplace
Plasterer and drywall
Electrician
Heating engineer
Hardwood floor layer and finisher
Linoleum, tile, and carpet layer
Tinsmith
Sidewalk and cement finishers
Air conditioning
Telephone and cable television
Painter and decorator
Landscaping
Swimming pool
Fencing—wood, metal, decorative, brick, stone, or
    other
Final cleanup—windows, doors, floors: vacuum and
    polish

Cleanup of grounds
Blacktop, gravel, concrete, or other

Overheads

Proportion of general office overhead to each job
Individual job overheads
Superintendent, job runner, and foreman
Sinking fund appropriations
Builders' loan fees
Mortgage fees
Federal taxes

Handover of Property

Professional fees
Attorney
Fixed fees for registrations
Notarization fees
Real estate
Accountancy
Accrued interest charges
Maintenance clause

Extras

You must get signed agreements for all extra work done over and above original drawings and specifications. This is most important.

Profit

You may be surprised to know how many bids are submitted where an allowance for profit has been forgotten. *Take care!*

Note well that this list is not exhaustive and should be used as a check against the take-off. It must be kept up to date. On large buildings there must be a checklist. The plans and specifications should always be very carefully examined for new types of materials and new methods of construction. *You will never have a complete guide list.* Discipline yourself not to forget to *take off all items.*

9.8 WEEPING TILE

Authorities state that where water may accumulate at the outside footings of basement or cellar walls, a continuous row of weeping tiles $0'\text{-}4''$ diameter, covered with not less than 0'-6" of coarse gravel or broken stone, shall be placed around such footings on undisturbed ground. They shall be laid with open joints (about ⅜"), which shall be covered with strips of rotproof saturated felt or equivalent. All runs of weeping tile shall be drained to the city main or to a soak-away area. See weeping tile, Fig. 9.1, p. 190.

Problem

Estimate the lin ft of weeping tile required for the plan shown by Fig. 8.6, page 166. Allow for elbows, one yoke, 20'-0" run to the soak-away area, and for the fabric to cover the joints.

Example

Step 1:

Take the basic perimeter:	200 lin ft.
Add for two right-side recess walls (2 × 5'-0"):	10
Add for eight outside corners:	16
Add for run to the soak-away area:	20
	246 lin ft.

Add for eleven elbows
Add for 100 sq ft of polyethylene for joints

Step 2:   Alternative Method. Using an architect's office-use steel tape, which is scaled ⅛" on one side and ¼" on the other, measure the outside footing perimeter. Pay out the tape wall-for-wall as each wall length is added plus 20'-0" for the run to the soak-away area. Make the final accumulated reading. Add for elbows, yoke, and joint fabric covering material.

Problems

1. Using a work-up sheet, make a neat estimate of the material required for a weeping-tile drainage system for the plans shown in Figs. 8.1 through 8.4.

Item	Fig. 8.1	Fig. 8.2	Fig. 8.3	Fig. 8.4
Weeping tile				
Elbows				
Yoke				
Rotproof fabric				

2. Estimate the time required to place the drainage sys-

tem for each of the plans in problem 1, above. Allow that two men working in restricted areas around the footings can place 500 lin ft of weeping tile in an 8-hr day at a rate of 30–35 lin ft per man-hr.

3. Estimate the number of cu yd of coarse gravel or broken stone required to cover the weeping tile to a depth of 0'-6" and 1'-6" wide. Assuming that the gravel is dumped close to the working area, a man should wheel and place about 1 cu yd of coarse gravel per hr. *Time your own jobs. There are no records as valuable as your own.*

4. Using local labor and material prices, make a complete labor and materials cost estimate for the complete drainage system for the plan in Fig. 8.4.

Weeping tile		
Elbows		
Fabric		
Labor		
TOTAL COST		

9.9 DAMP-PROOFING AND WATERPROOFING MASONRY BELOW GRADE

Degree of Protection

Where hydrostatic pressure occurs, waterproofing shall be applied to all exterior foundation walls and floors below grade. If hydrostatic pressure does not occur, damp-proofing shall be applied to all exterior foundation walls below grade.

Waterproofing

Waterproofing shall consist of a system of membrane waterproofing applied at the exterior surfaces of the foundation walls enclosing the basement or cellar area. Basement or cellar floors shall be protected by applying a system of membrane waterproofing between two pourings of concrete, each of which shall be at least 0'-3" thick. This membrane shall be carefully mopped to form a complete seal with the membrane of the exterior walls, which shall enter the basement or cellar area between the footings and the foundation.

Damp-proofing

Basement and cellar walls to be damp-proofed shall

be treated on the exterior from finished grade to outside edge of footing as follows:

(a) *Masonry-unit walls:* The first course of masonry shall be laid in a full bed of mortar with end joints, including end cavities of the blocks, and completely filled with mortar. This first course shall then be parged and the parging covered over the footing before further masonry units are laid. Over the parging apply at least one heavy, continuous coat of bituminous emulsion with or without mineral colloid, asphalt, cutback, undiluted hot tar, hot asphalt, or other acceptable compound.

(b) *Poured-concrete walls:* Apply at least one heavy continuous coat of undiluted hot tar, hot asphalt, bituminous emulsion with or without mineral colloid, or other acceptable compound.

Damp-proofing

There are numerous emulsions available for spraying or hand-painting foundation walls below grade. When estimating the number of sq ft to be treated, take the perimeter of the walls in sq ft and multiply by the depth in ft. Most emulsions are marketed by the 5-gal drum. The directions should be carefully followed and a note made of the sq-ft coverage per drum. The cost of application will depend upon the method used. For a single job it will be cheaper to apply by hand. Using a roofer's hand brush and painting into the crevices of the concrete, one man should apply the first coat at the rate of about 500 sq ft per 8-hr day. Succeeding coats will go faster and the emulsion will have a better coverage. The second coat would require about 6 to 7 hrs per 500 sq ft and the third coat about 4 to 5 hrs.

The coverage per hr for the first coat: 60–70 sq ft
The coverage per hr for the second coat: 80–90 sq ft
The coverage per hr for the third coat: 120–150 sq ft

Read the specifications for the type and number of coats required.

Waterproofing

Concrete specifications often call for a waterproof concrete for basement and cellar walls. This may be achieved by the design mix of the concrete to be used. If an additive to the concrete is specified, this must be allowed for on the estimate.

Membrane waterproofing by built-up fabric applied with hot pitch or tar may be called for on perpendicular walls or for floors or roofs. This would require the use

of a subtrade, for which the estimator would obtain the best price from a reputable company.

Waterproof mortar would be applied by plasterers. For quantities of cement and sand required per cu yd of mortar, see p. 288. A plasterer working in the confined space of an excavation should apply 60–70 sq yd of waterproofing mortar per 8-hr day. If the depth is greater than 5'-0", allow 40–60 sq ft per 8-hr day.

9.10 INSTALLATIONS WITH SALES APPEAL FOR SPECULATING BUILDERS

To remain competitive, a contractor must be adaptable to modern techniques and innovations and be prepared to install many services, such as:

(a) Air conditioning.

(b) Automatic push-button self-cleaning oven.

(c) Gas ranges and refrigerators.

(d) Automatic rotisserie and electrically operated thermometer.

(e) Double stainless steel kitchen sink with government-approved waste-disposal unit (sometimes called a garburetor).

(f) Automatic electric dishwasher.

(g) Fully formed formica counter tops and breakfast bars.

(h) Quick recovery large capacity water heater.

(i) Laundry room with installed washer and dryer, and an ironing area with ample electric outlets including a 220-volt outlet for dryer; outlets in this area for telephone and television; and ample storage for linen.

(j) Total electric or other heating.

(k) Door chimes.

(l) Chandeliers.

(m) The whole residence to be prewired for cable television, telephone in strategic areas, door chimes, and thermostatic controls.

(n) All rooms and areas to be provided with vacuum-cleaner outlets to carry dust directly to one large bag located in the garage; the bag may only require emptying every six months.

(o) Silent switches.

(p) Good plumbing fixtures, architectural metalware.

(q) Ample mirrors and cabinets in the bathroom.

(r) The latest and most attractive in metal ironwork and carpets.

(s) Burglar alarm.

Try drafting a newspaper advertisement for your houses before you build them. Then compare it for value, appeal, and location with current advertising for similar houses. This will fix in your mind the type of competition you are to face.

9.11 LAND APPRAISAL

Land means its geographical location and everything that is attached to it, such as buildings, fences, out-croppings of rock, or even a sundial attached to it by concrete.

An estimator of residential construction, especially if he is an entrepeneur, must have a wide knowledge of the industry, and he should know where, when, and what to build that will have sales appeal.

There are a number of reasons for appraising property:

(a) for the sale or purchase of land

(b) for financing a mortgage

(c) for assessing municipal taxes

(d) for disposition of property between beneficiaries under a will.

(e) For the takeover of the property by the government for the benefit of the public. This is called *the right of eminent domain,* and the government is entitled to do this, providing it gives fair compensation to the owner.

Land Appraisal

(a) by comparing recent sales of similar property in the same district

(b) by adjusting values against trends of the environment in the district

(c) by judging intended future development in the district by the municipality;

The appraiser should be a very experienced realtor and know the district intimately.

Let us consider a property that is being appraised for a savings and loan institution, which lends money on mortgages. The appraiser's report to the savings and loan company lists the value of the property at $100,000. The company will probably approve a loan of 60 percent of the appraised value, or $60,000. This is done in case the mortgagee does not honor his commitment to make regular, agreed repayments on the loan plus interest. The mortgagor then has the right to initiate foreclosure proceedings (forced sale of the property), and in this way he would receive the amount of capital and interest outstanding at the time of foreclosure (liquidation).

9.12 A GUIDE LIST OF SUBTRADES

The following list is presented as a guide to the different specialized subtrades. The list should be kept up to date and referred to when designing residential units. Remember that firm legal bids must be obtained from every subtrade before work is commenced. *Under no circumstances should word-of-mouth agreements be accepted.*

1. land levelling and grading
2. water drainage and control
3. excavating
4. ditching and trenching
5. concrete contractors and finishers
6. concrete: prestressed and post tensioned
7. concrete forms
8. concrete block
9. reinforced steel erectors
10. septic tank and field specialists
11. plumbing
12. electric and power
13. heating
14. air conditioning
15. telephone and television linesmen
16. tinsmiths
17. masons
18. swimming-pool installers
19. roofing specialists
20. metalwork
21. drywallers
22. plastering and stucco wire
23. acoustic ceiling
24. chimney builder
25. doors and windows
26. glass block, glass and glazing
27. mirrors
28. chandeliers
29. tiles earthenware, ceramic, plastic
30. insulators
31. I-beams: wood solid, wood built up
32. flooring: hardwood, linoleum, tiles, broadloom, indoor-outdoor carpeting
33. hardware suppliers
34. painting and decorating
35. blacktop, gravel, and other types of driveways

36. janitorial services

37. landscapers

38. sale and closing procedures

Problem

1. Call or phone your local builders' supply merchant (or search your A.I.A. file) and get all the information you can on the cost and sq-ft coverage of damp-proofing agents. Select any one type and make a materials and labor cost estimate to damp-proof the walls (by hand) 5'-0" below grade for the plan in Fig. 8.4.

2. What is the cost of a paint-spraying unit in your locality?

3. What is the cost of the rental of a paint-spraying unit in your locality?

QUESTIONS

1. Name the units in which the following estimates are made:

   (a) weeping tiles _____
   (b) gravel _____
   (c) concrete footing forms _____
   (d) damp-proofing _____
   (e) waterproofing _____
   (f) concrete footing keyway _____
   (g) concrete wall, form ties _____
   (h) parging _____
   (i) anchor bolts _____
   (j) header or rim joist _____
   (k) outside wall plates _____
   (l) building paper _____
   (m) water table or drip cap _____
   (n) flashing over the drip cap _____
   (o) stucco wire _____
   (p) stucco _____
   (q) frame wall sheathing _____
   (r) vapor barrier _____
   (s) mud sill _____
   (t) concrete floors _____
   (t) concrete floors _____
   (u) wood subfloor _____
   (v) insulation _____
   (w) rigid insulation _____
   (x) roof sheathing _____
   (y) roof shingles _____

2. Name five ways in which the floor-plan area of a house may be used in estimating quantities.

3. Name three ways in which the sloped area of any roof may be used in estimating quantities.

4. List ten preliminary building operations that must be accounted for when estimating.

5. List ten main headings for a house take-off.

6. Fill in, by printing, the parts (indicated by arrows) of the typical wall section shown in Fig. 9.2.

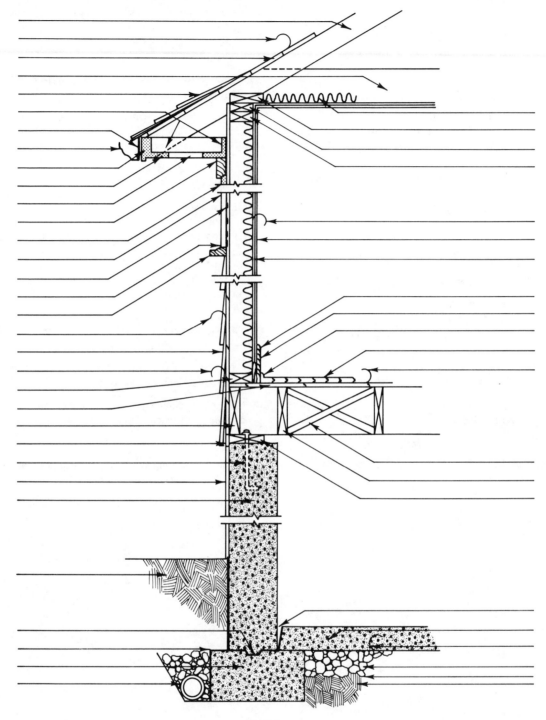

Fig. 9.2

10

EARTH WORK AND LAND LEVELING

This chapter on land leveling is designed for the small building contractor and contains sufficient information for the builder to take off the cut and fill on land leveling for housing or a small shopping center, without recourse to the services of a professional engineer. Where project housing or large land development is required, the services of an engineer should be obtained.

10.1 A GLOSSARY OF LAND-GRADING TERMS

Angle of Repose: The sides of an excavation that are dressed back at an angle so that small rocks will not fall onto the floor; such an angle of repose will obviate the need to support the sides of perpendicularly cut walls.

Basement: A basement has more than one-half of its vertical height above the finished grade of the land.

Batter Boards: Horizontal boards nailed to stout stakes and erected at the corners of excavations, over which taut strings or wires are stretched to indicate the building lines for walls and footings; see page 6. Sec. 1.12 and Fig. 10.1.

Bench Mark: A predetermined point used as a reference when building. The symbol is 〒; the center of the horizontal line is the reference.

Cellar: A cellar has more than one-half of its height below the grade of the land.

Compaction: Fill that is spread and compacted with heavy machinery.

Cut: Land that is cut (excavated) and hauled away from a parcel of land to reduce it to the desired elevation.

Fill: Land that is brought to a higher elevation by hauling in surplus cut from other areas.

Frost Line: The depth to which frost penetrates the ground in any given area; for official data see the local authority.

Grade: The present or proposed grade of a parcel of land as shown on the topographical drawings.

Grade Line: A predetermined line indicating the proposed finished elevation of the ground for an area such as a parking lot or the area around a new building.

Grade Striker: The title given by some authorities to the person whose duty it is to place official stakes on city property, indicating the depth of the sewer below the existing grade and showing the future grade of the sidewalk and of the center of the proposed finished road.

Grid: Baselines intersecting at right angles as used on survey maps.

Interpolate: To determine an intermediate (average) between two stations such as 112.7′ and 105.1′. Thus: 112.7′ minus 105.1′, divided by 2, gives an average reading of 108.9′ for land that must be graded to a bench mark reading of 100′; the amount to be cut away averages 8.9′. *Note that in surveying, measurements are described in feet and decimals of feet.*

Interstices: The voids between pieces of rock (or grains of any solid). The larger the pieces, the greater the collective volume of voids between the lumps. *This is an important fact to remember.*

Parcel of Land: Any area of land registered as one unit, such as a building lot. The area may be several acres, which may later be subdivided and registered as smaller parcels.

Sanitary Fill: This is a system of filling low-lying areas. First the topsoil of the area to be filled is removed and stock is piled on the site, then household refuse is deposited and compacted to a depth of several feet. This is followed by compacting clean, inert material such as crushed rock, boulders, sand, or broken brick to a few feet in depth. The whole area is thus sandwiched until such time as the topsoil is replaced; within a few years' time the area is perfectly sanitary, and light structures or a public park may be built on it.

Station: Any point on a parcel of land from which a survey reading is taken.

Test Holes: Holes drilled or dug on a proposed building site to determine the nature of the subsoil and to find the natural water table.

Undisturbed Ground: Uncut and undisturbed natural subsoil upon which the footings (underpinning) of structures stand, where the soil is able to withstand an imposed load without any settling.

10.2 CUT AND FILL

Cut refers to ground that has to be brought to a lower level by cutting and hauling, and fill refers to land that has to be brought to a higher level by filling and compacting.

It is imperative for the estimator to know the present and future grades of any project against which he is going to bid. In cities, towns, and villages, these levels may be obtained from the city engineers' department. The estimator should visit the site to determine if there are trees to be removed or demolitions to be made. An allowance may have to be made on the estimate for protecting adjoining properties during operations.

In some areas, test holes may have to be drilled or dug to ascertain the nature of the ground; also read the specifications for soil reports.

In urban developments, the proposed grades of the sidewalk and the center of the road can be obtained from the local authority. In most cases (but not all) the land in residential districts will be above grade. Where curves on expressways and freeways are planned, the grade will be lower on the inside curve of the road.

10.3 SANITARY FILL

Many authorities bring low-lying land areas to a desired grade by use of sanitary fill in the depression. First, the topsoil is stockpiled on site, then household refuse is compacted in the depression to a depth of several feet; this is followed with several feet of clean gravel or other approved, clean, compacted fill. This sandwiching is continued until the replaced topsoil brings the whole area to the desired grade.

Within a few years the area is perfectly sanitary and ready for building or other uses. City planning departments have plans available for study by the public which show all reclaimed lands under their jurisdiction. You must be fully aware of ground conditions for any area upon which you intend to place a bid. Estimating volumes of earth is not difficult, but encountering unusual ground conditions—from rock to water—is always a risk.

The estimator must always keep in mind that the plans and elevations of buildings show the *finished grades.* A house plan may be suitable for erection in any state or province where the contours of the land are unknown to the designer. If the land is low, it may be necessary to haul in fill, which may have to be purchased. When compacted with heavy machinery, some kinds of loam and clay will require 20 to 25 percent more earth than was occupied by the same earth in its natural state. This must be estimated. Conversely, the volume of excavated earth will swell anywhere from 6 to 10 percent for sand and up to 50 percent for rock.

10.4 BATTER BOARDS

Batter boards are frames placed adjacent to (but on the outside corners of) proposed excavations, over which a taut mason's line (or wire for larger jobs) is strung for delineating building lines when excavations are completed. See Fig. 10.1.

Let us assume that we have established the front building line stakes for a residence having a rectangular plan of 30'-0" × 48'-0". See Fig. 10.2. Note carefully that for small offsets from an otherwise rectangular plan, a large wooden square may be used.

To lay out the batter boards, proceed as follows:

Step 1: Using the methods of triangulation described, lay out the four corner stakes for the above-mentioned residence as in Fig. 10.2.

Step 2: Remember the importance of checking by steel tape all diagonals in field layout. They should be equal.

It once came to my notice that the building lines for a rectangular house 30'-0" × 90'-0" were 2'-0" out of square. The layout had been done using a transit level, but the diagonals were not checked for equality of length (diagonals must always be checked). The error was not discovered until the carpenters tried to lay the subfloor with 4'-0" × 8'-0" sheets of plywood. *Be very careful!* You cannot rush this kind of work.

Step 3: With two men working together, one sights over stakes A and B (in Fig. 10.3) to align stake B, which the second man places about 3'-0" from stake B.

Step 4: This operation may be repeated with one man now sighting over stakes B and C, while the other places stake C (about 3'-0" from B), and so on, until two stakes have been placed adjacent to each original corner stake. See Fig. 10.4.

Step 5: The original stakes A, B, C, and D may now be removed, leaving the area free for the excavator. Some small jobs may be accomplished using stakes only. See Fig. 10.5, which shows the original corner stakes removed and the mason's lines drawn over the outer stakes. Remove the lines for excavating, then replace them, dropping a plumb line from each of their intersections to reestablish the stakes in the bottom of the excavation. See Fig. 10.1.

Step 6: It will be seen from Fig. 10.6 that a more durable job of establishing the building lines is possible by using batter boards on the outside corners of proposed excavations. Remember that the further away that batter boards are placed from an excavation, the longer the arms must be. If the batter boards are placed 3'-0" clear from the proposed excavation, the arms must be at least 4'-0" in length.

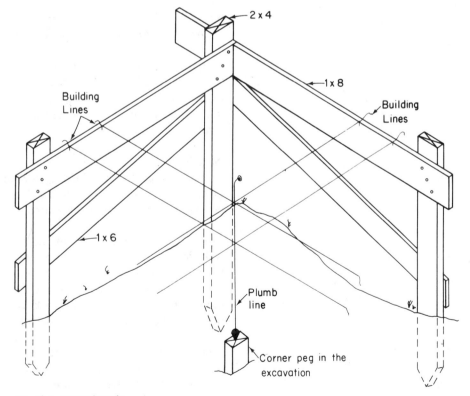

Fig. 10.1. Batter boards.

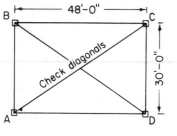

Fig. 10.2

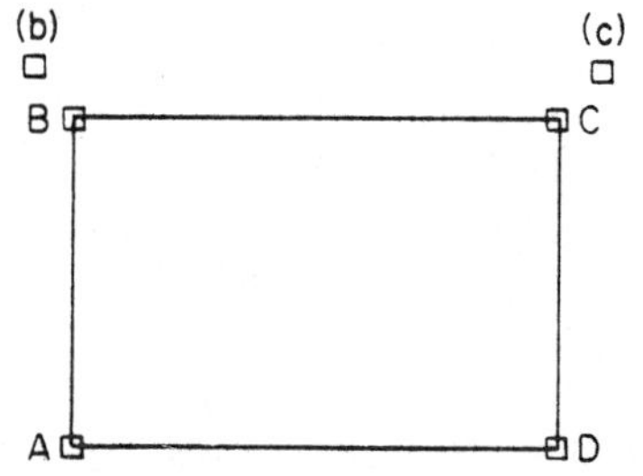

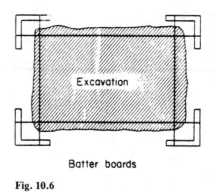

Batter boards

Fig. 10.6

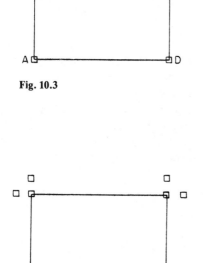

Fig. 10.3

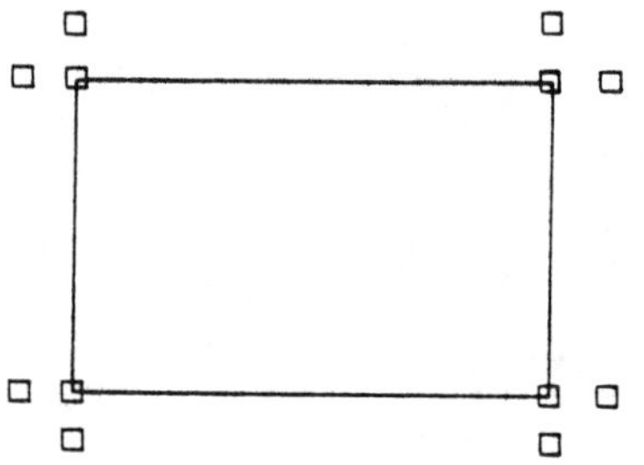

Fig. 10.4

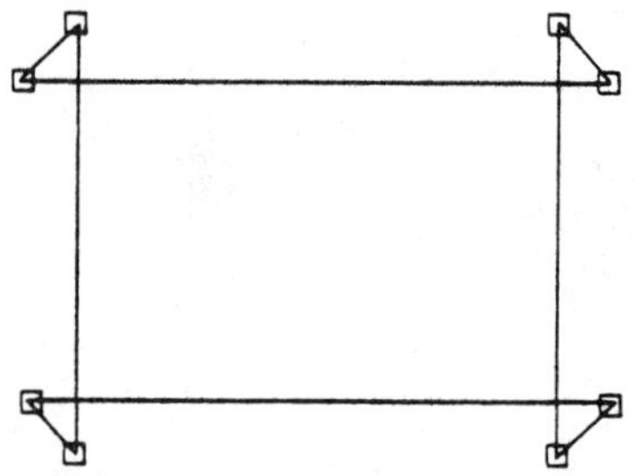

Fig. 10.5

Step 7: Set the arms of the first batter board with its top edges level to each other, and placed at a predetermined height, say, the top of the basement wall. The three other batter boards must be placed and leveled to the first.

Step 8: To level from one batter board to another, use a long straightedge of wood (say, 14'-0" in length) and a spirit level, the longer the better (see Fig. 10-7). Place one end of the straightedge on the first batter board; level the straightedge with the spirit level as shown in Fig. 10-7 and set a stake at the other end of the straightedge, exactly level with the batter board. Turn the straightedge and spirit level, end for end, and recheck the work. The bubble of the spirit level should read the same both ways. Then level from one stake to another until the next batter board is leveled from the first.

It is recommended that the first level be made to the center of the area to be excavated. Then the accuracy of the height of this stake allows it to be used as a reference to radiate all the remaining batter boards. The fewer times the straightedge has to be used, the less the risk of error.

Step 9: A saw kerf may be cut into the tops of the arms of each batter board indicating the building lines. On large jobs, the footing lines and wall widths may also be cut on the arms of the batter boards.

Step 10: It is important that the excavation be made well clear of the building lines to give room for men to work between the basement underpinning and the walls of the excavations. Some jurisdictions have stipulated minimum clearance for such working areas. Get a copy of the local building code from your local authority.

Step 11: When the area has been excavated, the building lines may be established in the excavation by dropping a plumb line from each corner intersection of the building lines above. See Fig. 10.8; also Fig. 10.5.

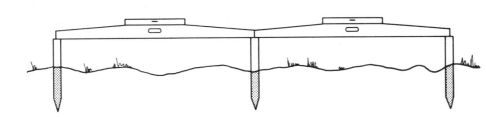

Use straightedge and spirit level for levelling

Fig. 10.7

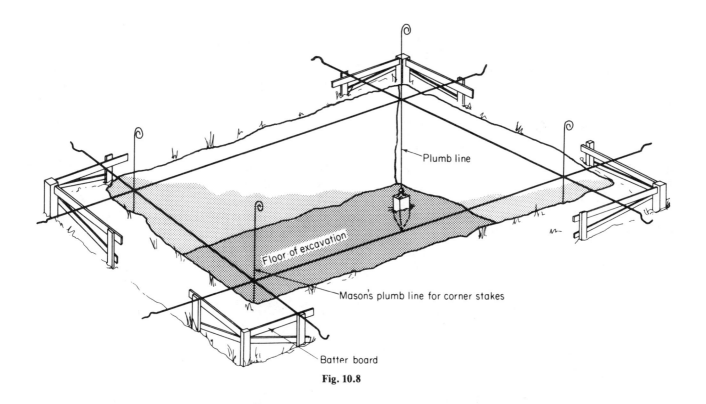

Plumb line

Floor of excavation

Mason's plumb line for corner stakes

Batter board

Fig. 10.8

10.5 SWELL PERCENTAGE OF CUT EARTH

The actual amount of swell for any given cut is determined by soil tests or by studying the soil reports given in the specifications. *Soil* means earth. Topsoil is expensive and is estimated separately for either buying or selling.

Type of Earth Cut	Swell Percentage
Sand and gravel	6 to 20%
Loam and clay	20 to 25%
Rock	30 to 50%

It would require an 8-cu-yd truck to haul an excavated 6 cu yd of cut clay. Think of the estimating haulage problem where 6 cu yd of cut clay will swell to 8 cu yd of loose earth, which, with heavy machinery, will compact to about 5 cu yd of fill.

10.6 BENCH MARK OR DATA

The grade to which land has to be cut and/or filled is taken from a bench mark having this symbol: ⊼ . The center of the horizontal bar is the bench-mark level or datum line. For small jobs this symbol is usually cut onto a stout stake, which is erected near the perimeter of the site and fenced around for protection during operations. The center of the bar is identified as 100.0 '; for example, 112.2 ' equals 12.2 ' of cut, and 89.2 ' means the land is 100.0 ' − 89.2 ' and requires 10.8 ' of fill.

Many bridges and large public buildings bear a bench mark showing the height above sea level; other levels may be taken from these to determine sewer levels, street drainage, and so on. A manhole cover in the center of a road may be taken as an initial reference. Internationally, sea level is the recognized datum.

10.7 DECIMAL EQUIVALENTS OF INCHES IN FEET

In this chapter all land dimensions are given in feet and decimals of a foot:

Inches	Feet	Inches	Feet
1	0.083	7	0.583
2	0.1667	8	0.667
3	0.25	9	0.75
4	0.333	10	0.833
5	0.417	11	0.917
6	0.5	12	1.0

Remember that when estimating quantities, computations are not worked out to exact mathematical quantities.

10.8 ESTIMATING CUT

Example Problem: 1

Estimate the number of cu yd to be cut from a square parcel of land 50.0' × 50.0', with the station elevations as shown above the bench mark in Fig. 10.9.

Step 1: From the given bench mark, estimate the average height of the stations:

$$\frac{7.1 + 6.3 + 9.8 + 8.4}{4} = 7.9'\text{ average height}$$

Step 2: Multiply the area of the land in sq ft by the average height of the stations in ft: 2500 × 7.9 = 19,750 cu ft. Since excavating is reckoned in cu yd, the estimate would be 731 cu yd.

Make a scaled sectional drawing in line *d–c*. Use graph paper. See Sec. 10.13.

Example Problem: 2

From the given bench mark, estimate the number of cu yd of earth to be cut and hauled from two separate parcels of land dimensioned as shown in Fig.10.10.

Note that in the previous example, *A* has already been estimated as 731 cu yd, as in Step 2, Example 1.

The average height of the stations in lot *B* is 7'-0" and the estimated area is 648 cu yd. The total cut of lots *A* and *B* is 731 + 648 = 1,379 cu yd. The total number of stations taken into consideration for the two lots is eight, four for each lot.

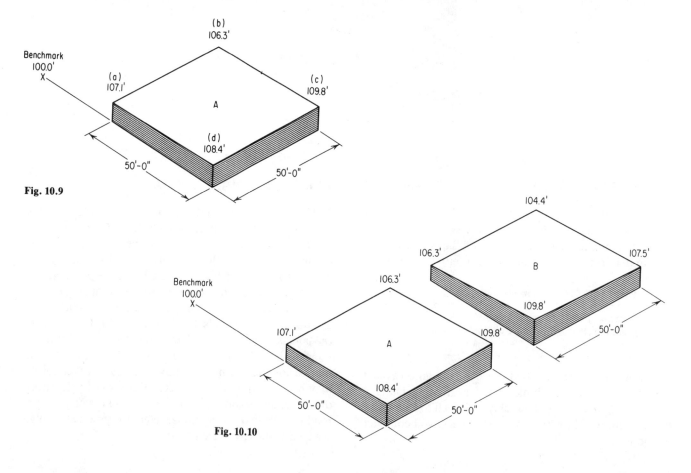

Fig. 10.9

Fig. 10.10

Remember that estimating *is* estimating and does not necessarily require mathematical exactitude.

Examine Fig. 10.11, where both lots are placed end-on-end.

It will be seen that the adjoining lot stations—that is, 6.3 and 9.8—were taken twice; once for parcel *A* and once for parcel *B*.

Since eight stations must be taken into account, set out the work as follows:

Station Readings	Times Taken	Totals
7.1	1	7.1
6.3	1	12.6
4.4	1	4.4
7.5	1	7.5
9.8	2	19.6
8.4	1	8.4
	8	59.6

Average cut is

$$\frac{59.6}{8} = 7.45,$$

and

$$\frac{7.45 \times 5,000}{27} = 1,379 \text{ cu yd.}$$

From the given bench mark, using six station elevations only, make a calculation of the cut required for the land shown in Fig. 10.11. Subtract the result of calculations using six elevations from the result of calculations using eight elevations. Satisfy yourself that it is more accurate to take into account eight elevations than six. Remember this is a very small parcel of land on an even plane.

Always be sure to check the number of elevations taken into account.

Rule: Take the number of station elevations in adjoining grids which have the same shape and perimeter and multiply by the number of grids. This equals the number of elevations to be taken into account.

Problem: Example 3

From the given bench mark, estimate the number of cu yd of cut on a parcel of land with the dimensions and stations as shown in Fig. 10.12. Read the rule again. Note that the area of Fig. 10.12 is the same as in Fig. 10.11. The station elevations are the same, the bench mark is the same, *but the grids are not the same dimensions.* Take each grid estimate separately, add them together, and then subtract the differences of cuts for Figs. 10.11 and 10.12.

Make a scaled sectional drawing on line *d–c,* Fig. 10.11. In colored ink, impose on this section a section of line *d–c* of Fig. 10.12 (see Fig. 10.13).

Satisfy yourself on the validity of the rule.

Problem

Estimate the number of cu yd of earth to be cut from a parcel of land with the dimensions, stations, and bench mark as shown in Fig. 10.14.

Reread the rule and set out your work as shown in Example 2. How many times are you going to take station reading 110.7?

Problem: Example 4

(a) Estimate the number of cu yd of earth to be cut from a piece of land with the dimensions, stations,

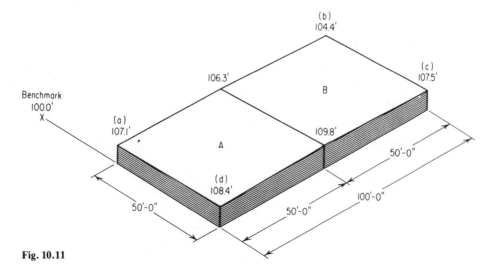

Fig. 10.11

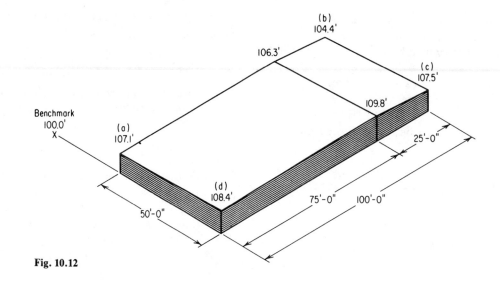

Fig. 10.12

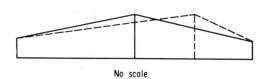

No scale

Fig. 10.13

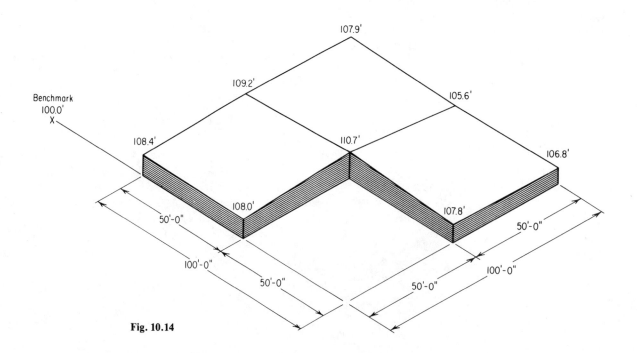

Fig. 10.14

and bench mark as shown in Fig. 10.15. Reread the rule. There are fifteen equal area grids and four station elevations to each grid. There are sixty stations to be reckoned, thus:

		Stations
(i) The four outside-corner stations are reckoned once:	$4 \times 1 =$	4
(ii) The outside adjoining equal area and perimeter grid stations are reckoned twice:	$12 \times 2 =$	24
(iii) The internal adjoining equal area and perimeter grid stations are reckoned four times:	$8 \times 4 =$	32
		60

For this problem, use the work-up sheets similar to those shown on p. 161, and set out your work as on the cut-and-fill estimating sheet on page 194.

(b) Allowing for a swell of 15 percent for cut earth, how many 8-cu-yd truckloads of borrow (cut earth) will have to be hauled from the site?

When setting out your work note that:

Station reading grid A –1 will be taken once
Station reading grid A –2 will be taken twice
Station reading grid B –2 will be taken four times

10.9 INTERPOLATION OF DATUM LINES

Problem

A square parcel of land $50.0' \times 50.0'$ with the station elevations and datum as shown in Fig. 10.16a is to be graded to datum 100.0. The earth is loam and clay.

Estimate:

(a) the number of cu yd of compacted fill required—allow 20 percent for machine compaction;

(b) the number of cu yd of cut;

(c) the number of 8-cu-yd truckloads of loose earth to be hauled away after allowing for fill—allow for 25 percent swell for cut earth.

Example Using the Graphic Method of Interpolation

Step 1: Make a scaled drawing of the area (say $2''$ to $50.0'$). Since part of the area is below datum, a datum level may be interpolated between *a–d* and *d–c* (see Fig. 10.16a).

Step 2: From *a* on line *a–b* (Fig. 10.16b), lay off 18 units of any convenient scale. as at *a–x*. *Turn Fig. 10.16b so that a–d is a base line.* The 18 units at *a* represent 18 ft above datum and the 19.2 units at *d* represent 19.2 ft below datum of $100.0'$.

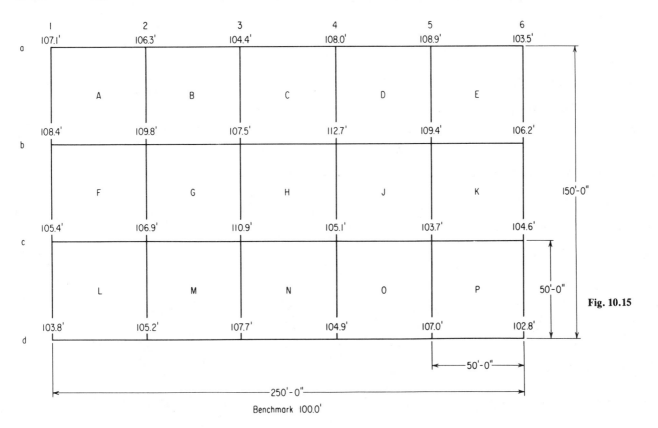

Fig. 10.15

Step 3: From *d* of line *d–c* (using the same scale) lay off 19.2 units, as at *d–y₁*, Fig. 10.16c.

Step 4: Join *x–y* (Fig. 10.16b,) which bisects *a–d* at the datum 100.0 '.

Note: Try this by first using units of ¹⁄₁₆ ", then by using units of ⅛ ", and then by making a sectional scale drawing on line *a–d*. *You must understand this operation before proceeding further.*

Step 5: From *d* of line *a–d* (Fig. 10.16d) project a line (*d–y₂*) of 19.2 units.

Step 6: From *c* on line *c–b* (Fig. 10.16c) lay off 6 units, as at *c–z*.

Step 7: Join *y₁–z*, which bisects *c–d* at the datum 100.0 '.

Step 8: Draw a line showing the approximate datum line dividing the end of the cut and the beginning of the fill (see Fig. 10.16d).

Step 9: The dimensions of the shaded triangle may be scaled off to the nearest 1.0 ' each way from *d* (Fig. 10.16d), 26.0 ' on line *d–a*, and 38.0 ' on line *d–c*.

Step 10: *Area of fill:*(Triangle

Triangle $\dfrac{26 \times 38}{2} = 494$ sq ft

Step 11: *Area of cut:* Total area of the parcel of land less the fill:

$$(50 \times 50) - \dfrac{26 \times 38}{2} = 2{,}006 \text{ sq ft area of cut}$$

Step 12: *Cubic yards of fill:*

(a) Area of fill multiplied by the average depth of three stations below datum.

(b) Three stations: $\dfrac{100 + 80.8 + 100}{3} = 93.6.$

(c) Average depth below datum is $100.0 - 93.6 = 6.4'.$

(d) Cu yd of fill is $\dfrac{\text{area} \times \text{ depth of fill}}{27}$ plus 25% (25 percent is for compaction of clay).

(e) $\dfrac{494 \times 6.4 \times 125}{27 \times 100} = 146$ cu yd est.

Step 13: *Cubic yards of cut:*

(a) From the total area of land, subtract the area requiring fill:

$$(50 \times 50) - \dfrac{(26 \times 38)}{2} = 2{,}500 - 494 = 2{,}006 \text{ sq ft}$$

for the area of the trapezoid.

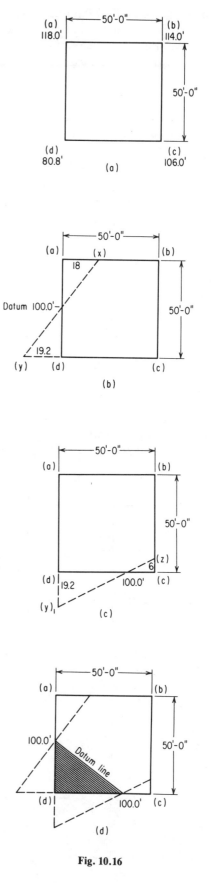

Fig. 10.16

(b) Multiply the area of the trapezoid by the average height of five stations (this assumes that the land is on a fairly even plane):

$$\frac{118 + 114 + 106 + 100 + 100}{5} = 107.6$$

(c) Average height above datum is $107.6 - 100 = 7.6$.

(d) $\dfrac{\text{Area of trapezoid} \times \text{average depth of cut in ft}}{27}$

$$= \text{cu yd}$$

(e) $\dfrac{2,006 \times 7.6}{27} = 570$ cu yd est. *Note very carefully that for rugged ground the trapezoid would have to be further subdivided into triangles.*

(f) 570 cu yd of this cut earth will swell 25 percent:

$$\frac{570 \times 125}{100} = 713 \text{ cu yd}$$

Step 14: *Number of 8-cu-yd truckloads to be hauled:*

(a) From the total cu yd of cut including swell, subtract the total amount of fill required (including compaction): $713 - 140 = 573$ cu yd.

(b) 573 cu yd of swelled earth to be hauled in 8-cu-yd trucks would require 72 truckloads.

(c) Estimate the time required for each truck. This may be done by actual timing of some trial runs through traffic.

Note that this is a very small piece of ground. Remember that there are more hazards in estimating earthwork of all kinds than in any other phase of construction.

Example Using the Mathematical Method for Interpolation

Examine Fig. 10.17. At some determinable point between stations 118.0 ' and 80.0 ', the slope of the land bisects *a–d* at grade 100.0 ', and we know from observation that part of the land must be cut and part filled.

Take the following steps to interpolate mathematically for 100.0 ':

Step 1: Using a metric scale, draw a plan 50 cm × 5 cm and a profile (1 cm to 1.0 ') on line *a–d* as in Fig. 10.17.

Step 2: Since station *a* is 18.0 ' above bench mark 100.0 ' and station *d* is 19.2 ' below, the total vertical height between station leves *a* and *d* is 37.2 ' and the length of the base is 50.0 '.

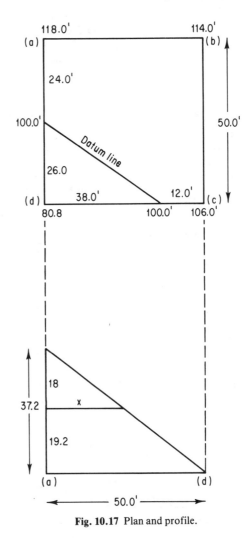

Fig. 10.17 Plan and profile.

Step 3: The profile shows how two similar triangles are formed. The first has a perpendicular height of 37.2 ' and a base of 50.0 '. The second has a perpendicular height of 18.0 ' (for cut) and the base is unknown.

Step 4: Therefore

$$18.0\text{ '}: s \text{ as } 37.2\text{ '}:50.0\text{ '}$$

$$\frac{18 \times 50}{37.2} = 24.2 \text{ or, } 24.0\text{ '} \quad \text{(scale off on plan)}$$

Step 5: *When interpolating mathematically, always be sure to scale off the point of interpolation (on plan) from the higher station reading, as in Fig. 10.17.*

Step 6: In a similar manner, interpolate mathematically for $00.0 '$ the bench mark on line *c–d* of Fig. 10.17.

Step 7: Station *c* is 6.0 ' above bench mark 100.0 ' and station *d* is 19.2 ' below. The vertical height between the two stations is 25.2 '.

JOHN DOE CONSTRUCTION COMPANY LTD

CUT AND FILL ESTIMATING SHEET

Job_____   Date_____

1	2	3	4	5	6	7
	Cut				Fill	
Grid	Area Sq Ft	Average Cut	Volume Cu Yds	Area	Average Fill	Volume Cu Yds
a	2500	17.5	1620			
B	2401	8.2	730	99	1.67	6
C						
D						
E						
F						
G						
H						
J						
K						
	TOTAL CUT				TOTAL FILL	

Step 8: Then

$$6.0\,' : x \text{ as } 25.2\,' : 50.0\,'$$

$$\frac{6 \times 50}{25.2} = 11.9 \text{ or } 12.0\,'$$

Step 9: Again always be sure to scale off (on plan) from the higher reading to the point of interpolation.

Step 10: Now see on the plan where both interpolated points are scaled off from the higher readings, as at 24.0' from point *a* and 12.0' from point *c* on plan.

Step 11: The datum line indicates where the cut ends and the fill begins; the station reading 100.0', 80.8', and 100.0' shows the area requiring fill. Note that the area of the triangle is $\dfrac{26 \times 38}{2} = 494$. (See the *graphic* method of interpolation at Step 10, p. 000.)

Problem

Estimate the cu yd of fill required. Estimate the cu yd of cut necessary for Fig. 10.17.

10.10 GRADES: CUT AND FILL

Let us examine Fig. 10.18. At the corner of each grid (using a bench mark of 100.0') plot the grade for cut or fill. For small jobs the contractor may have to make his own survey, which will become a charge on the estimate under "job overhead expenses."

Step 1: Write the existing elevations at the right-hand corner of each grid, as at *a, b, c,* and *d* in Fig. 10.18a.

Step 2: Write the future grade of 100.0' to the left of each existing grade, as shown in Fig. 10.18b.

Step 3: Deduct the higher reading from the lower, and write the cuts under the future grades and the fill under the present grades (see Fig. 10.18C) thus:

Future grade	Present grade
cut	fill

See the examples at the top of Fig. 10.18,C, where *c* = cut and *f* = fill; thus:

100.0'	103.9'		100.0'	94.6'
c3.9				f5.4

It is recommended in actual practice that you write all the cuts in red and all the fills in black, which facilitates abstracting to estimating sheets.

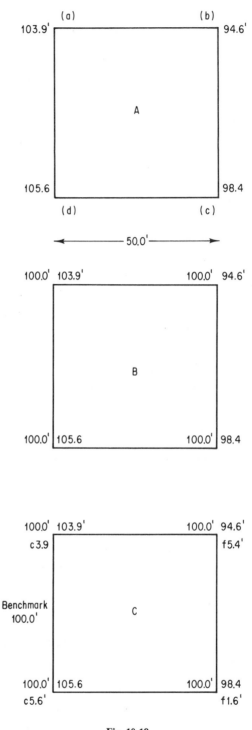

Fig. 10.18

Problem

A parcel of land (Fig. 10.19) 250.0' × 100.0' is to be graded to a datum of 100.0'. The existing grades are shown at the right-hand corner of each grid. The new grades (100.0') are shown to the left of existing grades.

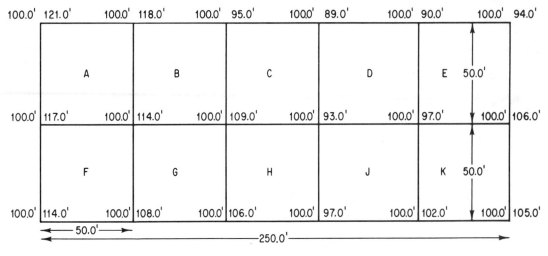

Fig. 10.19

Determine the *approximate* neutral 100.0 ′ between cuts and fills as follows:

Step 1: Draw a plan of the land shown in Fig. 10.19, making it 10″ × 4″ with ten grids each 2″ × 2″. Letter the grids as shown.

Step 2: Carefully examine all the grid-station elevations and get a feeling for the existing contours of the land. *It is most important that you sense the land "in relief" before starting to estimate it.*

Step 3: Examine grid *B.* A grade of 100.0 ′ must occur between stations 118.0 ′ and 95.0 ′, and it must be nearer to 95.0 ′ than 118.0 ′. Lightly mark the *approximate* point of intersection between these two stations where 100.0 ′ occurs.

Step 4: Still on grid *B,* lightly mark the *approximate* spot between stations 95.0 ′ and 109.0′ below. Draw a faint line across the grid indicating where the cut ends and the fill begins. This will form a triangle requiring fill with station readings 100.0 ′, 95.0 ′, and 100.0 ′.

Step 5: Repeat for top and bottom lines of grid *H,* drawing the faint line for the 100.0 ′ level. Grid *H* will be divided into two trapezoids.

Step 6: Repeat for grids, *E, K,* and *J.*

Step 7: Now that you have an educated assessment as to where the 100.0 ′ levels occur for cut and fill, interpolate for their exact positions by the graphic method and prove your findings are correct by the mathematical method.

Problems

(a) Estimate the area of cut for grid *B.*

(b) Estimate the area for fill for grid *B* and add the

areas for cut and fill; they should equal the total area of grid *B.*

(c) Estimate the area of fill for trapezoid *H.*

(d) Estimate the area of cut for trapezoid *H.*

(e) Estimate the areas for cut and fill for grids *E, K,* and *J* and check.

10.11 CUT AND FILL ESTIMATING SHEET

Let us carefully examine the following *pro forma* sheet, which is used for estimating cut and fill. This sheet should be 8½″ × 11″.

Step 1: Column 1 shows the grid letter.

Step 2: Column 2 shows the area being considered.

Step 3: Column 3 shows the average cut for the area being considered. Some grids will be all cut, some all fill, and others part cut and part fill.

Step 4: Column 4 shows the volume of cu yd of cut.

Step 5: Column 5 shows the area of fill. Some grids may be all or part fill.

Step 6: Column 6 shows the average depth of fill (if required).

Step 7: Column 7 shows the volume of *compacted* cu yd of fill (if required).

Step 8: Add the totals for cut and fill.

Using the Cut-and-Fill Estimating Sheet

Starting at grid *A* of Fig. 10.20, let us work up some examples of areas, averages of cuts and fills, and the volumes of cuts and fills, and enter them on the cut-and-

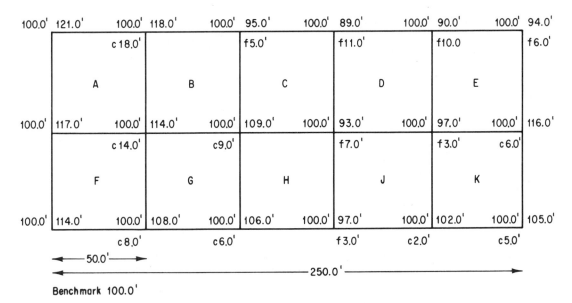

Fig. 10.20

fill estimating sheet as shown on p. 161. *Then you can finish the exercise.* The bench mark is 100.0'.

Step 1: Area of grid *A* is 50.0' × 50.0'. Enter 2,500 in column 2.

Step 2: The cut for grid *A* is the average of four station readings

$$\frac{121 + 118 + 114 + 117}{4} = 117.5$$

$$- 100 = 17.5 \text{ cut}$$

Enter in column 3.

Step 3: Volume of cu yd of cut for grid *A* is

$$\frac{17.5 \times 2,500}{27} = 1,620 \text{ (approx.)}$$

Enter 1,620 in column 4 and notice that the whole of this grid is cut.

Step 4: Interpolate on grid *B* (Fig. 10.20) for 100.0' between 118.0' and 95.0'. Make a freehand sketch of the section as shown. Then 18:*x* as 23:50 ≃ 39.0'

Note: You will never waste time by making freehand sketches of your problems. When interpolating mathematically, always scale off on the grid from the higher to the lower reading. This is very important.

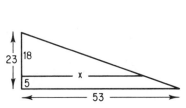

Step 5: Interpolate again on grid *B* for 100.0' between stations 95.0' and 109.0'. Then 9:*x* as 14:50 = 32.0'. *Scale 32.0' from the higher reading and draw in the triangle, which has an area of*

$$\frac{11 \times 18}{2} = 99 \text{ sq ft.}$$

The station readings are 100.0', 95.0', and 100.0', as shown at Step 6.

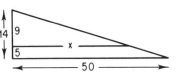

Step 6: Find the area for cut and for fill on grid *B*. Make a freehand sketch as shown and find the area of cut for grid *B* as follows:

$$(50 \times 50) - \frac{11 \times 18}{2}$$

$$= 2401.$$

Enter in column 2. Area of fill for grid *B* is 2,500 − 2,401 = 99.

Enter 99 in column 5. Check that the combined areas of cut and fill equal the total area of the grid:

2,401 + 99 = 2,500.

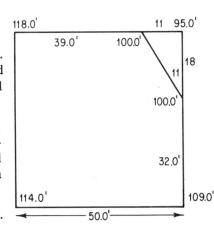

Step 7: The cut for grid *B* is the average of five station readings:

$$\frac{118 \; + \; 100 + \; 100 \; + \; 109 \; + \; 114}{5} \; = \; 108.2' \; - \; 100.0'$$
$$= \; 8.2'.$$

Enter 8.2′ in column 3 on page 000.

Step 8: To find the volume of cu yd of cut for brid *B,* multiply the entry in column 2 by the entry in column 3 and divide by 27, thus

$$\frac{2401 \; \times \; 8.2}{27} \; = \; 729$$

or nearly 730. Enter 730 for grid *B* in column 4.

Step 9: The depth of fill for grid *B* is the average of *three stations only,* thus

$$\frac{100 \; + \; 95 \; + \; 100}{3} \; = \; 98.3$$
$$\text{Then } 100 \; - \; 98.33 \; = \; 1.67$$

See the sketch at Step 6. Enter 1.67 average fill in column 6 on page 194.

Step 10: To find the volume of cu yd of *compacted* fill for grid *B,* multiply column 5 entry by column 6 and divide by 27; thus

$$\frac{99 \; \times \; 1.67}{27} \; = \; 6$$

Enter 6 in column 7.

Be aware that estimating cut and fill takes time. Pause for a moment and think of all the heavy equipment that will be operating on the land you have estimated and remember that patient, careful work now should mean profit later.

10.12 COMPACTION AND SWELL FACTORS IN CUT AND FILL

For this exercise, assume that earth is compacted by heavy machinery to 15 percent below its natural capacity, and that similar earth when cut swells 20 percent. That is to say that 85 cu yd of compacted earth will require 120 cu yd of cut earth.

Let us assume that 15,000 cu yd of earth are to be cut on site and compacted in a low-lying area of the same site, and that in addition it will require a further 20,000 cu yd of *compacted* fill (also with a swell factor of 20 percent) to complete the job.

Step 1: One cu yd of *compacted* earth will occupy 85 percent of its natural capacity. Then 1 cu yd of *compacted* earth will occupy only 0.85 of a cu yd.

Step 2: One cu yd of cut earth will swell 20 percent. Then 1 cu yd of cut earth will swell to 1.20 cu yd.

Step 3: If 85 cu yd of *compacted* earth need 120 cu yd of cut earth, then 1 cu yd of *compacted* earth needs almost

$$\frac{120}{85} \; = \; 1.41 \text{ cu yd}$$

Step 4: Therefore, 15,000 cu yd of earth swelling 20 percent will have a capacity of 15,000 + 3,000 = 18,000 cu yd.

Step 5: This 18,000 cu yd of swollen earth will compact to *85 percent of its original (15,000-cu-yd) capacity* and will equal

$$\frac{15,000 \; \times \; 85}{100} \; = \; 12,750 \text{ cu yd}$$

Step 6: Let us agree that 15,000 cu yd of earth to be cut on site will occupy 12,750 cu yd *when compacted.*

Step 7: Then we still need the difference between 15,000 (original in-site volume) and 12,750 cu yd of earth to be compacted, plus 20,000 cu yds of earth to be hauled onto the site and compacted. Then we need 15,000 − 12,750 + 20,000 = 22,250 cu yd of compacted earth to complete the job. Reread the paragraph preceding Step 1.

Step 8: The total cu yd of loose fill to be compacted (this includes that cut on site and that to be hauled in) is 22,250 × 1.41 = 31,373 (*see* Step 3).

Step 9: Using 8-cu-yd-capacity trucks, it would require 31,373 ÷ 8 = 3,921—say, 3,925 truckloads.

Estimate:

How many 8-cu-yd truckloads of earth must be hauled onto a building site where 18,000 cu yd are to be cut and compacted on site, and an additional 14,000 cu yd of compacted earth will be required to complete the job? Assume that earth compacts 12 percent and all the cut and hauled loose earth swells 16 percent.

10.13 CUT AND FILL: SAND MODEL

Using a scale of 6″ to 50′-0″ for each grid and a scale of 3/16″ to 1′-0″ for all station readings above or below bench mark 100.0′, make a ¼″ plywood-frame model of all the sections of the grids in Fig. 10.20.

Step 1: Prepare five pieces of ¼″ plywood 42″ long and 4″ wide.

Step 2: Gauge a line 2½″ from the bottom of each piece of plywood representing a bench mark of 100.0″.

Step 3: Using the station readings as shown in Fig. 10.22 (bottom line), lay off (above the gauged line on the plywood) a profile of the contours of the land for the longitudinal grid section and letter the grids *F, G, H, J,* and *K* as shown in Fig. 10.21.

Step 4: In a similar manner lay off and letter the other two longitudinal sections.

Step 5: Lay off six transverse (two-grid wide) sections.

Step 6: Cut the plywood on all the profile lines; assemble the frame as shown in Fig. 10.22, and fill the frame with sand to the level of all the grid contour lines.

Step 7: Examine the model and imagine that you are in charge of grading this land to a 100-0′ bench mark. Ask yourself the following questions:

(a) How much earth must be hauled from the site?

(b) Where would be the best place to start cutting the land?

(c) From which area would it be best to start to haul the cut?

(d) Where is the cut to be *legally* dumped?

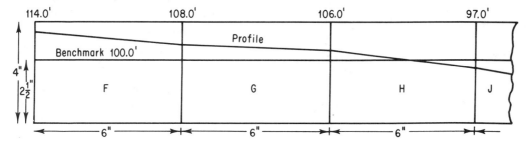

Fig. 10.21 Part of longitudinal section of plywood for sand model.

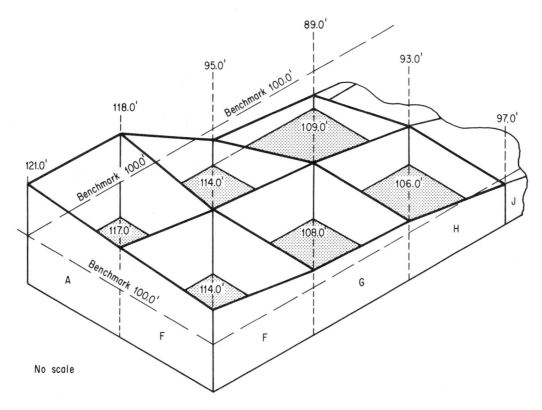

Fig. 10.22 Plywood frame for sand model.

(e) How long will it take a truck to make the round trip from cut to dump and return?

(f) What kind of machinery would you recommend to use on this job?

(g) If it is two miles from site to dump, how many trucks will be required to keep an empty truck always in line for filling without waiting and without employing more trucks than are necessary?

10.14 ESTIMATING CUT AND FILL FOR A PARKING LOT

Problem

A parcel of land (Fig. 10.23) 300.0' × 150.0' is to be graded for parking outside a small shopping center. The proposed grades are shown to the left of the present grades (see the legend). The finished grade will slope the land from the left and right sides to the center, and from the front to the back of the parking lot. Make a large-scale plan of the area and show all the grids with present and proposed elevations.

Estimate:

(a) the number of cu yd of machine-compacted fill required—allow for 20 percent more earth for compaction than in its natural state before being cut on site;

(b) the number of cu yd of cut;

(c) the number of 8-cu-yd truckloads of cut earth to be hauled away after allowing for fill—allow for 25 percent swell on the cut earth.

Grades, Cut, and Fill:

(a) The present grade is written above and to the right of each grid.

(b) The future grade is written to the left of each present grade.

(c) Deduct the higher reading from the lower reading and write the cuts under the future grades and the fill under the present grade. Thus:

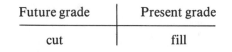

Future grade	Present grade
cut	fill

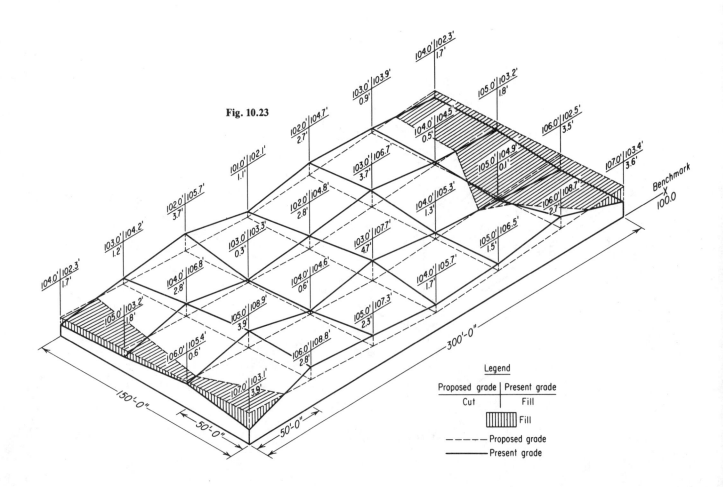

Fig. 10.23

Example, top left of plan:

104.0	102.3
	f 1.7

and the next to the right:

103.0	104.2
*c*1.2	

(d) Some estimators prefer to write in *black* the present grades and fill, and in *red* the future grades. In any case, be careful.

Draw the contours showing where the cut ends and the fill begins, then work out the estimate.

QUESTIONS

1. Define the following terms in connection with land grading:
   (a) cut
   (b) fill
   (c) swell on cut earth
   (d) compaction on filled earth
   (e) datum line
   (f) bench mark
   (g) a station

2. Whose responsibility is it to provide a bench mark on a parcel of land for leveling?

3. In what units of measurement are levels read and recorded in engineering practice?

4. Define sanitary fill.

5. Draw the symbol used for a datum line.

6. Examine the parcel of land to the right and interpolate for 100.0′.

7. The parcel of land shown below is to be cut and leveled to a bench mark grade of 100.0′. The percentage allowance of swell for cut earth is 22%. Using the work-up sheets provided (and complete a cut-and-fill estimate sheet similar to page 194, estimate the number of 8-cu-yd truckloads of earth that will have to be hauled away from the site. *Ans:*

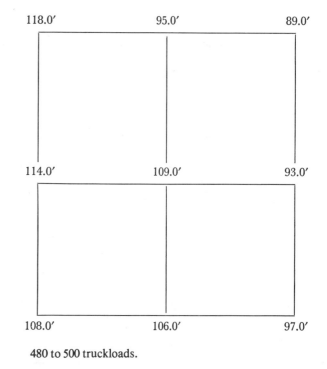

480 to 500 truckloads.

8. Define a batter board.

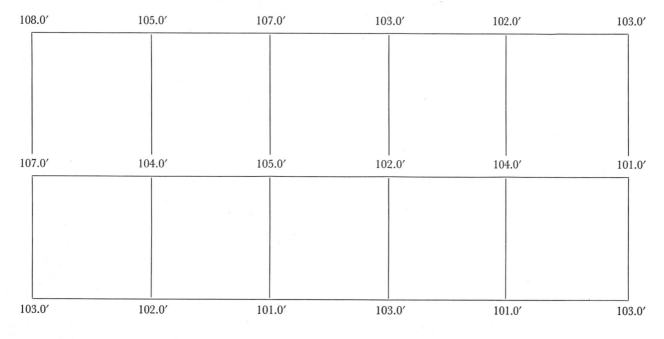

Answers to Problems

Fig. 10.9 731 cu yds

Fig. 10.10 731 and 648 cu yds

Fig. 10.11 eight stations = 1379

six stations = 1343

36 cu yds difference

Fig. 10.14 2328 cu yds

Fig. 10.15 10,070 cu yds and about 1510 truckloads

Fig. 10.15 requires about 320 truckloads

Fig. 10.16

WORK-UP SHEET   DATE *28 March*

SHEET No. *15* OF _____

BUILDING	LOCATION	ARCHITECT	ESTIMATOR
Rosemont School	University Blvd.	J. Doe & W. Wilson	J. Smith

Rough Carpentry

① 82/1x4 – 16'-0" @ $90.00 per M
  82
  64
  328
  492
  12)5248   .437
  437⅓ F.b.m.   90
                  $39.330 $ 39.33

② 76/2x3 – 12'-0" @ $82.00 per M
  76   .456
  6   82
  456 F.b.m.   912
                  3648
                  $37.392 $ 37.39

③ 56/1x10 – 14'-0" @ $94.00 M
  560   .653
  14   94
  2240   2612
  560.   5877.
  12)7840 $ 61.382 $ 61.38
  653⅓ F.b.m.

④ 44/2x10 – 16'-0" @ $86.00 M
  44   1.173
  320   86
  880   7038
  132..   9384.
  12)14080 $100.878 $100.88
  1173⅓ F.b.m.

⑤ 160/2x6 – 8'-0" @ $84.00 M
  160   1.280
  8   84
  1280 F.b.m.   5120
                  10240
                  $107.520 $107.52

11
COMMON SOURCES OF ERROR IN ESTIMATING

This chapter is a reminder to pay special attention to some of the most common errors that occur in estimating. They have come to my attention over a period of years and not only beginning estimators but experienced estimators can be guilty of them. The things that we are the most casual about are the things that we do most often.

11.1 MAIN SOURCES OF ERROR WITH FIGURES

(a) Probably the main source of error in all estimating is in forming of casual, misshapen figures. The figures 4, 7, and 9 quite frequently are mistaken for one of the others. The figures 5 and 8 are even more often mistaken for one another. The whole basis of successful estimating lies in the ability of the contracting company to make successful bids at their own price. This presupposes that estimates will be accurate and free from casual mistakes.

(b) The more simple the arithmetical problems—such as addition, subtraction, multiplication, and division—the easier it is to make errors. With machine accounting a careful check must be made of the typescript result. The machine will only deliver the correct computations if the correct information has been fed into it. There are no complicated mathematical problems in estimating, but disciplined accuracy and careful cross-checking are required.

(c) In all businesses a great source of error is in copying of figures from one paper to another. For example, a dimension such as 627'-0" is very easily written with a transposition of figures as 672'-0", with of course a difference of 45'-0". Imagine this lin-ft measure multiplied by a width of 80'-0". The difference would be 3,600 sq. ft. Assume that it was for an excavation to a depth of 3'-6". The difference would be 12,600 cu ft or 466 cu yd, or (allowing for 25 percent swell on cut earth) it would amount to about seventy four 8-cu-yd truck-loads of earth.

(d) If you use a slide rule, you must set your own decimal point. This is such a frequent source of error that some companies will not permit their estimators to use slide rules. Instead, they are given good electronic calculators. A great advantage of a calculator is that the printout may be stapled to the work-up sheets for anyone to recheck.

11.2 MAIN SOURCES OF ERROR IN ESTIMATING

(a) Not visiting the site. The estimator must see the shape of the land and the topographical features. He must check the distance from the railway and from the main road and consider how the vehicles may get to and from the site in all types of weather.

(b) Having too cursory a knowledge of earthwork. Some survey experience (or academic study) would be a great help in this field. If the estimator has had no experience with earthwork, an engineer must be consulted.

(c) Miscalculating haulage costs, distances, types of roads, and traffic, and forgetting to check the round trip for a dump truck from, say, city center to dump and return. Make the trip yourself through traffic and time it.

(d) Forgetting the cost of labor and materials to build adequate runways for wheelbarrows and buggies.

Every project is different, and a survey should be made for estimating lumber, fasteners, and carpenter's and helper's time. Much of this lumber can only be used once.

(e) Forgetting the cost of making and maintaining a good truck road on and off the property. The quality of this rough road will vary according to the length of time that it will be in use. The bigger and longer the job, the better the road should be.

(f) Cost of scaffold and its maintenance. An untidy supervisor can waste a great deal of time and expense on this one item. *Workmen will be just as untidy as they are permitted to be.*

(g) Using inadequate or poorly maintained equipment. There are certain times when things should be done; for example, immediately after placing concrete, the equipment should be thoroughly cleaned and oiled before the men leave the job.

(h) Over- and underestimating. When any item is incorrectly estimated, the cost of the item, the labor to install it, the tradesmen's pay, the helpers' pay, their fringe benefits, the company overhead expenses and the profit on all these items are also under- or overestimated.

(i) It is a cardinal rule among estimators never to destroy any calculations until they are proved to be incorrect. Nearly everyone has had the experience of throwing some calculations away only to try to retrieve them from the wastebasked later on. The estimator should never have to ponder over his original thinking. The work-up sheets should clearly show the original method.

(j) Errors in estimating the time required for a man or a gang or a machine to complete a work unit.

(k) Taking the liberty of making shortcuts when taking off quantities. Materials must be taken off in the strict order in which they are used. Materials will later be ordered against the estimate. Only those things should be delivered to the job that can be used immediately. There may be little storage room on the building site, or you may even be renting storage room.

(l) Forgetting such items as bridging, girths, and headers in rough carpentry wall framing.

(m) Omission of materials, labor, plant, overhead, and profits.

(n) Forgetting to check all estimated items against the specifications and your own guide list of main headings.

(o) Overestimating waste.

(p) Forgetting to make notes of unusual items or errors in drawings or specifications that are given to the job runner. Correcting physical conditions that could be avoided through good liaison within the contractor's staff is expensive.

(q) Forgetting to study soil reports for new jobs and not knowing whether the sides of excavations will be self-supporting or if shoring or extra excavation will be required to dress the banks back to an angle of repose.

(r) Failure to secure bond and insurance coverage in time to meet bid closing times. Make arrangements for coverage as soon as drawings and specifications are received.

(s) Incorrectly adding columns of dimensioned figures (see Sec. 6.12).

(t) Forgetting to include every estimate sheet total in the final summary of all sheets. Sheets must be numbered—e.g., "Sheet 1 of 75" through "Sheet 75 of 75"—then checked to see that there are 75 sheets.

(u) Remember gas and mileage for trucks and equipment.

(v) *Remember that the brain tires quickly, and quickly recovers; short burts of mental energy are recommended.*

(w) Using the wrong unit of measure when working up the quantities; such as lin ft for sq ft, or cu ft for cu yd. The difference between 81 cu ft and 81 cu yd of concrete is 78 cu yd; at $20 per cu yd this makes an error of $1,560.

(x) Finally remember to have a well-seasoned "hunch man" *to look for gross errors in the final estimate.* As an example: looking at $400'-'' \times 53'-0'' \times 8'-0''$, his mental reaction might be $4 \times 4 = 16$ plus four zeros plus something, i.e., $400 \times 400 \times 8 = 160,000$ (see rapid calculations, Chap. 6). It certainly could not be 16,000 plus; actually, it is 169,000. He would also look over the final additions and check that subcontractors' prices have been transferred correctly. He would make sure that, say, $13,377.70 had not been transferred as $1,337.70. Be careful!

11.3 TO AVOID ERRORS ON GENERAL ESTIMATE SHEETS

(a) Every sheet must be titled, dated, and numbered.

(b) Wherever possible the basic data such as floor, wall, and roof areas should be shown, and also the perimeter. One some jobs this basic data will be used over and over again. It must be correct.

(c) The general estimate sheets must contain sufficient information for them to be passed through to the cost department for a final work-up and cost accounting.

(d) It is an error to crowd the work on the general estimate sheets. Leave at least two clear lines between each main heading.

(e) There is no set rule for the use of the extension columns other than to say that they are provided to enable the paper to be set out methodically and easily read. See the Sections on earthwork (Chap. 10) and brickwork (Chap. 16).

11.4 DIMENSIONS

It is very important to write dimensions correctly. For example,

Area: $50'-0'' \times 36'-0''$, not $50' \times 36'$

Area: $50'-9'' \times 36'-6''$, not $50.75' \times 36.5'$, and not $50\frac{3}{4}' \times 36\frac{1}{2}'$

Excavations: $54'-0'' \times 40'-0'' \times 4'-9''$, not $54' \times 40' \times 4.75'$

End Section $2'-0'' \times 0'-8''$ (not $2' \times 8''$); this item must be written with figures as two feet zero inches times zero feet eight inches.

Lumber:

$12/2 \times 4$—$10'-0''$	In each case writing bd ft
$14/2 \times 6$—$12'-0''$	(shown struck out) is un-
$18/2 \times 8$—$16'-0''$	necessary.

Estimating sheets should contain nothing other than that which is impossible to leave out and also maintains clarity.

QUESTIONS

1. What is one of the prime causes of errors in estimating?

2. List fifteen main sources of error in estimating.

3. Write the following dimensions as they would appear on a general estimate sheet:
   (a) An area fifty feet, four inches by thirty-six feet, six inches
   (b) An excavation sixty-seven feet by thirty-seven feet and four feet, nine inches deep
   (c) An end section of a concrete footing five feet by nine inches
   (d) Forty pieces of lumber, two by ten and sixteen feet long
   (e) Four sheets of plywood three-quarters of an inch thick, solid one side, four feet by eight feet
   (f) Two hundred and sixty-three lineal feet of two-by-two
   (g) Enough No. 210 lb asphalt shingles to cover one thousand three hundred square feet of roof area
   (h) Three hundred and thirty-three pounds of three-and-a-half-inch common nails

12

PRELIMINARY, DETAIL AND UNIT ESTIMATES; PROGRESS SCHEDULES AND COST-SUMMARY SHEETS

This chapter discusses five different documents with which the estimator must be familiar. They are:

1. The preliminary and approximate estimate. This is a guide to the probable cost within, say, 5 to 10 percent of the new building.

2. The detail estimate as used by the contractor in arriving at his final proposal tender. This gives in great detail the cost of each item, the labor, the depreciation or rental of equipment, and also the overheads and profit. The detail estimate is then summarized for the final bid price to be tendered.

3. The unit estimate as prepared by quantity surveyors. This type of estimate is bid by the cost of each single unit times the number of units. Assume on the quantity surveyor's estimate that a certain job will require 5,200 cu yd of specified and placed concrete. The contractors would then estimate their individual cost

for the production and placing of 1 cu yd of such concrete. The bid for this item would then be 5,200 times the cost of 1 cu yd. In this manner all the items would be broken down into units and the contractors would know exactly what they are to bid on without having to work up the quantities themselves.

4. The progress schedule, on which is shown the anticipated progress of the work from the starting deadline date to the final handover of the completed project. It is usual for the architect to call for a progress schedule to be prepared and submitted by the contractor about one week before the actual building operations begin. It is from this schedule that the architect, the general contractor, and the subcontractors know what headway has been made against the original proposed progress schedule.

5. The cost-summary sheets (which are used by large companies), on which every item that is to be built into the new project is code-numbered and priced both for the cost of the item and the time it takes a man or a crew to install it. (See page 216.)

   This code number originates with the cost accounting department, where a constant check on prices for materials and labor is maintained. The same code number would appear on the daily time sheets that have to be completed on the job by the workmen, the timekeeper, or the foreman himself. It is from these time sheets that the actual cost of each coded item is taken and compared with the original estimate for such coded items.

12.1 PRELIMINARY ESTIMATES

An approximate estimate of cost for a new building (within, say, 5 to 10 percent either way) is used by architects in their first consultations with clients. A similar estimate is used by contractors to guide their thinking in the first approach to the probable cost of a new building. There are several ways of determining within reason the cost of a proposed building.

12.2 THE SQUARE-FOOT METHOD

This method may be used where a comparable building has recently been built on similar ground conditions. First, the sq-ft area of a completed building is taken and the cost per sq ft is reckoned by dividing the total cost by the number of sq ft. Second, the sq-ft cost of the newly completed building is multiplied by the sq-ft area of the proposed new building. This gives a quick approximate cost. Assume that a recently built warehouse cost $12 per sq ft and a proposed 3,200-sq-ft warehouse is to be built in the same general area and of similar construction;

the preliminary estimate of cost would be $12 × 3,200, which is $38,400. This method is used by both the architect and the contractor as a guide.

12.3 THE CUBIC-FOOT METHOD

This method is used by the architect and the contractor in much the same manner as the sq-ft method, for particular types of structures. As an example, assume that a school building with air conditioning is to be built in a tropical climate. Once the cu-ft cost of a school built to similar drawings and specifications has been established, it is reasonable to suggest that the proposed new school will cost very nearly the same amount per cu ft.

12.4 THE COMPARATIVE-APPRAISAL METHOD

This method is used by both architects and contractors to get a near-cost price for a proposed new building by comparing its size, design, specifications, and soil conditions with a similar new building.

Assume that the following information is known about the existing building:

(a) It is a recently completed *single-story elementary school* with library, infirmary, gymnasium, stage, and storage and administrative space. It comprises 31,310 sq ft. *Note: This is an elementary school and could only be compared with such. A high school, college, technical school, or university would cost more because of special labs, and so on.*

(b) It has fourteen classrooms and is designed for 420 student stations, with an average of 30 students per classroom.

(c) The total cost was $431,207.
    From this information we can derive the following:

   (i) *The cost per sq ft* is total cost divided by the number of sq ft: this is 431,207 ÷ 31,310 = $13.77 per sq ft.

   (ii) *The cost per classroom* is total cost divided by the number of classrooms: this is 431,207 ÷ 14 = $30,800 per classroom.

   (iii) *The cost per student station* is total cost divided by the number of student stations. This is 431,207 ÷ 420 = $1,027.00 per student station.

Let us make an appraisal of cost for a proposed new *single-story elementary school* comprising 27,500 sq ft, with twelve classrooms designed for 360 student stations, to be built to similar design, specifications, and ground conditions as the fourteen-room school. We may proceed as follows:

(a) *Assume the cost is $13.77* per sq ft, as for the fourteen-room school.

(b) *The total cost* will be the sq-ft area times the cost per sq ft = 27,500 × $13.77 = $369,720 total cost.

(c) *The cost per classroom* will be nearly the same as for the other school, which was $30,800 per classroom. Then 12 × 30,800 = $369,600 total cost.

(d) The cost per student station will be near (maybe more than) that of the fourteen-room school—say, $1,026 × 360 = $369,360 total cost.

(e) Another very quick guide would be to say that if a fourteen-room school cost $431,207, then a twelve-room school with everything else being equal would cost $^{12}\!/_{14}$ of $431,207 = $369,606. It can be seen that the guide cost in paragraphs (b), (c), (d), and (e) are very close to each other. But, other considerations would have to be taken into account—such as: Would total expenses be more or less for a smaller school? Would using heavy equipment be more economical on a bigger job? Would temporary buildings and even the cost of estimating the job be more or less on a smaller job? There are other considerations too, but all these appraisal methods are educated guesses and have value as guides.

In the case of a motel, the approximate cost may be appraised by the type and size of each proposed unit against the actual cost of similar completed types and sizes built under similar conditions.

All these methods should be used simply as a guide in the initial thinking. In the case of the architect, he can advise his client of the type and size of the building to expect for a given outlay. In the case of the estimator, he can advise his manager of the approximate cost. The manager, in consultation with the chief executive officers of the company, would be able to assess their financial obligations if they made a successful bid. There is a limit as to how far each company may become financially involved, as well as to whether or not materials, equipment, and labor will, indeed, be available at the correct time.

Every contracting company only wants to obtain contracts at their own price and build in their own available time. Usually, it is considered that a general contractor for an average new city building is financially involved for about 15 to 20 percent of the total cost. The rest is taken up by subcontractors. It does not require many contracts, even at this rate, for the general contractor to become deeply involved financially.

There is no data for you comparable to that of your own company. It is imperative that a cost analysis of every completed job be made and the records brought up to date. Toward the top left-hand corner of the general estimate sheet is a place for the "Subject" to be written. The subject would range from residences to schools, hotels, motels, garages, warehouses, and so on. The completed estimates would then be filed under the appropriate subject heading; the analysis of each job would also be filed under the correct subject heading. An estimator cannot have too many guides to regulate his thinking.

When an analysis of the cost of each job is taken by the sq-ft method, the cu-ft method, and the comparative-appraisal method, any one or, better still, a combination and average of all these methods may be used to arrive at a very reasonable educated guess of the cost of a new proposed building. However, the detail estimate is the document against which the final bid will be made.

12.5 THE DETAIL ESTIMATE

After the preliminary estimates have been made, the estimator will proceed with the detail estimates. He would use the preliminary estimates only as a guide for a comparison of price with his detail estimate. Assume that the preliminary estimates indicated a bid price of $180,000 for a building. It would be obvious that something had been omitted if the detail estimate only amounted to, say, $140,000. In this case a careful check would be made to see that every general estimate sheet total had been included in the summary or that the general overhead expenses or the profit had not been forgotten. There may have been a combination of errors.

It is always against the detail estimate that the contractor's proposal bid is finally made. To prepare the detail estimate, the estimator should proceed in the following manner:

Step 1: Check both the drawings and the specifications to be quite sure that all the drawings and the latest amendments and addendums are included. The drawings will be in the following order:

(a) Architectural drawings prefixed with the letter *A.* Assume that there are seventeen architectural drawings. They should be numbered "Sheet A-1 of 17" and so on, up to "Sheet A-17 of 17."

(b) Structural drawings, prefixed with *S,* are numbered "S-1 of . . ." and so on.

(c) Mechanical drawings, prefixed with *M,* are numbered "M-1 of . . .," and so on.

Step 2: Scan through all the drawings to get a general

idea of the scope of the work. Some estimators would now make a sketch view of the proposed building from four different angles—say, from:

(a) front and right side   (c) rear and right side
(b) front and left side   (d) rear and left side

This will fix in the mind of the estimator the shape of the building and the placing of the main openings in the walls of the building.

Step 3: The next drawing to look over would be the plan views. Find the front entrance and imagine walking through the building room by room. They are numbered on the plan and shown on the room schedule of the architectural drawings. This will give the estimator the general feel of the building. Follow the stairways from the basement to the top of the building. Unless an estimator can visualize the proposed new building and feel himself walking around inside it, it is very doubtful that he could hope to make a success of estimating it.

Step 4: Examine the elevations to get a general idea of the main structural fabric of the proposed work.

Step 5: Examine the typical wall sections for an appreciation of the types of walls, floor assemblies, roof assemblies, and so on.

Step 6: Look through the structural drawings. In these day of preformed concrete, there will be much for the estimator and job superintendent to absorb in the structural drawings.

Step 7: Read through the mechanical drawings and keep in mind that some units may have to be placed in rooms before some walls are built. *This is important.*

By this time the estimator will have formed his first concept of the scope of the work to be done.

Step 8: *Very carefully study (not just read) the specifications and type a series of notes of all unusual items or features.* No estimator is very concerned with ordinary routine construction. It is the *unusual* methods of construction and the introduction of new materials that will require very careful consideration and consultations with colleagues for an estimate of labor costs. With the introduction of new materials, someone has to be the first to assess labor costs. After labor costs have been estimated, new items should be processed back through the job superintendent, who should give special attention to the actual labor time required for installation by units of the material. The superintendent's reports would then be given very careful study by the costing department, who would compare their original estimated costs for installation against the actual costs. When the study is complete,

a new card would be made out for the item concerned, which would be filed in the cost indexes of the costing department for future pricing of similar units (see Sec. 12.11).

Step 9: To speed the preparation of the final bid, several persons may be delegated specific duties such as:

(a) getting costs of insurance, bonds, licenses, and all necessary permits;

(b) preliminary investigations of site, soil conditions, topography, datums, survey stakes, roads, distances from city, and availability of company mobile offices and temporary buildings;

(c) assuring availability of labor and job runner;

(d) pricing special specified materials or equipment, and costing interstate or interprovincial taxes and transportation costs for them.

(e) Obtaining bids from subcontractors.

By this time the estimator should be ready to take off all the quantities of materials required for the building. Every item, piece by piece, must be accounted for in the strict order in which the building is to be erected. The estimator must remember that if the bid is successful, the procurement department will use the estimate for the ordering of the materials. If the estimator has taken the liberty to skip from point to point on the take-off, it is probable that materials will arrive on the job before they can be used, with the result of possible damage on the job or at least cramping the superintendent for storage space. No materials should ever arrive on the job much sooner than they can be used.

Note: Some companies give their junior estimators excellent training by making them job sponsors. This means that the job sponsor would assist the estimator from the moment the drawings and specifications are applied for until the tender has been accepted. He would then become the right-hand man of the job superintendent. This is excellent training, and it is as beneficial for the company as for the man.

Step 10: The estimator should have a quiet, comfortable, well-lighted, and ventilated place in the office. His desk should be long, with dropleaf extensions for holding drawings and papers. On his left he should have a suitable library of reference books, and within his reach he should also have the A.I.A. file. On his right (on casters) should be a good calculating machine and a typewriter. Before him he should have:

(a) the drawings

(b) the specifications

(c) a guide list of main headings for a similar building

(d) the local and national building code

He should now proceed with conscientious, disciplined, and methodical neatness throughout the whole take-off. *Note: It is a cardinal rule with many companies that where any written figure is called in question by any other person in any department, such a figure will be referred back to its originator. As an example, assume that the figure 8 in $1846.10 was so poorly formed that there was a question as to whether or not it was intended to be a 5, making $1546.10 instead of $1846.10. Then it would be referred right back to the originator of the work. UNDER NO CIRCUMSTANCES MAY ANY FIGURES BE WRITTEN OVER.*

12.6 PROGRESS SCHEDULES

Two types of progress schedules are shown (see page 213). The first is a typical progress schedule as may be called for by an architect. This would have to be submitted about one week before building operations commence. The architect would use this schedule for both his own and the client's information. The architect would be able to see, at any time during the building operations, the actual progress of the work against the assessed time. He would be in a position to advise his client of the financial obligations to the contractor at the agreed progress payment scheduled times.

The contractor can make good use of his copy of the progress schedule in the following ways:

(a) He marks on the schedule in a different color the actual progress of the work, under the estimated progress.

(b) He can see at a glance if more men and/or machines are necessary on the job to maintain the scheduled progress. It must be remembered that a large contractor will have several jobs going at any one time, with a separate progress schedule for each. He may see the advisability of moving men or equipment from one job to another, or whether to recruit or lay men off.

(c) He can send extracts of the schedule to subtrades so that they will know when they are expected on the job **and quite definitely when they are to be off the job.** If a subtrade falls behind on its schedule, this could throw the whole schedule and completion date off, with possible penalties for late delivery.

Housing Project Daily Progress Report

The second daily progress report (p. 213) was devised (with minor exceptions) and used by my friend and colleague Mr. H. B. Smith of the Southern Alberta

THE MANSFIELD ENGINEERING AND CONSTRUCTION CO. LTD.

PROGRESS SCHEDULE

Project 18 Suite Apartment for Mr. J. Doe — Date: April 24
Starting Deadline: May 1   Scheduled Completion: September 3

	%	May	June	July	Aug	Sept
Preliminaries	3/4	—				
Survey	1/2	—				
Cut and fill	2					
Fencing	1/2	—				
Excavating	2 1/2	—				
Ditching	1/2		—			
Rough plumbing (subtrade)	5	—				
Concrete footings	3 1/2		—			
Concrete walls	5 1/2		—			
Backfill	1/4		—			
Rough carpentry, basement	4		—			
First-floor assembly	3		—			
Electrician (subtrade)	4			—		
Heating engineer (subtrade)	4 3/4			—	—	—
Plumbing (subtrade)	8			—		
Insulation and soundproofing	1/2			—		
Brickwork, first floor	2			—		
Second-floor assembly	4			—		
Rough carpentry, second floor	4			—		
Brickwork, second floor	2 1/2			—		
Third-floor assembly	5				—	
Rough carpentry, third floor	4 1/2				—	
Brickwork, third floor	3				—	
Rough carpentry, third floor	5				—	
Ceiling, third floor	3				—	
Flat roof assembly	4				—	
Bonded roof (subtrade)	2 1/2					—
Glazing (subtrade)	4 1/2					—
Interior finish	7					—
Painting and decorating (subtrade)	3 1/2					—
Finish floor (subtrade)	2					—
Sidewalks	1 1/2					—
Landscaping	2 1/2					—
Final Clean-up	1/4					—
Final Handover	100					—

Institute of Technology, to whom I extend my sincere thanks for permission to reprint it here.

There is no hard-and-fast rule as to how a progress schedule should be drawn up. However, there must be some kind of progress schedule in order to keep close control over the job.

(d) He will know when he may expect progress payments for completed work. Such payments may be expected monthly on most jobs, upon issuance of an architect's certificate to the owner, who may then instruct the bank to release to the contractor the amount of cash shown on the certificate. The progress schedule percentage column shows a breakdown of the cost per main item.

(e) Some contracts stipulate the amount of daily liquidated damages against the contractor for late delivery of a structure. Some contractors may feel more competent to control costs than time. For a time study see the critical-path method (CPM) in this chapter.

Notice that on the first line the job numbers are shown. In this particular example, they refer to each separate house as numbered on the contractor's projected overall plan. They are in no way associated with the addresses of the houses.

Building permits and applications for utility services on this type of project may be processed at the rate of ten or twelve at a time, in which case the date when each application was granted would be entered under each house number.

The progressive inspections are heavily underlined. These are very important dates, since builders' or mortgage progressive loans would not be issued before inspections are completed.

The first item on "Sheet 2 of 3" shows "Progress payment applied for" and, underneath, "Progress payment received." "Sheet 3 of 3" shows the final disposal of the property. You must make your own progress schedules.

12.7 THE GANTT BAR CHART PROGRESS SCHEDULE

This chart is named after its originator, Henry L. Gantt, a pioneer in modern management who invented it about the time of World War I. It is still used extensively in industry. Its essential feature is the presentation of projected events in their relation to time. It enables management to compensate for deviations from the planned course by the introduction of more men, machines, overtime, and so on at critical times.

In the construction industry the individual, or better still, the individuals responsible for making a Gantt Bar Chart (usually and hereafter called a progress schedule) must be practical men who can draw from their personal (and recorded) experiences the time it will take for a man, a crew of men, a single piece of equipment, or several pieces of equipment to perform any unit of work.

With this knowledge the following steps may be taken to produce a progress schedule:

Step 1: Examine the drawings, specifications, and *completed estimate sheets* for the job to be scheduled. Enter the starting date, the finishing date, and the number of working days at the head of the progress schedule as shown on the specimen on page 211. Divide the calendar days allowed for the completion of the project into five-day working weeks and deduct public holidays falling in that period of time. Assume that the job is to start on 1 June and end on 31 October of the same year as shown on the specimen. This is an elapsed time of 143 calendar days. Allowing for no work on Saturdays, Sundays, and public holidays, the actual days will be reduced to about 104 *working days.*

It is imperative to know the actual working days available so that the correct numbers of men and pieces of equipment may be assembled at the right times. Remember the penalty clause for late delivery of a building. If a building is late for delivery, everybody who has any investment in it is a loser. Assume that a person has $50,000.00 tied up (say retainage in the case of the builder) for two months longer than originally scheduled. This represents a direct loss, for the investor could be earning a yield on its employment elsewhere. Also keep in mind that if the general contractor has other jobs going at the same time, his men and equipment will be delayed in arriving at another job site.

It is important for a general contractor to maintain a good record for meeting completion dates. It could influence an architect in awarding a contract to a contractor with such a record, even when his bid is marginally higher than another.

Step 2: Using the drawings, specifications, and estimate sheets as guides, make a list of all the progressive main operations *and subtrades*. Head the list as shown on the specimen. Every project will have different divisions of operations, and no two individuals (working separately) would be likely to break the divisions down exactly alike.

Step 3: Using the estimate sheets as a guide for costs, assess the percentage of the total cost of each specific item listed, as at item numbers 1 and 2, "Preliminary and site operations" and "Cut and fill."

THE MANSFIELD ENGINEERING AND CONSTRUCTION CO. LTD.

DAILY PROGRESS REPORT

Sheet 1 of 3

Project No. _____   Date_____

Starting Deadline _____   Scheduled Completion _____

JOB NUMBER:	1	2	3	4	5	6	7	8	9	10	11	12	13	14	15	16	17	18	19	20	21	22	23	24
1. Permit																								
2. Layout																								
3. Utilities—gas																								
4. —water																								
5. —sewer																								
6. —telephone																								
7. —cable television																								
8. Excavation																								
9. City and mortgage inspection no. 1																								
10. Formwork for footings																								
11. Concrete footings																								
12. Posts, forms, beams, and floor joists																								
13. Concrete walls																								
14. Floor sheathing																								
15. City and mortgage inspection no. 2																								
16. Framing walls																								
17. Masonry unit walls																								
18. Chimney																								
19. Framing roof																								
20. Walls and roof sheathing																								
21. Roofing complete																								
22. City and mortgage inspection no. 3																								
23. Rough plumbing																								
24. Rough wiring																								
25. Rough heating																								
26. Rough telephone																								

THE MANSFIELD ENGINEERING AND CONSTRUCTION CO. LTD.

DAILY PROGRESS REPORT

Sheet 2 of 3

Project No._____   Date_____

Starting Deadline _____   Scheduled Completion _____

JOB NUMBER:	1	2	3	4	5	6	7	8	9	10	11	12	13	14	15	16	17	18	19	20	21	22	23	24
27. No. 1 progress payment applied for																								
28. No. 1 progress payment received																								
29. Exterior trim—windows/doors																								
30. Insulation																								
31. City and mortgage inspection no. 4																								
32. Gyprocking/lathing																								
33. Basement floor																								
34. Taping/plastering																								
35. Sheet metal/heating																								
36. City and mortgage inspection no. 5																								
37. No. 2 progress payment applied for																								
38. No. 2 progress payment received																								
39. Exterior wall finish																								
40. Meter applications																								
41. Hardwood floors laid																								
42. Interior finish																								
43. Painting																								
44. Floor sanding																								
45. Linoleum																								
46. Sidewalk and steps																								
47. Backfill																								
48. Finish—plumbing																								
49. —wiring																								
50. —heating																								
51. Lawn seeded																								
52. No. 3 progress payment applied for																								

THE MANSFIELD ENGINEERING AND CONSTRUCTION CO. LTD.

DAILY PROGRESS REPORT

Sheet 3 of 3

Project No. _____   Date _____

Starting Deadline _____   Scheduled Completion _____

JOB NUMBER:	1	2	3	4	5	6	7	8	9	10	11	12	13	14	15	16	17	18	19	20	21	22	23	24
53. No. 3 progress payment received																								
54. City and mortgage inspection no. 6																								
55. Sold proposal																								
56. Approval of sale																								
57. Purchaser's mortgage complete																								
58. Down payment received																								
59. Final inspection and holdback																								
60. Occupied																								
61. Garage																								

Step 4: Any large "change order" (see page 87) for either the prime contractor or any of the subcontractors will affect not only the final cost of the job but may also seriously affect the completion date. The recording and filing of these documents cannot be overemphasized. *They are very important documents* (see Section 12.7). A special book should be kept to record all change orders, and notices of proposed changes must be sent to all parties. In the preparation (for the architect) of costs for proposed changes, not only the prime contractors fee but also those of the subcontractors must be taken into account. Also overhead expenses and profit must be added, as well as a statement of the number of extra working days that such changes will add to the completion date of the project. A general contractor should consider very carefully what effect a delay in finishing one project may have upon completing another concurrently under construction by him. If change orders on one job may result in a claim against him for liquidated damages on the other project, should he not add these damages to his change order estimate? All interested parties must be kept informed regarding change orders, and notice of acceptance or rejection by the architect must be immediately passed on to all other parties to the proposed change.

Step 5: At the bottom of the specimen progress schedule the 100 percent total value of the project is shown as $244,680.00. Preliminary operations account for 6.2 percent of the total cost; 6.2 percent of $244,680.00 is $15,170.20. This figure is entered as shown on the chart under the column headed "Value," and so on throughout all of the itemized operations.

Remember that subtrades are not bound to move onto the job any earlier than they are scheduled. In short, a great deal depends upon the job superintendent to keep the work moving, at the same time maintaining amicable relations with his own work crew, the union business agent, subcontractors, and building suppliers not to mention inspectors, architects, engineers, workmen's compensation board representatives, and so on. It is one thing to prepare a good progress schedule, but quite another to manage people and things to conform to it.

Step 6: Carefully assess the time required to complete each of the main operations listed, and draw a bar line on the progress schedule showing the starting and completion date for each individual operation (see the specimen). Examine item 1, "Preliminary and site operations," which will include establishing the boundaries of the job site and setting up a datum, removing existing obstructions, assembly of temporary buildings, utilities, equipment, materials, and so on.

The percentage of work completed (including materials on the ground) by the end of each succeeding month is entered on the appropriate bar line, up to the comple-

MEANSCO
FORM 220

CONTRACT CHANGE ORDER

FROM:

_____

_____

_____

_____

TO:

_____

_____

_____

_____

CHANGE ORDER NO. _____

DATE _____

PROJECT _____

LOCATION _____

JOB NO. _____

ORIGINAL CONTRACT AMOUNT	$				
TOTAL PREVIOUS CONTRACT CHANGES					
TOTAL BEFORE THIS CHANGE ORDER					
AMOUNT OF THIS CHANGE ORDER					
REVISED CONTRACT TO DATE					

GENTLEMEN:

This CHANGE ORDER includes all Material, Labor and equipment necessary to complete the following work and to adjust the total contract as indicated;

☐ the work below to be paid for at actual cost of Labor, Materials and Equipment plus

_____ percent (_____ %)

☐ the work below to be completed for the sum of _____ dollars ($ _____)

_____

_____

_____

_____

_____

_____

_____

_____

_____

_____

_____

_____

_____

_____

_____

_____

CHANGES APPROVED

_____

By _____

_____

By _____

The work covered by this order shall be performed under the same Terms and Conditions as that included in the original contract unless stated otherwise above.

Signed _____

By _____

R. S. MEANS CO., INC., DUXBURY, MASS. 02332

tion of the job. Item 1 will not be totally completed until the whole job is finished; item 2 will be largely completed by the end of July when backfill is replaced. From the daily progress reports made by the job superintendent (and analyzed by the cost department), it will be seen how actual job progress is being made against the projected progress. The office copy of the progress schedule may be marked under each bar (in a different colored ink) with the actual progress, and remedial action taken to correct deviations.

Step 7: From an examination of the schedule, it may be seen that the following percentages of work (plus the value of materials on the ground) are projected to have been completed by the end of June:

Preliminary and job site operations	60% of $15,170.20 =	$ 9,102.00
Cut and fill	90% of 6,117.00 =	5,505.00
Layout and ditching	100% of 3,247.50 =	3,247.00
Formwork	100% of 4,893.60 =	4,894.00
Concrete and reinforcing	80% of 8,141.50 =	6,513.00
Plumbing	10% of 2,153.20 =	215.00
		$29,476.00

Step 8: Remember that there will be a 10% holdback by the architect of all progress payments on this job. Thus the first check (assuming the work is kept up to time) would be in the sum of $23,414.00 less 10 percent = $26,528. Holdback (retainage) differs on government contracts and also on some others, so read the conditions of contract carefully!

It should also be observed that by the end of the first month, although much work will have been accomplished and much material may be on the ground, the first progress payment cannot be expected from the owner (through the architect's certificate for payment) before five or six weeks after the start of the work. The contractor must finance all payrolls, insurances, and so on during this period of time.

Step 9: In a similar manner, a further analysis may be made of progress payments for the months of July, August, September, and finally October.

Step 10: The subcontractors must be given an extract of the schedule showing when they are due on the job and specifically when they are due to leave the job. If one subcontractor is late on leaving the job, the whole schedule for the job completion date may be affected, because other trades will be held up as a consequence.

Step 11: From these figures the accountant may make a financial progress schedule. Such a schedule is critical, especially where a contractor has several jobs on the go at one time with commitments for payrolls, subcontractors, material suppliers, and so on. Such a statement could be used by a contractor when negotiating for new loans from financiers; remember that the retainage is actual money already earned and only subject to change for contingencies (things that may or may not happen).

Step 12: When the job is in operation, any differences between projected and actual performance may be seen on the chart, and corrective action may be taken by the introduction of more men, overtime, more machines, hastening of materials from suppliers, and so on. It is imperative that the job be kept moving and be terminated on time.

Step 13: Be aware that unless the architect calls for separate bids from any or all of the subcontractors, *the total responsibility for the completed project is that of the prime contractor.*

When conditions in the construction industry are buoyant, there is always the difficulty of recruiting good men for a "limited" time. If, however, a company has a subsidiary or itself operates a remodelling company, men may be temporarily withdrawn from such operations to boost the general contractor's main project. The union, too, may be helpful in finding men.

Finally, remember that subcontractors will want to submit bids to a general contractor who consistently has a job ready for them on scheduled time and from whom regular contracted progress payments are received.

12.8 THE CRITICAL-PATH METHOD (CPM)

Part of a dictionary definition of the word *critical* is "quick to find fault." The CPM was devised for the construction industry to assess the minimum time required to erect a structure, and for planning the critical consecutive jobs to be done to accomplish this end. See the following Critical-Path Flow Chart (Fig. 12.1) and observe how the arrowed flow indicates the progress of the units of work which follow each other. Those items are critical that have to be completed before other items may be started; and when such items are given a time limit for completion, we are *quick to find fault* if they fall behind schedule.

The term *project* denotes the whole structure. The term *activity* denotes any identifiable job that takes time to complete within the project. Every activity occurs before, after, or at the same time as another activity or activities. Study the chart and note that when the formwork is being made and erected, it is an identifiable activity; it occurs after the layout, but before the activity of placing concrete footings; it also occurs at the same time that reinforcing is being set in the formwork;

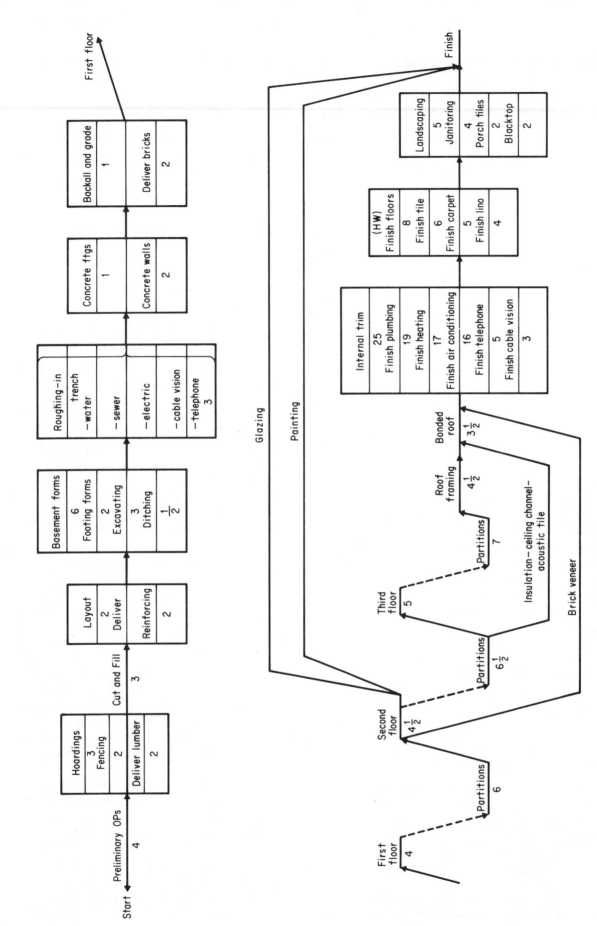

Fig. 12.1 Critical path flow chart. All units are in days.

furthermore, while the forms are being made, excavating may proceed. The critical activity in the above operations is formwork, because other activities cannot proceed until it is finished. This is an activity that must be carefully watched so that it may be completed in the time scheduled.

It is suggested that, to make up a simple CPM flow chart, you proceed as follows:

Step 1: Type a list of all identifiable activities for the project. Examine the example as a guide.

Step 2: Cut out all typed activities and place them in a single straight line on a flat surface.

Step 3: Arrange the activities in sequential operational order using the following reasoning:

(a) This activity occurs before that. (See cut and fill and layout.)

(b) That occurs after this. (See backfill after concrete walls.)

(c) This happens at the same time as that. (See footing forms, wall forms, excavating, and ditching.)

(d) This occurs at the same time as all these things. (See glazing and plumbing, which occur during lots of other activities.)

Rearrange all activities until you have satisfied yourself that they are in sequential order of events. The term *event* denotes a point in time between the finish of one activity and the start of another.

Step 4: On a long piece of paper (a good quality piece of kitchen shelving paper will do) make a freehand graphical arrowed flow chart similar to the one shown. *Note carefully that the length of the arrows is immaterial* and they may be straight, bent, or curved.

Step 5: Underline each activity, and underneath it write the duration of time required to complete it. The term *duration* is an experienced judgement of the time required to complete an activity.

Step 6: When you are completely satisfied that all the activities (together with their durations) are in correct order, make a final CPM chart. *Be aware that the making of a* CPM *chart* takes time and patience, and can only be attempted against the background of someone's experience in the field.

You will notice the dotted lines on the chart between first, second, and third floors and the following partitions. These lines indicate that some of the carpenters will leave the floor framing to start working on the partitions before the whole floor assembly is completed.

You will find the experience of making a CPM chart interesting and rewarding, and it will fix in your own mind the best order of operations; you will also be better able to communicate with others about the job.

Using the CPM Chart

Step 1: To find the total duration of the project, add together the number of days required for all activities that appear on the top arrowed paths. See column (a) on CPM Analysis Chart below.

Step 2: Add up all the days required for all activities appearing on the arrowed path, but including the second-row arrows instead of the first—column (b).

Step 3: Add up all the days required for activities appearing on the arrowed path, but including the third-row arrows instead of the first—column (c).

Step 4: Add up all the days required for all activities appearing on the arrowed path, but including the fourth-row arrows instead of the first—column (d).

Step 5: Compare the total times for each arrowed path in columns (a), (b), (c), and (d). *The longest path is the critical path.* Anything that can speed any of the activities on this path will shorten the total time required for the project. Assuming that, by using more men and equipment or by working overtime, the longest critical path was shortened to fewer days than the second longest path, then the next longest path would become critical and an effort would be made to speed it up. This is how we examine the CPM chart for critical events. Where more men, machines, or equipment are used to shorten a critical path, it is reflected in direct costs; but indirect costs will go down by completing the project in a shorter period of time.

Computing CPM Time

Time is measured by a five-day week, excluding holidays. If a building takes 75 working days to complete, this is $75 \div 5 = 15$ weeks plus public holidays.

To calculate on the CPM chart when brickwork will commence we make what is termed a *forward pass.* This is done by adding together all the preceding time activities and relating this total to a calendar date (allowing for public holidays), which yields the date for the bricklaying to commence. The subtrades must know when they are to arrive on the job, and they must know quite definitely when they must be finished with their part. A delay in any subtrade can very seriously affect the whole course of the progress of the project. *You will notice that on the* CPM *chart the delivery of bricks is*

ANALYSIS CHART CPM

Item	(a) Activity		(b) Activity		(c) Activity		(d) Activity	
1	Preliminary ops	4	Preliminary ops	4	Preliminary ops	4	Preliminary ops	4
2	Deliver lumber	2	Deliver lumber	2	Deliver lumber	2	Deliver lumber	2
3	Hoardings	3	Fencing	2	Fencing	2	Fencing	2
4	Cut and fill	3	Cut and fill	3	Cut and fill	3	Cut and fill	3
5	Layout	2	Layout	2	Layout	2	Layout	2
6	Deliver rein	2	Deliver rein	2	Deliver rein	2	Deliver rein	2
7	Basement forms	6	Footing forms	2	Excavating	3	Ditching	$\frac{1}{2}$
8	Roughing-in	3	Roughing-in	3	Roughing-in	3	Roughing-in	3
9	Concrete ftgs	1	Concrete walls	2	Concrete walls	2	Concrete walls	2
10	Backfill	1	Backfill	1	Backfill	1	Backfill	1
11	Deliver bricks	3	Deliver bricks	3	Deliver bricks	3	Deliver bricks	3
12	First floor	4	First floor	4	First floor	4	First floor	4
13	Partitions	6	Partitions	6	Partitions	6	Partitions	6
14	Second floor	$4\frac{1}{2}$	Second floor	$4\frac{1}{2}$	Second floor	$4\frac{1}{2}$	Second floor	$4\frac{1}{2}$
15	Partitions	$6\frac{1}{2}$	Partitions	$6\frac{1}{2}$	Partitions	$6\frac{1}{2}$	Partitions	$6\frac{1}{2}$
16	Third floor	5	Third floor	5	Third floor	5	Third floor	5
17	Partitions	7	Partitions	7	Partitions	7	Partitions	7
18	Roof framing	$4\frac{1}{2}$	Roof framing	$4\frac{1}{2}$	Roof framing	$4\frac{1}{2}$	Roof framing	$4\frac{1}{2}$
19	Bonded roof	$3\frac{1}{2}$	Bonded roof	$3\frac{1}{2}$	Bonded roof	$3\frac{1}{2}$	Bonded roof	$3\frac{1}{2}$
20	Interior trim	25	Finish plumbing	19	Finish heating	17	Finish telephone	5
21	Finish floors	8	Finish tile	6	Finish carpet	5	Finish lino	4
22	Landscaping	5	Janitoring	4	Porch tiles	2	Blacktop	2
	Days:	109	Days:	96	Days:	92	Days:	$76\frac{1}{2}$

Note: Items 12–16 have similar activities for each floor level; the extra time is taken up by the raising of material and placing at progressively greater heights.

shown at the time of backfill. It is important to show delivery times for important items, including mechanical items.

To calculate on the CPM chart when the carpet layer should commence work, we make what is termed a *backward pass.* This is done by adding together all the timed activities (including that of the carpet layer), starting with his alloted duration time until the finish of the project, and subtracting this number of days from the total number of working days for the whole project. Remember to transcribe this total of working days into calendar dates.

For housing units and small jobs you may make a CPM chart yourself, but for large, complex jobs you may, *according to the terms in the specifications,* have to make a computerized CPM chart. There are specialists in this field who will set up a program and revise it every month or so. The rental service for a computer will range from $60 to $360 per hour, but most jobs can be processed in less than 1 hour.

12.9 QUANTITY SURVEYING

In some places, architects employ for a fee the services of registered (or chartered) quantity surveyors who take off the quantities of units required by the contractor in the execution of the required work. As an example, they would examine the plans and specifications and make the estimate of, say, the number of cu yd of concrete required (by type). The contractors then would all be bidding against this stated number of cu yd and their price would be so much per cu yd placed. This price would include the contractor's overhead expenses, the maintenance or rental of equipment, the fringe benefits for workmen, and the profit.

Quantity surveyors jealously guard their organization by requiring applicants for membership either to pass one of their entrance examinations or to submit acceptable evidence of proficiency to their board of directors. They have a standard system of measurement and all measurements, sizes, and quantities are taken as net. The contractor is to allow for all cutting and waste in his prices.

Under this sytem the contractors would be given:

(a) drawings

(b) the contract document, including the tender specification and the bill of quantities (all under one cover).

The following pages show specimen examples of extracts of specifications and quantity surveyors' sheets. These are issued by the architect along with the drawings to general contractors for proposal bids.

The index may read as follows:

INDEX

	Page
Notice to contractors	2
Tender	4
Particular conditions of contract	5
Specification	6
Bills of quantities	31
Appendix	125
Schedule of prime costs and provisional sums and items	128

Some extracts from bills of quantities follow. Under "Bill No. 1—Preliminary and General," some of the contractor's estimated costs would be made up under such headings as:

	Item	$	cts
Insurance	A	250	00
Cut-away and make good	B	420	00
Temporary utilities	C	130	00
Temporary buildings, and so on	D	600	00

Each page of Bill No. 1 would now be totaled and carried to an abstract at the end of all the pages comprising Bill No. 1. When abstracts have been made of all the individual bills in the bills of quantities, the abstracts are then carried to a final summary at the end of the cost estimate. (Examine Bill No. 2, p. 223.)

Excavator, Concretor, and Bricklayer

A specification of the scope of the work to be performed by each trade is shown directly under each bill. Then the actual bill of quantities for each trade follows.

Specimen Preamble to a Schedule of Quantities

Schedule of Quantities for the erection of Proposed New (School) the (Beaumont Primary) in accordance with the Contract Drawings and Specification prepared by and to be executed under the superintendence and to the entire satisfaction of: (Johnson, Olsen and Associates), Architects.

DATED
30 January 19____   Signed_____

This schedule of quantities contains 125 pages in consecutive order. The contractor is requested to check the number of pages, and should any of the pages be missing or duplicated, or the figures or writing indistinct, or the descriptions ambiguous, the contractor is to apply to the surveyor at once and have the errors rectified. Any errors arising from the contractor failing to do this will not be admitted after the submission of the tender.

Contractor's Note

In the event that the tenderer is not able to obtain materials and firm quotations for items measured and specified in these contract documents, he is to report in writing to the architects immediately, or not later than three days to the closing date of the tenders.

Should the tenderer fail to do this, his tender will be taken as firm for the items as described and he will be responsible for the supply of the same at the rates tendered.

Bill No. 1
Preliminary and General

A. FORM OF AGREEMENT AND CONDITIONS OF CONTRACT

Note: *Here followed the general conditions of contract, which in this case covered 15 pages. Any expenses to the contractor were carried out to the right-hand columns and then recorded on the "Abstract" of "Bill No. 1," which covered all the page totals. Finally the total amount of "Bill No. 1" was carried to the "Summary" at the end of the estimate.*

Abstract of Bil No. 1
Preliminary and General

	Item	$	cts
Total of page number	1	621	17
Ditto ditto	2	380	68
Ditto ditto	3	492	55
and so on			
Ditto ditto	15	74	49
		31,869	71

Total Amount of Bill No. 1 Carried to Summary (see p. 196).   * $31,869.71

Bill No. 2
Excavator, Concretor, and Bricklayer

A. *All foundations are provisional and subject to remeasurement as executed.*

Excavator

B. *Note:* Prices for all excavations are to include the bottoms of trenches being well rammed before concrete is laid.
C. All excavations have been measured net bulk as "in ground," and the contractor must allow for increase in bulk upon excavating and for making good any slips.
D. The excavations are measured the net width of the concrete.
E. Should the contractor require any extra working space, he must allow here, or include in his prices, such extra sum as he may consider necessary.

F. The contractor is to visit the site and from his own observations ascertain the nature of the ground and must include in his prices all costs for excavating in same as no extra will be allowed owing to his failure to do so.

	Unit	Quantity	$	cts
G. Clear the whole area of the site to be built upon and also clear for approximately 10'0" all around the building, all rubbish, debris, shrubs, and vegetable matter and, in general, roughly level the site and prepare for building upon including excavating 6" deep to remove all grass roots and deposit top soil on site for filling or spread over site as directed. (Approximate area 3806 square yards.) *Note: The cost per sq yd must be supplied by the contractor.*	sq yds	3806	1903	00
H. Excavate in all kinds of ground met with, not exceeding 6'0" deep, small part return, fill in and ram, remainder haul and deposit on site for filling (if approved) or spread over site as directed to the following:				
I. Reducing levels	cu yds	447	223	50
J. Surface trenches	cu yds	769	461	40
K. Column bases	cu yds	11	8	80
L. Bases to isolated piers and sleeper piers	cu yds	18	16	20
M. Steps and flower boxes	cu yds	23	34	50
N. *And so on* Carried to Abstract of Bill No. 2 *				
Total amount of Bill No. 2, p. 196.			17,462	89

Example 1

	Unit	Quantity	$	cts
I. Reducing levels *Note: The cost per cu yd must be supplied by the contractor.*	cu yds	447	223	50

Example 2

	Unit	Quantity	$	cts
J. Surface trenches *Note: The cost per cu yd must be supplied by the contractor.*	cu yds	769	461	40

With the quantity-survey method, every contractor is estimating against a given number of units of quantity for each trade. In short, each contractor is bidding his own efficiency in purchasing materials, erecting, and administering the work against all other bidders.

When all the pages of quantities comprising Bill No. 2 have been completed, they are carried to an abstract

Abstract of Bill No. 2
Excavator, Concretor, and Bricklayer

		Item	$	cts
Total of page number	*	17	17,462	89
Ditto　　　　ditto		18	7,348	19
Ditto　　　　ditto		35	6,237	41
			28,473	68

Total Amount of Bill No. 2 Carried to Summary　　　　$28,473.68

at the end of the pages of Bill No. 2. Afterward, the total of the abstract is carried to a final summary at the end of the cost estimate.

The summary at the end of all the bills and abstracts would appear thus:

SUMMARY

		Bill	Page	$	cts
Preliminary and general Bill No. 1	*	1	15	31,869	71
Excavator, concretor, and bricklayer Bill No. 2		2	36	28,473	68
Rough carpenter Bill No. 3 *and so on*		3	68	9,432	49
			Total	178,466	36

12.10 UNIT ESTIMATES

The unit system of estimating (under the quantity-survey method) is based by the contractor on the consistency of previous projects correctly estimated by units of work and cost of materials. This means that every job that has been completed by the contractor must be analyzed to find the actual cost of the work against the estimated cost of the work.

Consider the cost sheet:

This would not necessarily mean that all future concrete will cost the company $11.90 per cu yd to place, but it does mean that because of unusual weather conditions, a new factor has been added for concrete to be placed in the month of April each year.

In this manner, every item would be dealt with and the cost records brought up to date accordingly. This is highly specialized work, and the only way in which companies can hope to remain in business is by rigorous cost analysis.

COST SHEET

THE MANSFIELD ENGINEERING AND CONSTRUCTION CO. LTD.

Rosemont School　　　　　　　　Date 20 April 19____

Est'd Cost	Actual Cost	Overest.	Underest.	Reason
Placing concrete walls $11.50	$11.90	.40		Unusual frost in April

12.11 COST SUMMARY SHEETS

On the following pages a code-numbered cost summary sheet for a school is shown. As an example only the first page is ruled. *The columns are used in the following manner:*

(a) Each item is given its code number.　　　　column 1

(b) Each item is named.　　　　column 2

(c) The unit in which each item is ordered and estimated—for example, cu yd of concrete for foundations or sq-ft units of concrete for floors, walks, and so on.　　column 3

(d) The actual cost for proposal bidding.　　column 4

The code number would be used by:

(a) the cost department for working up proposal bids

(b) the procurement department, when ordering materials.

(c) the timekeeper or foreman, when making out the workmen's weekly time sheets, both for hand and machine labor.

The unit cost would include labor (plus Social Security benefits), materials, and plant. Using the quantity surveyor's estimated quantities of each unit, the total cost would equal the number of estimated units times the cost of each unit. *It can be appreciated that the cost per unit must be correct or the error will equal the error for one unit times the number of units.* The cost of labor originates on the actual job. The time sheets must be made out accurately; remember that costing requires a remorseless review and research.

There must be a cost summary sheet for every different type of building. Those for a school would be very different in many respects from those of a hotel, although some code numbers would be the same for both buildings.

THE MANSFIELD ENGINEERING AND CONSTRUCTION CO. LTD.

COST SUMMARY SHEET

Subject: Schools　Sheet no. 1　　　　Date _____

Code　　Item	Unit	Unit Cost
100 *Preliminaries*		
101 Visit the site		
102 Temporary buildings		
103 Temporary services		
Gas		
Light and power		
Telephone		
Water		
104 Test holes		
105 Building permits		
106 Insurance, fire		
Performance bond		
107 Supervision		
Superintendent		
Foreman		
Timekeeper		
Maintenance record		
Diary		
Watchman		
Clean-up man		
108 Final clean-up (subtrade)		

Continuation Sheets Follow:

The following items would also appear on the No. 2 Cost Summary Sheet, and so on.

109　Cut-away and make good
110　Performance bond
111　Storage
112　Fencing
113　Advertising sign
114　Equipment rentals

115　Own plant—all equipment
116　Legal—temporary easement
117　Storage area lighting
118　Finance charges
119　Heating and winter protection

200 Survey

201　Site layout
202　Batter boards
　　　Building lines
203　Datum—bench marks

300 Cut and Fill

301　Clear trees
302　Demolish and remove old buildings
303　Remove topsoil and storage
304　Grading

400 Excavations

401　Mass excavation
402　Hand excavation
403　Shoring and fencing
404　Pumps
405　Backfill—power
406　Compaction—testing
407　Gravel fill
408　Trenching
409　Drain tile
　　　Paper
　　　Coarse fill
410　Backfill—hand
411　Ground detrosting for winter work
412　Haulage
413　Temporary road

500 Concrete

501　Footings
　　　(a)　Forms
　　　(b)　Ties
　　　(c)　Stripping and recondition
502　Walls
　　　(a)　Forms
　　　(b)　Ties
　　　(c)　Stripping and recondition
503　Columns (form clamps)
504　Beams
505　Floors
　　　Main
　　　Coal bunker
　　　Boiler room
506　Stairs
507　Sidewalks
508　Ramps

509 Floor drain
510 Special concrete
      Front entrance
511 Lintels
512 Curing
513 Testing
514 Expansion joints
515 Cold-weather placing
      additives
516 Sleeves—inserts
517 Waterproofing
518 Screeds
519 Anchor bolts
520 Vapor barrier
521 Carborundum rub
522 Parging
523 Metal pan
524 Trough deck
525 Steel Tex

600 Reinforcing

601 Rods
602 Stirrups
603 Mesh
604 Chairs
605 Tie wire
606 Temperature rods
607 Placing
608 Brick and block ties

700 Plumbing: Subtrade

800 Electrical: Subtrade

801 Bell system
802 Call system
803 Intercommunication system (includes telephone)

900 Heat: Subtrade

901 Fire sprinkler system

1000 Sheet Metal: Subtrade

1001 Flashing
1002 Eaves trough

1100 Structural Steel: Subtrade

1101 General contractor placing

1200 Mill Work: Subtrade

1300 Masonry

1301 Facing brick
1302 Backing brick
1303 Wyths

1304 Lightweight building block (concrete, tile, gypsum)
1305 Bearing partition blocks (heavy)
1306 Terra cotta
1307 Stone
1308 Precast lintels
1309 Precast coping
1310 Precast sills
1311 Ties, masonry
1312 Ladder reinforcing
1313 Cleaning brickwork
1314 Pointing
1315 Waterproofing
1316 Insulation, loose fill
1317 Scaffold
1318 Hoist
1319 Concrete fill—beams—columns
1320 Fire brick
1321 Flue linings
1322 Mortar

1400 Glass Block

1401 Caulking and expansion joists
1402 Expansion

1500 Precast Concrete: Subtrade

1501 Sills
1502 Coping stones
1503 Lintels

1600 Rough Carpentry

1601 Form work (See *Concrete 500*)
1602 Floor assembly
1603 Joists
1604 Header
1605 Bridging
1606 Subfloor
1607 Strapping
1608 Furring
1609 Framing
1610 Girths—backings
1611 Rough bucks
1612 Ceiling joists
1613 Ceiling bridging
1614 Rafters
1615 Collar ties
1616 Rough temporary stairs
1617 Drop ceilings
1618 Partitions
1619 Sheathing
1620 Plaster grounds

1700 Ornamental Ironwork: Subtrade

1800 Lathing: Subtrade

1801 Insulation: subtrade
1802 Vapor barrier
1803 Expanding metal (metal lath)
1804 Metal corner beads

1900 Plastering and Stucco: Subtrade

1901 Parging

2000 Boilers: Subtrade

2100 Glazier: Subtrade

2200 Roofing: Subtrade

2201 Shingles, asphalt
2202 Shingles, cedar
2203 Roll roofing
2204 Bonded roofing
2205 Metal roofing

2300 Finish Floors: Subtrade

2301 Hardwood
2302 Tile
2303 Asphalt
2304 Vinyl
2305 Carpet
2306 Linoleum
2307 Terrazzo
2308 Quarry tiles

2400 Hardware: Prime cost

2401 Foot scrapers
2402 Mat sinkings
2403 Miscellaneous metal

2500 Painting: Subtrade

2501 Priming (roof deck)

2600 Outside Finish

2601 Stucco
2602 Siding
2603 Window wall
2604 Curtain wall

2700 Landscaping: Subtrade

2701 Seeding
2702 Sodding
2703 Trees and shrubs

2800 Final Cleanup: Subtrade

2801 Cleaning glass: subtrade
2802 Janitor service

2900 Final Handover

2901 Legal

3000 Maintenance (one year)

Estimating Department

Columns 1, 2, and 3 of the general estimate sheet are the responsibility of the estimating department. This requires the services of building technologists, who have had many years of on-the-job experience in the field (see the example below).

Work-up and Extension Department

Columns 4, 5, and 6 are the responsibility of the workup and extension department. This requires the services of a competent clerical staff with some knowledge of either estimating or costing.

Costing Department

Columns 7, 8, 9, and 10 are the responsibility of the costing department. This is a most important department and requires the services of men of long experience, both in the field and in the estimating department.

Columns 7 and 8 are not very difficult, since they are simply a matter of making purchases in the best markets. This is done in large companies by the procurement department.

Referring to the general estimate sheet, the blocks cost $35 per 100, as shown in column 7. This figure is then multiplied by the total estimated quantity as shown in column 6. The total estimated material cost for this item is then entered in column 8, which, in this case, is $262.50, as shown.

Column 9 is the most intricate to assess. The cost department would look up their own indexes to find their own latest information as to how long it would take a mason and his helper to lay up 100 concrete blocks. *There are no published figures as reliable as your own company indexes. Keep them up to date.*

12.12 FIRE DAMAGE ASSESSING, RENOVATING, AND ALTERATION WORK

There is a constant need for competent fire damage assessors and renovators; if you are interested in this field, contact any or all of the fire insurance companies in your area and arrange to submit a bid on such work. The nature of the work requires extraordinarily careful

BUILDING	*Western Warehouses Limited*		ESTIMATE No. *69*
LOCATION	*149-19th Street S.E. Brompton*		SHEET No. *5 of 22*
ARCHITECTS	*Johnson & Olsen & Assoc.*	**GENERAL ESTIMATE**	ESTIMATOR *B. Smith*
			CHECKER *J.B.*
SUBJECT	*Warehouses*		DATE *April*

DESCRIPTION OF WORK	No. Pieces	DIMENSIONS			Extensions	Extensions	Total Estimated Quantity	Unit Price Mat'l	Total Estimated Material Cost	Unit Price Labor	Total Estimated Labor Cost
(1)	(2)	(3)			(4)	(5)	(6)	(7)	(8)	(9)	(10)
Concrete Blocks: 7⅝"×7⅝"×15⅝"											
South Wall:		17'-8"×40'-0"			707	707					
Outs:											
Cont. con. sill		0'-4"×40'-0"			13						
Con Beam over door		0'-10"×14'-0"			12						
Beam over window		0'-10"×11'-0"			9	34	663 sq. ft.				
663 sq. ft. @ 113 blocks per 100							750	35/100	26250	32/100	24000

surveying, not only for obvious fire damage, but also for the clearing of pockets of noxious gases trapped in undamaged framed walls. It also requires sound construction knowledge and an understanding of all subtrades.

The clearing away of charred and damaged materials is a dangerous and dirty job, but there is always someone willing to do it for a price. The estimate must cover the tearing down and hauling away of materials to a legitimate dump, thoroughly cleaning the affected area, and the installation of new work.

As with alteration work to undamaged buildings, this class of work should be given a good profit mark-up. Estimating costs and accident risks are much greater here than for new work.

QUESTIONS

1. Write a short explanation of the following three ways of making a preliminary estimate (say within 5% to 10%) of the probable cost of a new building:

   (a) the square foot method
   (b) the cubic foot method
   (c) the comparative appraisal method as for a school

2. How would an estimator use the data in Question 1?

3. How would the architect use the information obtained in Question 1?

4. Define revisions.

5. Define an addendum.

6. Define the following abbreviations used on the title block of architects' drawings:
   (a) A. 1 of 17
   (b) S. 1 of 4
   (c) M. 1 of 3

7. State what action you would take if you were working in an estimator's office and you had doubts about the value of a misshapen figure (written by someone other than yourself) in the dimension column of a general estimate sheet.

8. Define a progress report.

9. State how the following persons or departments would use a progress schedule:
   (a) architect
   (b) contractor
   (c) subtrades
   (d) superintendent
   (e) expediting department

10. Define quantity surveying.

11. Define cost analysis.

12. Define the use of cost summary sheets in heavy building construction.

13. Define the following terms as used in the critical path method flow charts for controlling the progress of the erection of a new structure:
   (a) project
   (b) activity
   (c) event
   (d) duration
   (e) forward pass
   (f) backward pass

FLOOR ASSEMBLIES

This chapter is designed to help you to estimate different types of floor assemblies. The principal kinds discussed range from wood joist and concrete slab on ground to intermediate floor assemblies, concluding with the method of estimating metal pans in concrete floors. The estimator gets most of his information about floor systems from the cross-sectional drawings. These he must study closely, along with the relevant sections and the specifications. As an estimator you must be alert to new methods of design, which require new techniques and methods in estimating both materials and construction costs.

13.1 SUPPORTING FIRST-FLOOR JOISTS

There are several methods of supporting first-floor joists above a basement or cellar floor.
Note: An accepted definition of a basement is that it

is more than halfway out of the ground; a cellar is more than halfway in the ground.

The first-floor joists may be supported on:

(a) brick or masonry columns supporting a wood or steel beam;

(b) monolithic columns supporting a wood or steel beam;

(c) standard steel pipes supporting a wood or steel beam;

(d) telescopic adjustable metal posts supporting a wood or steel beam;

(e) lally posts, which are metal cylinders filled with concrete (named after their originator);

(f) a wood-frame partition directly supporting wood joists.

Problem

Assume a 2 × 6 wood-frame wall with one sole plate and two top plates, studs 16″ OC, and one centrally placed girth that directly supports a wooden floor assembly for the plan shown in Fig. 8.1, p. 163. The clearance between the basement floor and the underside of the floor joists is to be 8′-3″ (see Fig. 13.1).

Example

Step 1: Check your local building code for:

(a) maximum spans for wood beams;

(b) minimum stud and plate sizes for frame wall supporting wood floor joists;

(c) maximum spans for wood joists.

Step 2: On the plan, measure the length of the 2 × 6 frame wall required. Plans are usually drawn to the scale of ¼″ to 1′-0″. Multiply the number of inches as read on the tape by 4; each ¼″ reading on the tape is equal to 1′-0″ on the job.

Step 3: Estimate the lin ft of plates required. The wall is 48′-0″ long (exclude the thickness of the concrete bearing walls). It requires one sole plate and two top plates.

Estimate three plates, each 48-0 lin ft, = 144 lin ft of 2 × 6.

Order in merchantable lengths of 2 × 6 = 9/2 × 6—16′-0″.

Step 4: When studs are placed 16″ OC, there are three studs in every 4 lin ft of wall plus one extra for the end:

$$\frac{48 \times 3}{4} + 1 = 37 \text{ studs}$$

The height of the wall is 8′-3″, including the plates and concrete base. The studs may be cut from 8′-0″ lengths.

Order 37/2 × 6—8′-0″ studs.

Step 5: Allow 48′-0″ lin ft of 2 × 6 for girths. *If you are unfamiliar with some technical names, use the glossary that begins on page 355 (see Fig. 13.1 for girths).*

Fig. 13.1 Floor assembly supported on 2 × 6 framing.

When estimating girths, discount the thickness of the studs.

Order 3/2 × 6—16'-0" girths.

Step 6: Summarize:

12/2 × 6—16'-0" plates and girths;

37/2 × 6—8'-0" studs.

Add for nails.

Add for anchor bolts to the concrete supporting the frame wall. See specifications for OC's.

Allow for labor. A carpenter should erect about 400 bd ft of framing per 8-hr day, or 1000 bd ft in 20 hrs.

Problem

Using local prices for labor and materials, estimate (on a work-up sheet) the cost of a wood-frame 2 × 6 bearing partition to the same specification as the example given, but for a wall 56'-0" long.

As an estimator there are many things in the building code that you should know without referring to the code. As an example: "Wood posts or wood-frame walls supporting girders or floors shall bear on solid concrete bases raised at least 0'-3" above the level of the finished concrete floor. Each base shall be at least 2" wider than the supported post and shall bear directly on the post footing."

13.2 FLOOR JOISTS

Floor joists may be supported on:

(a) the tops of beams

(b) the tops of supporting walls

(c) ledger boards spiked to the sides of wood beams

(d) metal joist hangers (see Fig. 13.2).

Problem

Estimate the Pacific hemlock lumber required for floor joists spaced 16" OC for the plan shown in Fig. 8.1. The supporting center wall is 2 × 6 wood-framed.

Example

Check the plans and specifications; also the local building code for the quality of lumber and the type of fastening devices to be used in the construction—for example, nails or metal joist hangers. *It is often cheaper to use expensive fasteners and save on expensive labor costs.*

Step 1: The building code states that 2 × 12 Pacific hemlock joists spaced 16" OC may span up to 18'-7".

Step 2: Allow three joists (16" OC) for every 4 ft of floor to be supported. The length of the plan in Fig. 8.1 is

$$48'-0" \frac{48 \times 3}{4} + 1 = 37 \text{ joists}$$

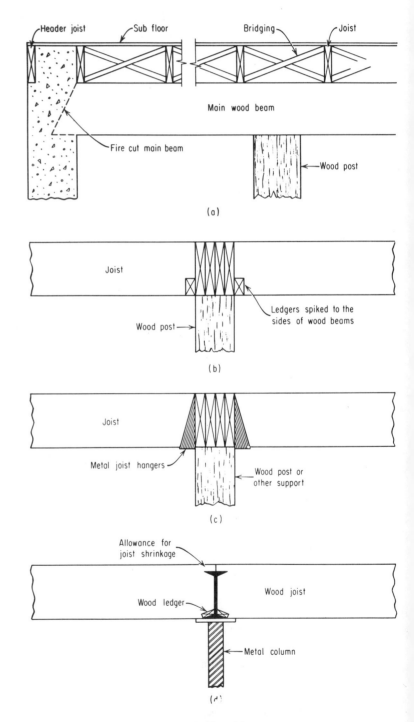

Fig. 13.2 Floor joists.

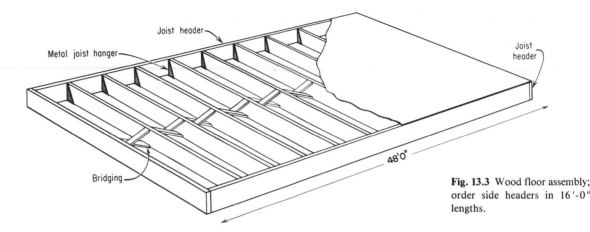

Fig. 13.3 Wood floor assembly; order side headers in 16'-0" lengths.

on each side of the supporting 2 × 6 frame wall. The total joists required is 74/2 × 12—20'-0" (the width of the floor is 38'-0").

Note: With most building codes it is mandatory that wood joists be doubled under all partitions more than 6'-0" in length than run parallel to the floor joists.

Step 3: Headers will be required to secure the ends of the floor joists. These are 48'-0" (lin ft) on either side of the building (see Fig. 13.3).

Order the headers in merchantable lengths: 96 lin ft of 2 × 12, which is equal to 6/2 × 12—16'-0" headers.

Step 4: *Summarize:*

74/2 × 12—18'-0" joists	2664
6/2 × 12—16'-0" headers	192
	2856 bd ft

Note: No allowance has been made on this estimate for doubling under partitions.

Allow for spikes, nails, and/or metal joist hangers.

Labor: A carpenter should erect about 100 bd ft of 2 × 12 joists per hr, or 1,000 bd ft in 10 hr with equal helper time.

In the example given, the total floor area (36'-0" × 48'-0") supported with 2 × 12 joists is 1,728 sq ft. The total amount of floor joists including headers (but excluding any doubling for bearing partitions above the floor) is 2,856 bd ft. Find the number of bd ft of 2 × 12 joists required for each sq ft of floor:

$$2,856 \div 1,728 = 1.65 \text{ bd ft (approx.)}$$

It will require 1.65 bd ft of 2 × 12 floor joists for every sq ft of floor supported; add for doubling under parallel partitions and for floor openings such as stairwells, chimneys, chutes, and so on. Use this figure as a check on your calculations of bd ft for 2 × 12 floors. *As an estimator you must always be thinking in this manner.*

Problem

Find the bd-ft joist factor per sq ft of floor area for the following wood floors:

(a) floor 24'-0" × 36'-0" with joists 16" OC supported on a central girder;

(b) floor 30'-0" × 48'-0" with joists 16" OC supported on a central girder.

You will have to read your local building code for the minimum-dimension joists for the spans given.

13.3 BRIDGING FOR WOOD JOISTS

There are several different types of bridging used to stiffen wood joists that support floors (see Fig. 13.4).

Note that:

(a) diagonal wood bridging may be in specified dimensions as 1 × 3, 1 × 4, 2 × 2, or 2 × 4;

(b) solid-wood bridging is usually mandatory under frame walls running at right angles to the joists;

(c) steel bridging of various gages to meet differing conditions may be specified.

An extract from a building code states "Joists shall be bridged with 1 × 3, 1 × 4, 2 × 2, or 2 × 4 diagonal cross bridging with 1 double row per span of 7'-0" to 12'-8" and two double rows in spans over 12'-8". Acceptable metal bridging may be used in lieu of wood bridging."

Problem

Estimate the lin ft of 2 × 2 bridging required for the plan in Fig. 8.1, p. 163.

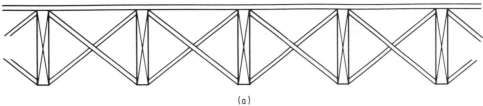

(a)
Cross bridging

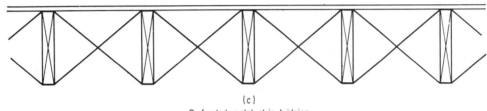

(b)
Solid wood bridging

(c)
Perforated metal strip bridging

Fig. 13.4

Example

Step 1: Read the building code. The span of the joists is 18'-0", and each span will require two double rows of cross bridging. Each row will be the length of the floor: 48'-0".

Step 2: Allow three times the length of the floor for each double row of bridging. It will require four double rows, each 48'-0" long:

$$48 \times 4 \times 3 = 576 \text{ lin ft of } 2 \times 2$$

Step 3: Order in lin ft or convert to bd ft. Bridging may be cut from cheaper stock than the joists. This method allows for cutting poor portions to waste.

Labor for Bridging for Wood Joists: A carpenter should machine-cut and then nail into place 100 lin ft of double-row wood cross bridging in about 6 hr. It will require about 40 lb of 2½" common nails for 1,000 lin ft of a double row of cross bridging.

Problem

Using local prices for labor and materials, estimate the cost to frame the floor joists complete with headers and 2 × 2 cross bridging for the house design on pp. 301-302.

13.4 COMBINATION TILE AND CONCRETE FLOORS ON GRADE

Clay tiles are made in various standard sizes of twelve inches in length. When laying concrete tile drained floors directly on the ground, the following method should be adopted:

(a) The slab must be above the surrounding land and care should be taken to assure that the ground slopes away from the building on all sides.

(b) The clay tiles should be laid on a sand or gravel fill— the thickness of the fill depends upon the building site. If it is a high, well-drained site, a fill of only 1" may suffice. Such a fill provides a firm cushion for the tile and helps to reduce capillary rise of moisture.

(c) A vapor barrier should be placed over the fill before the tile is placed.

(d) A minimum of 2" of rigid waterproof edge insulation should be used.

(e) When the floor slab is to be heated, make a close study of the specifications for insulation and installation specifications.

Problem

Using local prices, make a proposal bid to install a tile and concrete floor for the warehouse shown in Fig. 16.2. The specifications are as follows:

(a) Assume the area is rough-graded natural gravel.

(b) Allow for 2″ of sand over the total area.

(c) Allow for a 6-mil polythylene sheet over the sand.

(d) Allow for 4″ clay tile over the floor area.

(e) Allow for 2″ of rigid waterproof edge insulation around the perimeter of the clay tile and the concrete floor.

(f) Allow for a 3″-thick 3,000-psi-high, early-strength concrete-slab floor.

Labor: Allow for a mason and helper requiring 3 hr each per 100 sq ft of floor area prepared for receiving ready-mixed concrete.

Price List: Obtain price lists from your nearest local suppliers.

Code Index Card: Make a suitably documented code index card or cards for this floor.

Clay Tile and Concrete Floors

For these, calculate the total number of lin ft of tiles in the floor system. Each tile is 12″ long; reduce to number and order in correct size.

Metal Pan Forms in Concrete Floors

Read the specifications to see if the pans have to be left in place. Two gage types are manufactured, and the heavy gage may be used over and over again. Many companies specialize in supply, fix, and removal of metal pan at a price per sq ft of floor area. For lightweight metal pans that are to be left in place, the contractor may use his own workmen. The sizes of metal pans are as follows:

Depth	6″, 8″, 10″, and 12″
Width	20″ and 30″
Length	1′-0″, 2′-0″, and 3′-0″.

The cost of labor in placing metal pans is very low, but the forms to receive them must be set very carefully.

See Fig. 13-5 and Fig. 13-6 for the grid system of metal pans.

Reinforcing for concrete floors is a subtrade, unless the job is very small.

Concrete for Reinforced Concrete Floors

This is reckoned by the floor area in sq ft times the depth of the floor in ft. The final unit is in cu yd. Deduct for voids made by either clay tiles or metal pans, but make no allowance for displacement of concrete by reinforcing rods.

13.5 TYPICAL FLOOR-BEARING DETAILS FOR CONCRETE MASONRY WALLS

The estimator must be aware of the method by which a whole floor assembly is supported from the bearing walls. Study the three concrete masonry unit drawings for precast joists, clay tile, and reinforced-concrete floors shown in Fig. 13.7.

13.6 STRUCTURAL I-BEAMS

Structural I-beams are marketed by lb per lin ft. As an example, two sizes which have been taken from a manufacturer's table are shown below.

Secure a complete price list from your nearest structural-steel manufacturer. *Be sure to keep up to date with your A.I.A. or R.A.I.C. file; and it is imperative that, when any file has been taken from the cabinet, it be returned to its correct place as soon as it has served its function.*

WEIGHTS AND DIMENSIONS OF STRUCTURAL BEAMS			
Size, inches	Weight per lin foot	Web thickness, inches	Flange width, inches
3	5.7	0.170	2.330
15	. . . and so on . . . 35 | 0.330 . . . and larger . . .		5.500

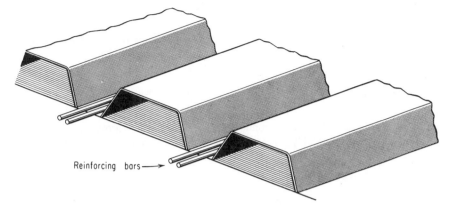

Fig. 13.5 One-way metal pans.

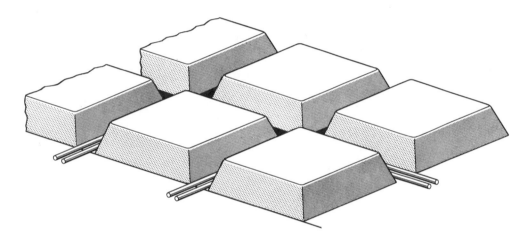

Fig. 13. 6 Grid system of metal pans.

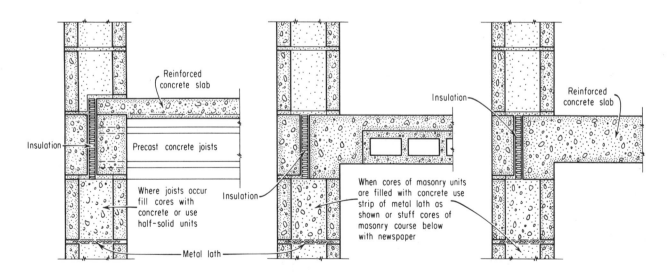

Fig. 13.7 Typical floor bearing details with concrete masonry walls
(Courtesy nation Concrete Masonry Association).

QUESTIONS

1. State an accepted definition of the difference between a basement and a cellar for housing units.

2. Make a neat sectional drawing of each of five different methods of supporting first-floor wood joists in residential construction.

3. Make a neat sectional drawing of each of three different types of bridging as used in a wood floor assembly.

4. Name two different kinds of wood that may be used for joists in wood floor assemblies in your locality.

5. What authority in your locality has jurisdiction over maximum allowable spans for wood joists?

6. Name six important considerations that must be taken into account when a tile and concrete floor is to be laid directly on the ground.

7. Make a neat sectional drawing of a one-way combination tile and reinforced concrete intermediate floor system.

8. Make a neat sectional drawing of a one-way metal pan and reinforced concrete intermediate floor system.

9. Make a neat perspective drawing of a grid system for a metal pan and reinforced concrete intermediate floor system.

14

CONCRETE: PROPORTIONING, MIXING, AND PLACING

This chapter is designed to give you an appreciation of estimating concrete, both ready-mixed and mixed on the job. Only ready-mixed "weight-designed concrete" must be ordered for delivery by the cu yd in the correct design and quantity from any ready-mixed concrete manufacturer. On-the-job "volume-mixed concrete" requires a knowledge of the components and proportions of dry-mix materials. It is with the latter method that the discussions in this chapter are largely concerned. The estimator is earnestly recommended to take a course in design and control of concrete mixtures. There are numerous excellent publications on the subject. Every year the Portland Cement Association makes publications available to the engineer, the builder, and the farmer in a language suitable to each.

The estimator must make himself familiar with concrete by reading the latest bulletins and also by on-the-job observations. His library should contain sufficient up-to-date information to meet his needs.

Sections 14.1 thru 14.9 have been excerpted from Design and Control of Concrete Mixtures, metric edition, and is reproduced here with the permission of the copyright holders.

Improved practices and techniques have added greatly to our ability to produce good concrete. It is the responsibility of those in charge of construction work to become familiar with these practices and techniques and to make sure that they are put into full use. The extra effort and care required are small in relation to the benefits.

14.1 MEASURING MATERIALS

To produce concrete of uniform quality, the ingredients must be measured accurately for each batch. Most specifications require that batching be done by mass rather than by volume because of the inaccuracies in measuring solid materials (especially damp sand) by volume. Use of the mass system for batching provides greater accuracy, simplicity, and flexibility. Flexibility is necessary because changes in aggregate moisture content require frequent adjustments in batch quantities of water aggregates. Water and liquid admixtures can be measured accurately by either volume or mass.

Specifications generally require that materials be measured in individual batches within the following percentages of accuracy:

Cement	1
Aggregates	2
Water	1
Admixtures	3

Equipment should be capable of measuring quantities within these tolerances for the smallest batch regularly used as well as for larger batches. The accuracy of batching equipment should be checked periodically and adjusted when necessary.

On small jobs where bagged cement is used, the batches should be of such size that only full bags are used unless fractional bags are accurately measured.

Air-entraining admixtures, calcium chloride, and other chemical admixtures should be charged into the mix as solutions and the liquid should be considered as part of the mixing water. Admixtures that cannot be added in solution may be measured by volume as directed by the manufacturer. Admixture dispensers should be checked daily since errors in admixture batching, particularly overdosages, can lead to serious problems in both fresh and hardened concrete.

14.2 MIXING CONCRETE

All concrete should be mixed thoroughly until it is uniform in appearance, with all ingredients evenly distributed. Mixers should not be loaded above their rated capacities and should be operated at approximately the speeds for which they were designed. Increased output should be obtained by a larger mixer or by additional mixers, not by speeding up or overloading the equipment on hand. If the blades of the mixer become worn or coated with hardened concrete, the mixing action will be less efficient. Badly worn blades should be replaced and hardened concrete should be removed periodically, preferably after each day's run of concrete.

If the concrete has been adequately mixed, samples taken from different portions of a batch will have essentially the same density, air content, slump, and coarse aggregate content. Maximum allowable differences in test results within a batch are given in appropriate American and Canadian standards.

14.3 READY MIXED CONCRETE

Ready mixed concrete is manufactured by three methods of mixing:

1. Central-mixed concrete is mixed completely in a stationary mixer and is delivered either in a truck agitator, a truck mixer operating at agitating speed, or a special nonagitating truck.

2. Shrink-mixed concrete is mixed partially in a stationary mixer and then completed in a truck mixer.

3. Truck-mixed concrete is mixed completely in a truck mixer.

CSA Standard A23.1-M, Clause 18.3.4.3, notes that when a truck mixer is used for complete mixing, 70 to 100 revolutions of the drum or blades at the rate of rotation designated by the manufacturer as *mixing speed* are usually required to produce the specified uniformity of concrete. No more than 100 revolutions at mixing speed should be used. All revolutions after 100 should be done at a rate of rotation designated by the manufacturer as *agitating speed*. Agitating speed is usually about 2 to 6 rpm, and mixing speed is generally about 6 to 18 rpm.

CSA A23.1-M, Clause 18.4.3, limits the time between batching and complete discharge to 2 hours. Mixers and agitators should be operated within the limits of volume and speed of rotation designated by the equipment manufacturer.

14.4 REMIXING CONCRETE

Fresh concrete that is left standing tends to stiffen before the cement has hydrated to its initial set. Such concrete may be used if, upon remixing, it becomes plastic enough to be consolidated in the forms. Under careful supervision a small increment of water may be added to delayed batches providing the following conditions are met (1) maximum allowable water-cement ratio is not exceeded; (2) maximum allowable slump is not exceeded; (3) maximum allowable mixing and agitating time (or drum revolutions) are not exceeded; and (4) concrete is remixed for at least half the minimum mixing time or number of revolutions.

Indiscriminate addition of water to make the concrete more fluid should not be allowed because this lowers the quality of the concrete. Remixed concrete will harden rapidly; therefore, concrete placed adjacent to or above remixed concrete may cause a cold joint.

14.5 TRANSPORTING AND HANDLING CONCRETE

Each step in transporting and handling concrete should be carefully controlled to maintain uniformity within the batch and from batch to batch so that the completed work is consistent throughout. It is essential to avoid separation of the coarse aggregate from the mortar or of water from the other ingredients.

Segregation at the point of discharge from the mixer can be minimized by providing a downpipe at the end of the chute so that the concrete will drop vertically into the center of the receiving bucket, hopper, car, or forms. Similar provisions should be made at the ends of all other chutes or conveyors.

All hoppers should be provided with a vertical drop at the discharge gate. When discharge is at an angle, the larger aggregate is thrown to the far side of the container being charged and the mortar is concentrated on the near side. Such segregation probably will not be corrected upon further handling of the concrete.

Many methods can be used to transport and handle concrete. They include the use of such diverse equipment as hoppers and chutes; buckets lifted by crane, tower hoist, or cableway; belt conveyors, railcars; trucks; pumps; wheelbarrow and carts operated over runways; boats or barges; and, in special cases, helicopters.

The method of transporting and handling concrete and the equipment used should not place a restriction on the consistency of the concrete. Consistency should be governed by the placing conditions. If these permit the use of a stiff mix, the equipment should be designed and arranged to facilitate transporting and handling such a mix. This may require larger chutes on steeper slopes, large discharge-gate openings, and modification of other features.

There have been continuous improvements in equipment for handling concrete. However, improved equipment is no guarantee of a uniform, high-quality job. A good mixture (with an adequate cement factor) and constant supervision are required to make sure that concrete fulfills expectations.

Chutes

Open-trough chutes should be of metal or metal-lined, preferably round-bottomed, and large enough to guard against overflow. Chute slopes neither flatter than 1 to 3 nor steeper than 1 to 2 are often recommended, but there is no objection to the use of steeper slopes for stiff mixes. The maximum or minimum slope should be determined by the condition of the concrete as discharged from the chute.

As stated previously, a downpipe should be provided at the end of the chute so that the concrete will drop vertically. The greatest objection to the use of chutes is that segregation occurs at discharge, but the use of a downpipe that is sufficiently long, at least 600 mm (24 in.), will eliminate or greatly reduce segregation.

Buckets

Buckets are made in different shapes and sizes, varying up to 9.0 mm³ (6.9 cu yd) for different applications. Some larger buckets, used primarily in massive work, have rectangular cross sections, but most buckets are circular. the concrete is released by opening a gate that forms the bottom of the bucket. for massive work the buckets often have straight sides with gates that open to the full area of the bottom. for most types of work, buckets having the lower part of their sides sloping to a smaller gate are usually preferred. Buckets preferred for small jobs have gates that can be regulated to control the flow of concrete and closed after only part of the concrete has been deposited. Gates may be operated manually or by hydraulic or pneumatic means.

Buckets are transported and handled by cranes, cableways, lift trucks, tower hoists, railcars, trucks, helicopters, boats, and barges. Regardless of the method used, care should be taken to prevent jarring and shaking; these may cause segregation, particularly if the concrete is relatively fluid.

14.6 DEPOSITING THE CONCRETE

Concrete should be placed as close as possible to its final position. In slab construction, placing should be started along the perimeter at one end of the work with each batch dumped against previously placed concrete. The concrete should not be dumped in separate piles and the piles then leveled and worked together. Nor should the concrete be deposited in big piles and then moved horizontally to its final position. These practices result in segregation because mortar tends to flow ahead of coarser material.

In general, concrete should be placed in horizontal layers of uniform thickness, each layer being thoroughly consolidated before the next is placed. Layers should be 150 to 500 mm (6 to 20 in.) thick for reinforced members, 375 to 500 mm (15 to 20 in.) thick for mass work—the thickness depending on the width between forms and the amount of reinforcement.

Concrete should not be moved horizontally over too long a distance within forms or in slabs. In some work, such as against sloping wingwalls or beneath openings in walls where concrete may have to be moved, the distance must be kept to a minimum. When concrete is moved horizontally in forms, water and mortar are forced ahead of the concrete, resulting in a poorer quality of concrete.

In walls the first batches in each lift should be placed at each end of the section; the placing should then progress toward the center. This method also should be used in placing beams and girders. In all cases, water should be prevented from collecting at the ends and corners of forms and along form faces. In sloping wingwalls water may collect along the sloping top surface, an area most vulnerable to weathering. However, if the top form boards of the sloping face are omitted, concrete can be placed directly in this section of wall. If necessary, boards forming the sloping surface may be placed as concreting progresses.

Drop chutes will prevent buildup of dried mortar on reinforcement and forms. If the placement can be completed before mortar dries, drop chutes may not be needed. The height of free fall of concrete need not be limited unless separation of coarse particles occurs, in which case a limit about 1 m (3 pt) may be adequate.

Concrete is sometimes placed through openings, referred to as windows, in the sides of tall narrow forms. When a chute discharges directly through the opening there is danger of segregation. A collecting hopper outside the opening permits the concrete to flow more smoothly through the opening and there is much less tendency to segregate.

When concrete is placed in tall forms at a fairly rapid rate, there is likely to be some bleeding of water to the top surface, especially with non-air-entrained concrete. Bleeding can be reduced by placing more slowly and by using concrete of a stiffer consistency. When practicable, the concrete should be placed to a level about 300 mm (12 in.) below the top of the form in high walls and an hour or so should be allowed for settling. to avoid formation of cold joints, concreting should be resumed before initial set occurs. It is good practice to overfill the form by 25 mm (1 in.) or so and to cut off the excess concrete after it has partly stiffened.

To avoid cracking due to settlement, concrete in columns and walls should be allowed to stand for at least two hours, and preferably overnight, before concrete is placed in slabs, beams, or girders framing into the columns and walls. Haunches and column capitals are considered as part of the floor or roof and should be placed integrally with them.

14.7 FINISHING CONCRETE SLABS

Concrete slabs can be finished in many ways, depending on the effect desired. Various colors and textures, such as an exposed-aggregate surface, may be called for. Some surfaces may require only striking off to proper contour and elevation, while in other cases a broomed, floated, or troweled finish may be specified.

One of the principal causes of surface defects in concrete slabs is finishing while bleed water is on the surface. *Any finishing operation performed on the surface of a concrete slab while bleed water is present will cause serious dusting or scaling.* The use of low-slump, air-entrained concrete with an adequate cement content and properly graded fine aggregate will minimize bleeding and will help ensure maintenance-free slabs.

Mixing, transporting, and handling of concrete for slabs should be carefully coordinated with the finishing operation. Concrete should not be placed on a subgrade or in forms more rapidly than it can be spread, struck off, consolidated, and bullfloated or darbied. In fact, concrete should not be spread over too large an area before strikeoff, nor should a large area be struck off and allowed to remain for a time before bullfloating or darbying.

14.8 STRIKING OFF

Screeding is a term often used to describe the process of striking off the excess concrete to bring the top surface to proper grade. The templet used is known as a straightedge; however, the lower edge may be straight or curved, depending on the surface requirements. When hand methods are used, the straightedge should be moved across the concrete with a sawing motion and advanced forward a short distance with each movement. There should be a surplus of concrete against the front face of the straightedge to fill in low areas as the tool passes over the slab. Allowing too great a surplus, however, may tend to leave high spots and hollows. Straightedges are sometimes equipped with vibrators that consolidate the concrete and assist in reducing the work of screeding. (Vibrating strike-offs are discussed under "Consolidating Concrete.")

Bullfloating or Darbying

To eliminate high and low spots and to embed large aggregate particles, a bullfloat or darby is used immediately after screeding. Generally, the bullfloat is used on areas too large to reach with a darby. For normal-weight, non-air-entrained concrete they should be magnesium.

Bullfloating or darbying must be completed before any excess bleed water accumulates on the surface. Care must be taken not to overwork the concrete; overworking will result in a less durable surface.

Although sometimes no further finishing is required, in most cases bullfloating or darbying is followed by one or more final finishing operations: edging, jointing, floating, troweling, and brooming. The final finishing operations should not begin until the water sheen is gone and the concrete will sustain foot pressure with only about 5-mm (¼ in.) indentation.

Edging and Jointing

Edging is required for isolation joints in most floor slabs and is common practice for outdoor slabs such as walks, drives, and patios. Edging produces a neat, rounded edge that prevents chipping or damage, especially when forms are removed. Edging also compacts and hardens the concrete surface next to the form where floating and troweling are less effective.

In the edging operation, the concrete should be cut away from the forms to a depth of 25 mm (1 in.), using a pointed mason trowel or a margin trowel. Then an edger should be held flat on the surface and run with the front slightly raised to prevent digging into the surface. Caution is necessary to prevent the edger from leaving too deep an impression. Edging may be required after each subsequent finishing operation.

Immediately after or during the edging operation, the slab should be jointed or grooved. Proper jointing practices can eliminate unsightly random cracks. Control joints can be made in the plastic concrete by using a hand groover or by inserting strips of wood, metal, or premoulded joint material into the concrete. Additional information on jointing is discussed under the heading "Making Joints in Floors and Walls."

Floating

After the concrete has been edged and jointed, it should be floated with a wood or metal hand float or with a finishing machine using float blades.

The purpose of floating is threefold: to embed aggregate particles just beneath the surface; to remove slight imperfections, humps, and voids; and to compact the concrete at the surface in prepration for succeeding finishing operations. The plastic concrete should not be overworked as this may cause an excess of water and fine material on the surface and result in subsequent surface defects.

Floating produces a relatively even (but not smooth) texture that has good slip resistance. Accordingly, it is often used as a final finish, especially for exterior slabs. Where such a finish is desired, it may be necessary to float the surface a second time.

Marks left by edgers and hand groovers are normally removed during floating unless they are desired for decorative purposes, in which case those tools should be rerun after final floating.

Troweling

Where a smooth, hard, dense surface is desired, floating is followed by steel troweling. No troweling should ever be done on a surface that has not been floated.

It is customary when hand-finishing large slabs to float and immediately trowel an area before moving the kneeboards. These operations should be delayed until after the concrete has hardened enough so that water and fine material are not brought to the surface. Too long a delay, of course, will result in a surface that is too hard to float and trowel. The tendency in a majority of cases, however, is to float and trowel the surface while the concrete is too soft and plastic. Premature floating and troweling may cause scaling, crazing, or dusting and will result in a surface with reduced wear resistance.

Spreading dry cement on a wet surface to take up excess bleed water is a bad practice that may cause crazing. When excess bleed water is on the surface, finishing operations should be delayed until the water either evaporates or is removed with a squeegee.

The first troweling may be sufficient to produce the desired surface free of defects. However, surface smoothness, density, and wear resistance can all be improved by timely additional trowelings. There should be a lapse of time between successive trowelings to permit the concrete to become harder. As the surface stiffens, each successive troweling should be made with smaller trowels. using progressively more tilt on the trowel blade. The final pass should make a ringing sound as the trowel moves over the hardening surface.

When the first troweling is done by machine, at least one additional troweling by hand is required to remove small irregularities. If necessary, tooled edges and joints should be rerun after troweling to maintain uniformity and true lines.

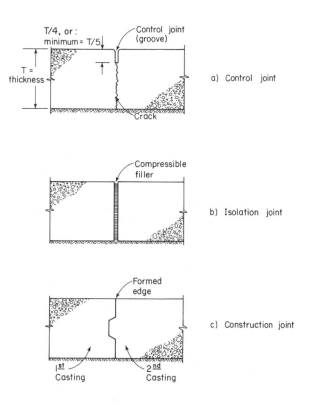

Fig. 14.1 The three basic types of joints used in concrete construction

14.9 MAKING JOINTS IN FLOORS AND WALLS

Three basic types of joints are commonly used in concrete construction (examples are illustrated in Fig. 14-1):

Isolation joints are used to separate different parts of a structure to permit both horizontal and vertical differential movements. For example, such joints are used around the perimeter of floors on ground and around columns and machine foundations.

Control joints provide for differential movement in the plane of a slab or wall. They are used to confine contraction caused by drying shrinkage. Control joints should be constructed so as to permit transfer of loads perpendicular to the plane of the slab or wall. If no

control joints are used in slabs on ground or in lightly reinforced walls, random cracks will occur when shrinkage produces tensile stresses in excess of the concrete's tensile strength.

14.10 WATER

Water as used in concrete should be fit to drink. When it is drawn from sloughs in the country, it should be tested for silt, bacteria, and so on—all of which would be injurious to the concrete.

The estimator must constantly keep in mind that water is an integral part of making concrete, and, as such, it may have to be purchased by piping from the city mains, or obtained by digging a well, by pumping from a river or slough, or by hauling it by water tanker. Any one of these methods will affect the cost and must be allowed for on the estimate. For many jobs, it is necessary to make a survey of the area. Even the simplest country job may require considerable lengths of ¾" hose to run a water supply from the nearest source of supply.

14.11 WATER-CEMENT RATIO

Mixed water and cement is known as cement paste. The whole concept of concrete making is based on the relationship of gallons of water to 1 sack of cement. The ratio of 5 gal of water to 1 sack of cement is known as a 5-gal paste. Similarly, the ratio of 6 or 7 gal of water to 1 sack of cement is known as 6- or 7-gal pastes, respectively. It would follow that the thicker (stronger) the paste, the stronger and more durable the concrete; and the thicker the paste the more expensive the concrete. With this in mind, the estimator would expect to find concrete mixed with a very strong paste to be used in the most psi-demanding places on a building. The concrete specifications always state the mix of concrete to be used in each section of the building. The estimator must examine the drawings and specifications very carefully to see where different psi mixes are required, since the cost of the concrete will vary with the different psi requirements.

When the cement paste is intimately mixed with the aggregates, every particle of sand and crushed rock must be completely coated with cement paste. Some architects specify the minimum allowable time for the mixing of every batch of concrete. This time must be given careful thought—especially since the number of men in the crew and the total time required to place the concrete will be governed by the mixture time allowed per batch. Assume a specification that calls for a 2-min mix for a 1-cu-yd delivery batch mixer as against another specification calling for a 1½-min mix for the same machine. The saving in crew time is 25 percent with the latter time.

14.12 STORAGE OF CEMENT

Since good cement is so dry and finely ground, it follows that it is very susceptible to dampness. If it goes lumpy while in storage and will not revert to its powered form when rolled in the sack, it should be discarded. The addition of moisture sets up a chemical reaction, and once this reaction has started the cement will not revert to its powdered form and so should be discarded entirely. On many jobs the inspector will insist that lumpy cement be hauled away from the building site.

Cement must be stored under very dry conditions. No amount of careful estimating can prevent a loss unless every precaution is taken to keep the cement (and indeed all building materials) secure against injurious hazards.

When it is necessary to haul cement onto the job some time before it will be used, it should be enveloped in polyethylene. First a wood platform should be provided about 0'-6" above the ground level. This may consist of a few pieces of 4 × 4 common lumber supporting any rough loose boarding such as subfloor material. Then the polyethylene sheet is placed over the platform and the stack of cement is enveloped by the sheet. *Remember that if you are using tarpaulins or polyethylene sheets, the cost will have to be allowed for on your estimate.*

14.13 BANK OR PIT-RUN AGGREGATES

Bank or pit-dug aggregate is the natural virgin product of sand and gravel as may be discovered in most places by digging a pit.

The common characteristics of this class of aggregate are:

(a) It usually contains far too high a ratio of fine-to-coarse aggregate (i.e., fine rock (sand) to rock over ¼ in. in size).

(b) It often is dirty and unfit for use in concrete without washing.

(c) It may contain organic or acid matter.

14.14 SURFACE AREA OF AGGREGATE

Example

Step 1: A 12" cube has a surface area of 12 × 12 × 6 sides = 864 sq in.

Step 2: A 6" cube has a surface area of 6 × 6 × 6 sides = 216 sq in.

Step 3: The cubic capacity of the 12" cube is equal to the sum total cubic capacity of eight 6" cubes.

Step 4: The surface area of eight 6" cubes is 6 × 6 × 6

sides, which is 216 sq in.; this times 8 cubes is a sum total surface area of 1,728 sq in.

It will be recalled that the determining factor in the making of concrete is the water-cement ratio (cement paste). The difference in quality between, say, a 4½-gal and a 7-gal paste is simply that in the latter case the yield of cement paste is 55 percent greater than in the former, and the latter is consequently a weaker paste and only suitable for restricted uses.

The finer the aggregate, the more surface area of aggregate to be covered.

14.15 DESIGN MIX BY VOLUME

In this chapter it is accepted that the most accurate method of designing concrete mixes is by weight; but very great quantities of concrete are placed every day in all countries by volume mix. This is especially true on small jobs which are too far from a ready-mix plant.

Very satisfactory concrete can be made with a volume mix, and there are many examples of concrete structures made this way that are still standing after many hundreds of years. As a notable example, the great dome of the Pantheon in Rome was built of concrete over 1800 years ago and is still standing.

Some building codes state, "If a proportion by volume of 1:2:4 (cement: sand: gravel) is mixed with just sufficient water to make a plastic mix, a 2000-psi concrete will result providing that the aggregates are good." The words "just sufficient water" imply that the sand may be wet and, therefore, just sufficient water would be less water than that required for very dry sand. Such a mix would be suitable for foundations for small light buildings, such as housing.

We must be aware that any of us could find ourselves in an undeveloped country having to do a job of building using such local labor, equipment, and materials as we find on hand. In any case, all estimators and builders must be very resourceful men.

14.16 ESTIMATING QUANTITIES OF DRY MATERIALS FOR A CONCRETE FLOOR AND FOR EACH BATCH OF THE MIXER TO BE USED

Example

You are referred to the completed work-up sheets after studying the following problem, which should be carefully followed step by step (see pp. 245 through 249).

Problem

A concrete floor 40'-0" × 42'-0" is to be placed with 0"-5" of finished concrete. The mix is to be by volume. Using a 3-cu-ft-capacity concrete mixer, a 5-gal paste and a 1:2¼:3 mix, estimate the following:

Allow for 1½ times as much material in the dry state as for wet cement.

1. the area of the floor;
2. the quantity of placed concrete required for the floor;
3. the total volume of materials in the dry state required to place the floor **(allow 1½ times the volume of dry materials to place the volume of concrete required)**;
4. the number of sacks of cement required to place the floor;
5. the number of cu ft of sand required to place the floor;
6. the number of cu ft of gravel required to place the floor;
7. the quantity of water required to place the floor;
8. the quantity of dry cement required per batch;
9. the quantity of sand required per batch;
10. the quantity of gravel required per batch;
11. the quantity of water required per batch;
12. the number of batches required to place the floor;
13. the time it would take to place the floor, allowing two min per batch and allowing at least one half-hour extra for starting and cleaning up afterwards;
14. the estimated cost to place the floor using the following price list:

Materials	Cement	$1.50 per sack	
	Sand	$3.40 per cu yd	take up to the
	Gravel	$3.85 per cu yd	next higher cu yd
Labor	1 man	$5.80 per hour	
	4 men	$4.40 per hour each	

In addition to these costs, an allowance must be made for lumber, screeds, and formwork, and possibly for constructing a wheelbarrow ramp. Make an allowance also for depreciation of plant, maintenance, and gas. If the project is on the outskirts of a town, you may have transportation costs for the crew.

14.17 MEASURING THE CORRECT VOLUMES PER BATCH

The resolved problem of the volumes of cement, dry sand, gravel, and water for a 5-gal paste and a 1:2¼:3 mix per batch for a 3-cu-ft-capacity mixer is $^{18}/_{25}$ sack of cement, $1^{31}/_{50}$ cu ft of sand, $2^{2}/_{25}$ cut ft of gravel, and $3^{3}/_{5}$ gal of water per batch. The standard American gal

weighs 8.377 lb and the Canadian Imperial gal weighs 10 lb. *See the following work-up sheets,* pp. 245 through 247 *and Table 14-1.*

Cement

To measure the correct amount of cement per batch, take an empty 5-gal drum and pour into it 68 lb of cement (63 lb of Canadian). This poundage in each case is $^{18}/_{25}$ of the sacks respectively. Calibrate a drum at the height to which it is to be filled with dry cement by indenting it with a nail set or marking it by a paint line. Keep this drum only for measuring the dry cement.

Sand

The volume of dry sand required will be $1^{31}/_{50}$ cu ft per batch. This may be measured by making a 1-cu-ft measuring box; use 1 full box measure plus $^{31}/_{50}$ (say, $^{3}/_{5}$) of another, or take $1^{31}/_{50}$ of the weight of 1 cu ft of sand ($1^{31}/_{50} \times 94$ lb). Calibrate a drum to measure the sand. One 5-gal drum may be used if calibrated to hold just one-half the quantity of required sand. Use two measures for this, since it would be too heavy for one man to be lifting $1^{31}/_{50}$ cu ft of sand all day. Sand weighs approximately 94 lb per cu ft. Be certain that the men do not guess at the quantities.

14.18 MEASURING THE CORRECT VOLUME OF DRY MATERIALS AND WATER PER BATCH

Problem

Assuming a concrete foundation is to be placed using an 11-S mixer with a specified 7-gal paste and a 1:3:5 mix, how much water, cement, sand, and gravel will be required per batch?

Example

Step 1: An 11-cu-ft side-delivery concrete mixer will require 1½ times its wet delivery capacity of dry materials to be fed into it per batch.

Step 2: $11 \times 1½ = 16½$ cu ft of dry materials to produce 11 cu ft of wet concrete.

Step 3: The mix is 1:3:5 with a 7-gal paste. The total units of dry materials per batch are 9, of which cement is one.

Step 4: $16½ \div 9 = 1^{5}/_{6}$ *sacks of cement per batch.*

Step 5: It will require 3 times as much sand as cement: $1^{5}/_{6} \times 3 = 5½$ *cut ft. of dry sand per batch.*

WORK-UP SHEET

DATE _____

SHEET No. _1_ OF _3_

BUILDING	LOCATION	ARCHITECT	ESTIMATOR
Concrete Floor	40'-0" x 42'-0" x 0-5" deep	Olsen Jensen & Assoc.	J. Smith

1. Area of floor | 40'-0" x 42'-0" = 1680 sq. ft. | 5 gal. paste 1-2¼-3 | 1680 sq ft

2. Concrete Volume | Place 0'-5" deep | $\frac{1680 \times 5}{12}$ = 700 cu. ft. | 700 cu. ft.

3. Total Dry Materials: Add ½ extra to fill the voids

700 × 1½ =   1050 cu. ft.   1050 cu. ft.

4. Cement: 5 gal. paste and a 1-2¼-3 mix

The total number of dry units required
is 1 + 2¼ + 3 = 6¼

1050 ÷ 6¼ = $\frac{\cancel{1050}}{\cancel{25}}^{42} \times 4$ =   168 cu. ft.   168 sacks cement

5. Sand: It requires 2¼ times as much
sand as cement
168 × 2¼ = $\frac{\cancel{168}}{\cancel{4}}^{42} \times 9$ = 378 cu. ft.   378 cu. ft. sand

6. Gravel: It requires 3 times as much gravel
as cement
168 × 3 =   504 cu. ft.   504 cu. ft. gravel

7. Water: It requires 5 gal. of water for
each sack of cement
168 × 5 =   840 gals.   840 gals. water

Check the total volume of dry materials required:

168 cu. ft. (sacks of cement)
378 cu. ft. sand
504 cu. ft. gravel

Total dry mix required 1050 See item No. 3 above

WORK-UP SHEET

DATE _____

SHEET No. _2_ OF _3_

BUILDING	LOCATION	ARCHITECT	ESTIMATOR
Concrete Floor	40'-0"×42'-0"×0'-5" deep	Olsen Jensen Assoc	J. Smith

Quantities per batch:

(X) A 3 cu. ft. capacity mixer required $4\frac{1}{2}$ cu. ft. of dry materials per batch. The mix is 5 gal paste and $1-2\frac{1}{4}-3$

8 Cement:

$4\frac{1}{2}$ divided by the number of units in the batch $= 6\frac{1}{4}$

$4\frac{1}{2} \div 6\frac{1}{4} = \frac{9}{2} \times \frac{4}{25} = \frac{18}{25}$ cu. ft. | $\frac{18}{25}$ cu. ft. cement

9 Sand:

It requires $2\frac{1}{4}$ times as much sand as cement $\frac{18}{25} \times 2\frac{1}{4} = \frac{18 \times 9}{25 \quad 4} = \frac{81}{50} = 1\frac{31}{50}$ cu. ft. | $1\frac{31}{50}$ cu. ft. sand

10. Gravel:

It requires 3 times as much gravel as cement $\frac{18}{25} \times 3 = \frac{54}{25} = 2\frac{4}{25}$ cu. ft. | $2\frac{4}{25}$ cu. ft. aggregate

11. Water:

It requires 5 gallons of water for one sack of cement, and it requires $\frac{18}{25}$ of 5 gals of water for each batch.

$\frac{18}{25} \times 5 = 3\frac{3}{5}$ gals of water per batch | $3\frac{3}{5}$ gals water

Check the total dry materials required per batch:

$\frac{18}{25}$ cement plus $1\frac{31}{50}$ sand plus $2\frac{4}{25}$ gravel

$= \frac{(36+31+8)}{50} + 3 = 4\frac{1}{2}$ cu. ft per batch

(X) See Quantities per batch above

12   Number of batches required to place the floor: The number of batches required is equal to the total volume of the placed concrete floor divided by the wet capacity of the mixer

$700 \div 3 = 233\frac{1}{3}$ say 234 | 234 batches

WORK-UP SHEET

BUILDING	LOCATION	ARCHITECT	ESTIMATOR
Concrete Floor	40'-0" x 42'-0" x 5" deep	Olsen Jensen Assoc.	J. Smith

13 Time required to place the floor:
234 batches @ 2 min. per batch = 468 min

468 ÷ 60 = 7 hrs. 48 min.
Say 9 hrs including start and clean up   9 hours

14 The estimated cost to place the floor:
Cement: 168 sacks @ $1.50 per sack. $252.00 cement

Sand: 378 cu. ft   27)378 (14 cu. yds
                         27
                        108

14 cu yds @ $3.40 per cu yd $47.60 $47.60 sand

Gravel: 504 cu. ft. 27)504 (18⅔ Say 19 cu yd.
                         27
                        234
                        216
                         .18

19 cu yds of gravel @ $3.85 per cu yd $73.15 Gravel

Labor: 1 man @ $5.80 per hr. = $5.80
        4 men @ $4.40 per hr = 17.60
                              23.40

9 crew hrs. @ $23.40 per hr $210.60 $210.60 Labor

	Summary:	Cement	252	00	
		Sand	47	60	
		Gravel	73	15	
			372	75	
		Labor	210	60	
			$583	35	$583.35 Cost

Note: This estimate does not cover
labor and materials for forms
runways, equipment, traveling
time, overheads and profit.
For overheads and profit add 25%

Table 14-1

Suggested Proportions of Water to Cement for Various Kinds of Concrete Work and Trial Mixes

Kinds of Work	Add U.S. gal of water to each sack batch if sand is:			Trial mixture			Materials per Cu yd of Concrete*		
					Aggregates			Aggregates	
	Very wet	Wet	Damp	Cement Sacks	Fine cu ft	Coarse cu ft	Cement Sacks	Fine cu ft	Coarse cu ft
5-gal paste for concrete subjected to severe wear, weather, or weak acid and alkali solutions									
Toppings for two-course work	3¾	4	4½	1	1¾ (Max size 1½″)	3	7¼	13	22
One-course industrial, creamery and dairy plant floors	3½	4	4½	1	2 (Max size 1½″)	2¼	8	16	18
Thin sections of dense, strong concrete	3½	4	4½	1	2 (Max size ⅜″)	1¾	9	18	16
6-gal paste for concrete to be watertight or subjected to moderate wear and weather									
Watertight floors such as industrial plant, dairy barn, basement, etc.	4¼	5	5½	1	2¼ (Max size 1½″)	3½	6	13½	21
Watertight basement walls. All watertight concrete for storage tanks, septic tanks, swimming pools, etc.	4	4¾	5½	1	2½ (Max size ¾″)	2½	6½	16½	16½
Concrete subjected to moderate wear or frost action, such as walks, driveways, tennis courts, etc. Reinforced structural beams, columns, slabs, etc.	4	4¾	5½	1	2½ (Max size ⅜″)	2	7½	18¾	15
7-gal paste for concrete not subjected to wear, weather, or water									
Foundation walls, footings, mass concrete, etc.	5	5½	6¼	1	2¾ (Max size 1½″)	4	5¼	14½	21
	4½	5½	6¼	1	3¼ (Max size ¾″)	3	5½	18	16½
	4½	5½	6¼	1	3¼ (Max size ⅜″)	2¼	6½	21	14¾

* Quantities are estimated on wet aggregates, using trial mixes and medium consistencies (5 in. slump).
Quantities will vary according to grading of aggregates and consistency of mix.
Change proportions of fine and coarse aggregate slightly if necessary to get a workable mix.
Quantities are approximate. No allowance has been made for waste.

NOTE: If concrete aggregates are sold in your locality by weight, you may assume for estimating purposes that a ton contains approximately 22 cu ft of sand or crushed stone; or about 20 cu ft of gravel. For information on local aggregates consult your building material dealer. (*Courtesy of the Portland Cement Association*)

(See American and Canadian gallons, page 224)

Step 6: It will require 5 times as much gravel as cement: $1\frac{5}{6} \times 5 = 9\frac{1}{6}$ *cu ft of gravel per batch.*

Step 7: Check the total cu ft of dry volumes required per batch: $1\frac{5}{6}$ cement, $5\frac{1}{2}$ sand, and $9\frac{1}{6}$ gravel added together equal $16\frac{1}{2}$ cu ft of dry materials (see Step 2).

Step 8: It requires 7 gal of water per sack of cement: $7 \times 1\frac{5}{6}$ sacks of cement equals $12\frac{5}{6}$ gal per batch.

14.19 SIZES OF CONCRETE MIXERS

Some of the standard types of concrete mixers are as follows:

Type of mixer	Cu ft of wet concrete capacity
3-S	3
6-S	6
11-S	11
16-S	16

Note: 3-S means 3 cu ft side delivery and so on.

14.20 AMERICAN AND THE OLD CANADIAN GALLON

Note: This problem has been worked out using the standard gallon of the United States, which is the old English wine gallon containing 231 cu in. and weighing 8.377 lb. The Canadian Imperial gallon contains 277.418 cu in. and weighs 10 lb. For Canadian use, in Step 8 *express* $12\frac{5}{6}$ *American gal in pounds and ounces and use the same weight of water per batch.*

Problem

How much water, cement, sand, and gravel will be required per batch of concrete when 5-gal paste and a $1:2\frac{1}{4}:3$ mix in a 16-S concrete mixer are used?

14.21 RUNWAYS AND RAMPS

Nearly all concrete jobs require runways, ramps, lumber for screeds, and so on. Every job has its own estimating problems, so that a survey of the area may be required.

The following points should be remembered:

(a) Estimate for the lumber, nails, ties, and labor needed to construct runways and ramps.

(b) Allow material for screeds; this lumber cannot be reused.

(c) Allow for scaffolds.

(d) Remember that formwork must be strong enough to withstand the concrete pressures so that it will not deflect under load.

(e) Floors will have to support wet concrete, power buggies, men, and tools.

(f) Estimate whether or not any of the lumber used in (a), (b), or (c) may be reused. Most specifications clearly state that all lumber on the job may be used for one purpose only.

14.22 MACHINE MIXING TIME

The machine mixing time is reckoned from about one min per batch for a small-batch mixer up to two min per batch for a large unit. Opinions vary, but architects may specify the minimum mixing time for each batch. Study this mixing time very carefully indeed, since the whole operation of placing concrete is entirely dependent upon the delivery time per cu yd from the batch mixer. The maximum allowable time for placing concrete is usually about forty five min after mixing.

14.23 WEIGHT OF CONCRETE

One cu ft of concrete weighs about 140 lb. If the first floor of a building is $40'-0'' \times 60'-0''$ and is placed to a depth of $0'-6''$ with concrete, the weight of supported concrete would be 168,000 lb or 84 tons. This does not include the weight of the forms, nor of the men, nor of the tools, nor of the loaded buggies during placing. The estimator must get the "feel" of a job when estimating, in the same way that the job superintendent must be aware of all the hazards inherent in the physical work to be performed.

No amount of careful estimating can offset inefficient job management. The chief ingredient of estimating and job management is the morale of all persons involved. The estimator and the job superintendent should have comparable job knowledge.

14.24 CURING

The object of curing is to keep the newly placed concrete from either drying out too fast or, even more important, to keep it from freezing. In the former case, it may have to be covered with polyethylene sheets; or it may be specified that the concrete must be cured by the application of a very fine water spray on the surface for seven to twenty eight days, according to the nature of the job. In the latter case, it may have to be protected with straw, burlap, tarpaulins, and so on. In all cases both labor and materials must be estimated.

Some of the agents for curing are:

(a) heat to repel freezing conditions;

(b) water spraying for warm conditions;

(c) continuous water-saturated covering of sand;

(d) wet burlap;

(e) spraying the flat surfaces with water and covering with balsawood panels;

(f) water-sprayed surfaces covered with cotton mats;

(g) water-sprayed surfaces covered with polyethylene sheets;

(h) sealing compounds

Most specifications call for exposed surfaces of newly placed concrete to be kept moist for a minimum of seven days; some require fourteen days.

Water is applied to troweled surfaces as soon as the concrete has set sufficiently so that the cement will not wash away. For untroweled surfaces, water may be sprayed on as soon as the forms are removed.

The beginning estimator is strongly urged to take a course on design and control of concrete mixtures.

14.25 DEMURRAGE

Great progress has been made, especially during this century, in the design and mix of concrete. In all our cities, concrete manufacturers will deliver (ready-mixed) designed concrete to meet specified requirements. There is a maximum free time allowed for unloading ready-mixed concrete from the delivery vehicle. After the allowed free time has elapsed, a demurrage charge is made for each extra portion of an hour delay. For small jobs and for delivery of ready-mixed concrete to ordinary householders, the merchants will deliver in units of 1 cu yd and to the next higher ⅓ cu yd. They will also place and finish the concrete for an inclusive price. *Mass concrete is charged by the cu yd, and finished concrete— as for concrete floors and sidewalks and so on—is charged by the sq ft placed and finished.*

QUESTIONS

1. Define a weight-designed concrete mix.

2. Define a volume-designed concrete mix.

3. What is the function of a field batching plant?

4. How should cement be stored?

5. Why should cement be stored as stated in the answer to Question 4?

6. Define water cement ratio in concrete manufacturing.

7. Define sand bulking.

8. How would a knowledge of sand bulking be of advantage to an estimator?

9. State two ways in which the amount of free water in sand would affect the design of a concrete mix.

10. Which would bulk more, fine sand or coarse sand?

11. What is the advantage of using fine "manufactured" aggregate over pit run sand when making concrete?

12. Why should aggregates be free from an excess of long slivers?

13. What is the ideal in shaped aggregates for cement?

14. How would you make a silt test on pit run sand?

15. What is the approximate weight of a sack of cement in your area?

16. What is the cubic capacity of a sack of cement in your area?

17. What is the approximate weight of 1 cu ft of concrete?

18. What is the approximate weight of 1 cu yd of gravel?

19. What is the weight of 1 gal of fresh water in your area?

20. List seven methods of curing concrete.

21. What is a demurrage charge?

CONCRETE FORMWORK AND REINFORCING STEEL

In this chapter you will learn the method of estimating wood formwork to receive wet concrete. This is followed by a discussion of the advantages of using semipermanent forms, their cost in materials, and labor for making, removing, and reconditioning. The chapter concludes with data for estimating reinforced steel in concrete, and the methods for this estimating.

15.1 CONCRETE FORMWORK: FOOTINGS

The amount of lumber required for concrete footings is reckoned by the amount of lumber touching wet concrete, plus stakes, spreader ties, nails, and waste.

Example

Estimate the amount of lumber required for the footing forms for the plan in Fig. 8.5, p. 164. The concrete footings are to be 1'-10" wide and 0'-6" deep (see Fig.

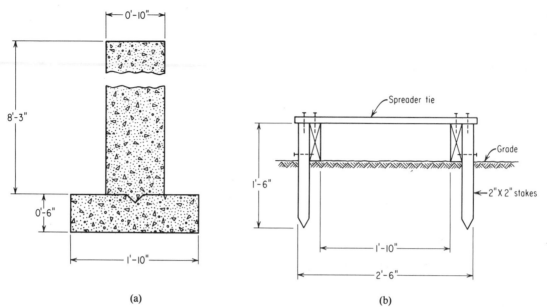

Fig. 15.1 (a) Concrete Footing and wall (b) Footing forms.

15.1). They are to be made with 2 × 6 stringers nailed to 2 × 2 stakes with 1 × 3 spreader ties 4'-0" *on center* (OC).

Step 1: Make a freehand sketch of the forms and name the parts. *You will never waste your time by making neat sketches when estimating.*

Step 2: Estimate the amount of lumber touching wet concrete. This is a prime consideration in all concrete formwork.

Step 3: Perimeter times the depth of the footing times 2 (two stringers) equals the estimated area in sq ft of the lumber touching wet concrete.

Step 4: The perimeter of Fig. 8.5 is 210 lin ft: 210'-0" × 0'-6" × 2 sides = 210 sq ft of lumber touching wet concrete.

Step 5: Add 5 to 10 percent for waste according to ground conditions. Assume 7 percent waste:

$$\frac{210 \times 107}{100} = 225 \text{ sq ft (approximately)}$$

$$225 \text{ sq ft of } 2 \times 6 = 450 \text{ bd ft}$$

$$28/2 \times 6 - 16'-0" \text{ is 448 bd ft.}$$

Step 6: Estimate the 18"-long 2 × 2 stakes 4'-0" OC. Perimeter divided by the OC of the stakes is ²¹⁰⁄₄—say, 53 pairs of stakes each 18" long: 53 × 2 × 1½ = 159 lin ft of 2 × 2. Allow 10 to 15 percent for waste according to ground conditions; assume 12 percent waste:

$$\frac{159 \times 112}{100} = 178 \text{ lin ft of } 2 \times 2.$$

Step 7: Estimate 1 × 3 spreader ties 4'-0" OC. It will require 53 spreader ties each 2'-6" long.

$$53 \times 2½ = 133 \text{ lin ft of } 1 \times 3$$

Assume 16 percent waste:

$$\frac{133 \times 116}{100} = 155 \text{ lin ft of } 1 \times 3$$

Step 8: Estimate the number of pounds of nails required; allow six 2½" common nails per 4'-0" OC of stakes: 53 × 6 = 318; add 10 percent for waste—say 350 nails for the job. Convert nails to pounds (see tables, p. 158): one hundred 2½" common nails = 1 lb.
Note: In foundation formwork there would always be pier footings, chimney footings, and so on. The drawing must be carefully read for all such items. If they are forgotten, there will be a loss in material cost, labor, and profit.

Step 9: Summarize:

Step 1: order 28/2 × 6—16'-0" stringers	448 bd ft
Step 2: order 178/lin ft of 2 × 2—say,	60 bd ft
Step 3: order 155/lin ft of 1 × 3—say,	40 bd ft
Step 4: order 3½ lb of 2½" common nails.	

Problem

Using local prices for labor and materials, estimate the cost of the wooden foundation formwork for the house design on pp. 302-303. Assume that the drawings call for a centrally placed concrete foundation footing supporting a 2 × 6 frame wall. Both the perimeter and the centrally placed foundations are 1'-10" × 0'-6".

The salvage value of the 2 × 6 stringers is equal to no more than three uses, at the most.

Labor: A carpenter and helper will erect about 275 lin ft of light-construction 2 × 6 foundation formwork (ready for placing concrete) in 8 hr. Allow 35–40 lin ft per hr.

Problem

Using local prices, estimate the cost to the nearest higher ⅓ cu yd of concrete required for the foundations in the problem on page 256. To estimate the cu ft of concrete for small footings, take the perimeter of the building times the width times the depth in ft. Convert to cu yd.

Problem

Examine the plans and specifications for any modern residence in your own locality and make a labor and materials estimate for:

(a) removal of topsoil to a depth of 0′-8″, stacked on site;

(b) excavations to depth required;

(c) footing forms;

(d) concrete foundations.

15.2 CONCRETE FORMWORK: BASEMENT WALLS

There are many types of semipermanent patented forms used in concrete work. The following example is for a contractor making his own forms for repeated use by his own men. The number of times that such forms may be used depends upon the organizing ability of the job superintendent. Some men would only get five uses out of them while others would use them up to fifty times. For the purpose of this exercise, it is assumed that the forms will be used thirty times.

Specifications

The specifications for a set of semipermanent ¾″ plywood forms framed with 2 × 4 dimension lumber is as follows:

(a) 80 forms 4′-0″ × 8′-0″ SIS ¾″ plywood (SIS means solid one side);

(b) 16 forms 2′-0″ × 8′-0″ SIS ¾″ plywood;

(c) 8 form 1′-0″ × 8′-0″ SIS ¾″ plywood;

(d) provide one top plate and one bottom plate for all

forms, and space the studs 16″ OC for the 4′-0″ × 8′-0″ forms; all others to have stud spacings of 12″ OC;

(e) construction-grade dimension lumber shall be used;

(f) frames shall be secured with 3½″ common nails;

(g) plywood shall be secured to frames with 2½″ common nails;

(h) sheet-metal strapping shall be secured to all outside corners of forms (see Fig. 15.2).

15.3 SEMIPERMANENT FORM: ESTIMATING

Use the current labor and materials costs in your locality and estimate the following problem.

Making Semipermanent Concrete Forms

Allow:

(a) 20 hr of carpenter's time per 1,000 sq ft of completed forms;

(b) 5 hr of carpenter's helper's time per 1,000 sq ft of completed forms;

(c) 10 lb of 3½″ common nails per 1,000 bd ft (for the 2 × 4);

(d) 20 lb of 2½″ common nails per 1,000 sq ft (for plywood);

(e) 20 sq ft of galvanized sheet cut for strapping the corners of the frames per 1,000 sq ft of frames.

15.4 STRIPPING AND RECONDITIONING OF FORMS

As soon as the forms are taken from the concrete, they should be wire-brushed, cleaned of all loose and clinging concrete, repaired, oiled, and level-stacked ready for shipping or carrying to the next job. Note carefully that no allowance is made in this exercise for shipping costs from job to job. This must be allowed for according to distances and conditions.

Allow:

(a) 3 hr of carpenter's time per 1,000 sq ft for reconditioning;

(b) 6 hr of carpenter's helper's time per 1,000 sq ft for reconditioning.

Erecting

Allow:

(a) 6 hr of carpenter's time per 1,000 sq ft of forms;

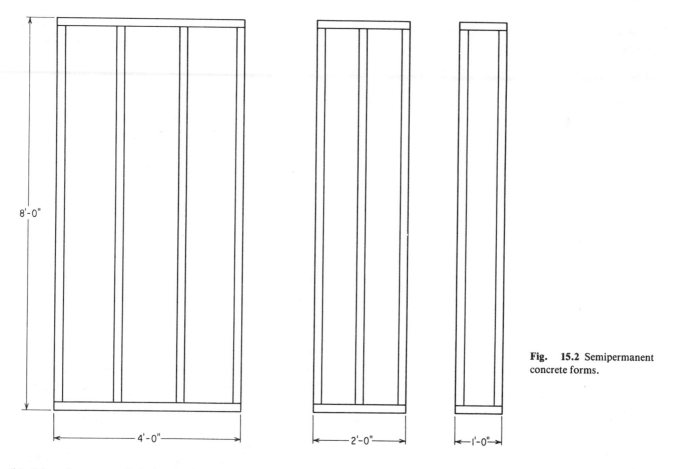

Fig. **15.2** Semipermanent concrete forms.

(b) 6 hr of carpenter's helper's time per 1,000 sq ft of forms;

(c) 200 form ties per 1,000 sq ft of outside wall area—unit of purchase is per 100 or 1,000.

When wire is used to tie the forms, purchase it by the roll. Use a minimum no. 8 gage wire.

Estimate:

(1.) the total labor and materials cost for making the complete set of forms;

(2.) the cost of using the forms for the tenth time—include the cost for stripping and reconditioning once;

(3.) the cost of using the forms for the twentieth time—include the cost for stripping and reconditioning once;

(4.) the cost of using the forms for the thirtieth time—include the cost for stripping and reconditioning once;

(5.) the cost per sq ft of using the forms 30 times—this estimate includes erecting once, stripping once and reconditioning, but does not include shipping from job to job.

When this type of information has been obtained and proven correct by analyzing a number of jobs, it is entered on a code-numbered card and filed in the estimating or cost-accounting office. A code-numbered card should be made out for every job operation and should show the time that a man, a gang, or a machine would take to accomplish an operation or part of an operation. Examine the two specimen cards shown.

Code no. 501	F
2 × 6 Footing forms Allow for using the stringers three times. Set up, remove, and clean 100 lin ft.	
Carpenter time	1 hr
Carpenter's helper's time	1 hr

Code no. 502	F
Semipermanent ¾″ plywood forms with 2 × 4 frames and reinforced metal corners. Using thirty times including erecting and reconditioning once. Cost per sq ft _____	

There is no particular size for these cards—they should, however, be in standard sizes.

Small Contractors use the card-index system for filing the actual time that it would take a man or gang to accomplish a given unit of work in 1 hr. If the workmen's rate per hour is increased, then the cost per unit of work done by a man or gang in 1 hr is similarly increased. Some cards (but not many) are made out with a priced unit—for example, the estimate of cost per sq ft for semipermanent concrete forms. Where this is done, the cards should be reviewed at least once a year.

Large Contractors use the card-index system in the cost-accounting department. Large engineering jobs will be estimated against a quantity estimate, which is provided by the architect and consulting engineers together with the drawings and specifications. With this method, the number of units of each item is stated. Assume that the quantity estimate had listed 559 cu yd of concrete. The contractor would include in each single unit of 1 cu yd of concrete the cost of the cement, sand, aggregate, additives, water, ramps, equipment (rented or owned), labor, fringe benefits for labor, workmen's compensation, job overheads, and profit. The actual cost of 1 cu yd of concrete would then be multiplied by the number of units of concrete required—in this case 559, and an error in one unit would be multiplied by the number of units.

All costing by either large or small companies requires a rigorous analysis. Once these costs are established and filed, the time required for future estimates is considerably reduced.

The code number shown at the top left of the card is given by the estimating or cost accounting department. (See code no. 500, page 259.)

The use of the code number is as follows:

(a) The foreman or job timekeeper should make out a daily or weekly time sheet for every man on the job. This should show the code number of the job on which each man has been working and the number of hours. This time period is analyzed by the estimators to keep their records up-to-date. If the job shows a loss, it would immediately be reflected in this analysis and the job superintendent would be so informed.

(b) Each type of material used on the job is ordered under its own code number. This is analyzed by the procurement section, and waste would be reflected immediately.

(c) An operation-by-operation analysis is made of the job until completion.

Most estimators claim that they know how much the materials for a building will cost. The real test is to know how much the labor for the building will cost.

15.5 CONCRETE WALLS: ESTIMATING

Estimate the number of cu yd of concrete required for the basement walls in Fig. 8.5, p. 164. The walls are to be 0″-10″ thick and 8′-0″ high.

Example

Step 1: Take the perimeter times the thickness and height of the wall in feet:

$$210 \times \frac{5}{6} \times 8 = 1,400 \text{ cu ft}$$

Step 2: Deduct four corners, which have been taken twice:

$$0'\text{-}10'' \times 0'\text{-}10'' \times 8 \times 4 \text{ corners}$$

$$\frac{5}{6} \times \frac{5}{6} \times 8 \times 4 = 23 \text{ cu ft (approximately)}$$

Step 3: $1,400 - 23 = 1,377$ cu ft. Reduce to 51 cu yd.

For small jobs (with offset walls) where all the walls are the same thickness and height, each internal corner will cancel out one external corner in cubic capacity of concrete required (see Fig. 15.3).

For large commercial jobs, it is usual for estimators to take off the quantities of concrete, formwork, and reinforcing at the same time: wall-for-wall, column-for-column, and beam-for-beam. This method lessens the risk of forgetting one of the items. Use the extension columns on the general estimate sheet and head them for concrete, forms, and reinforcing.

15.6 REINFORCING STEEL FOR CONCRETE

Your A.I.A. or R.A.I.C. file would not be complete without the publications made available by the research of the Concrete Reinforcing Steel Institute, by whose courtesy Table 15.1 is reprinted.

15.7 REINFORCING-STEEL CONTRACTORS' SPECIFICATIONS

The following clauses are typical extracts from the specifications for the reinforcing-steel contractor:

(a) Reinforcing steel shall be stored on racks or skids to protect it from dirt and to keep its fabricated form.

(b) All reinforcing steel shall be placed by experienced steel men and shall be wired in position and shall be approved by the architect or his representative before concrete is placed.

(c) Reinforcing bars shall be of medium-grade deformed steel suitable for a working stress of 20,000 psi and shall conform to ASTM and CESA standards.

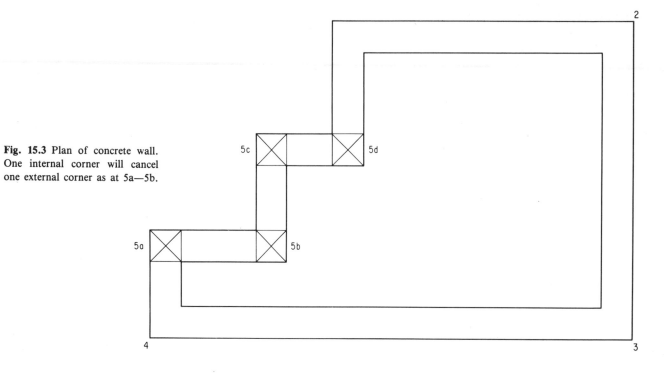

Fig. 15.3 Plan of concrete wall. One internal corner will cancel one external corner as at 5a—5b.

Table 15.1

Sizes and Weights of Reinforcing Steel Rods

ASTM STANDARD REINFORCING BARS				
BAR SIZE	WEIGHT	NOMINAL DIMENSIONS—ROUND SECTIONS		
DESIGNATION	POUNDS PER FOOT	DIAMETER INCHES	CROSS SEC- TIONAL AREA SQ. INCHES	PERIMETER INCHES
#3	.376	.375	.11	1.178
#4	.668	.500	.20	1.571
#5	1.043	.625	.31	1.963
#6	1.502	.750	.44	2.356
#7	2.044	.875	.60	2.749
#8	2.670	1.000	.79	3.142
#9	3.400	1.128	1.00	3.544
#10	4.303	1.270	1.27	3.990
#11	5.313	1.410	1.56	4.430
#14	7.65	1.693	2.25	5.32
#18	13.60	2.257	4.00	7.09

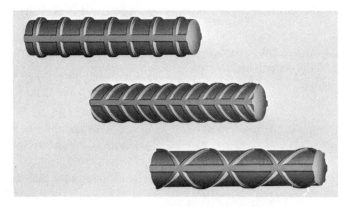

Fig. 15.4 Various types of deformed bars.

Note that ASTM *is American Society for Testing Materials and* CESA *is Canadian Engineering Standards Association.*

(d) Reinforcing steel shall be bent cold and shaped as shown or required, accurately spaced and located in forms and wired and secured against displacement before concrete is placed.

(e) Place reinforcing so that the distance from the face of the steel to the nearest face of the concrete is not less than 1 diameter nor in any case less than the following:

Footings	3″	Walls	2″
Columns	1½″	Slabs	¾″
Beams	1½″		

(f) Surface of bars shall be absolutely clean and free from mill scale, loose rust, oil, paint, and so on. Wire for tying shall be 18 U.S.S.G. annealed. *(Note:* U.S.S.G. *is United States Standard Gage.)*

(g) Bend horizontal wall steel around corners and continue 40 diameters of the bar. Support reinforcing on steel chairs and space with bar spacers. Support footing steel on brick or stone.

(h) Necessary splices not shown on the drawings shall be made by lapping and wiring adjacent bars. Splices and adjacent bars shall be lapped at least 24 bar diameters.

15.8 ESTIMATING REINFORCING-STEEL BARS

As stated in the foregoing table, reinforcing bars are determined by their designated numbers, but the weight of steel per lin ft for each bar number is still the same as it was by the old method of describing them by the diameter of the bars. Thus the old bar diameter of ¾″ is now known as no. 6 bar (six-eighths).

Figure 15.4 shows six types of deformed reinforcing bars. The deformations are to prevent slippage of the bars under tension in the concrete. Reinforcing bars are priced by weight.

Example

Step 1: List the reinforcing bars required by the denominated number of the bar.

Step 2: From the drawing, list the number of denominated bars that are of equal length.

Step 3: Extend the requirements of denominated bars to lin ft.

Step 4: Read the tables for the weight per lin ft of denominated bar required and extend to total weight. Thus:

$$\begin{array}{lll} \text{Bar no. 6} & 14/16'\text{-}0'' = & 224 \\ & 12/10'\text{-}0'' = & 120 \\ & 6/\ 8'\text{-}0'' = & \underline{\ 48} \\ & & \overline{392}\ \text{lin ft} \end{array}$$

Bar no. 6 weighs 1.502 lbs per lin ft
$392 \times 1.502 = 588.784$ lbs

Step 5: Check your price list for the local warehouse price of all denominated bars per lb. *Note that the smaller the bar, the higher the relative price.*

Not only is the intial cost of the smaller sizes of reinforcing bars greater than the large bars, the smaller bars also cost more to offload and to set up in forms. A man can carry and place a no. 9 bar 16′-0″ long as easily as he can carry and place a no. 3 rod of the same length.

$$1/16'\text{-}0''\ \text{no. 9 rod weights 54.4 lb}$$
$$1/16'\text{-}0''\ \text{mo. 3 rod weighs 6.016 lb}$$

In addition to this, the no. 3 bar is much easily damaged and bent in offloading than the no. 9 bar, which is quite rigid.

Reinforcing Bars: Setting and Tying

On small jobs, a contractor's own men should place and tie about 700 to 900 lb of reinforcing bars per man per 8-hr day. Most large jobs are let to subcontractors who will give an inclusive price to supply and fix.

15.9 CONCRETE COLUMN FOOTINGS SCHEDULE

On one of the drawings a schedule of column footings will be shown, which must be studied together with the specifications. An example of the extracts on a schedule follows:

Schedule of Reinforced Concrete Footings		
Footing no.	Footing size	Footing reinforcement
A-1 A-2 A-4	36″ × 36″ × 12″	6/no. 4 φ rods EWH 6″ cc
B-1 C-1	48″ × 30″ × 12″	no. 4 φ rods 6″ cc EWH

Notice that on the schedule each footing is numbered. These footings must be located and counted on the drawing to check that they do indeed comply with the total number and sizes required. Note the symbol ϕ for reinforcing rods; EWH means "each way horizontally," and cc means "center-to-center" (see Fig. 15.5).

The specifications must also be studied. An extract from the specifications for a simple warehouse follows.

15.10 FORM LUMBER SPECIFICATIONS

It is not usual for the formwork of concrete to be shown on the drawings, but a careful study should be made of the specifications that will give the minimum requirements of the type of formwork allowable. This must be complied with and may read as the following example:

1. All formwork to be of fir plywood, spruce, fir ship-lap, or boards.
2. All form lumber to be dressed on four sides.
3. All form lumber to be new when brought to the job.
4. All form lumber to be of sufficient strength to carry the load of wet concrete without deflection.
5. All form lumber may only be used once on the job.

Wood Forms

Lumber used in forms shall be dressed to uniform width and thickness and shall be free from loose knots or other defects. Joints in forms shall be horizontal or vertical. Lumber once used in forms (if permitted to be used more than once) shall have nails withdrawn and surfaces that will be in contact with concrete shall be thoroughly cleaned before being used again. Plywood good on one side should be used where plywood forms are called for. (Plywood GIS means "good one side.")

Design of Forms

(a) Forms shall be substantial and sufficiently tight to prevent leakage of mortar. They shall be properly braced and tied together so as to maintain their position and shape. If adequate foundations for shores cannot be secured, trussed supports shall be provided.

(b) Snap ties shall be used for internal ties so arranged that when the forms are removed no metal shall be within 1″ of any surface. Wire ties will be permitted

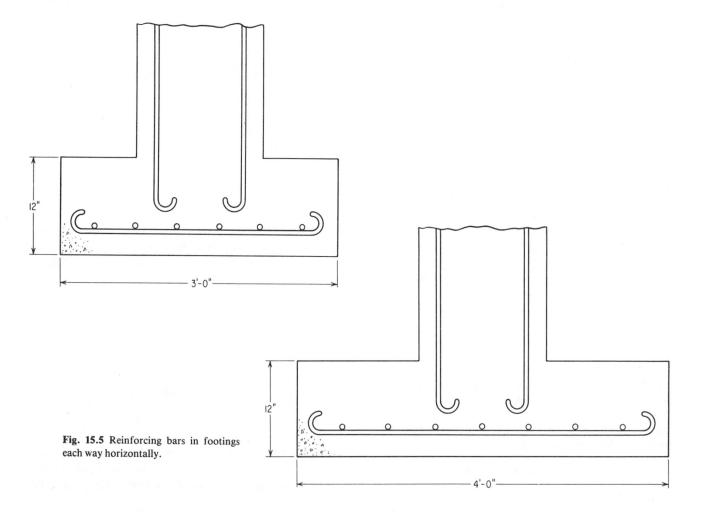

Fig. 15.5 Reinforcing bars in footings each way horizontally.

only on light and unimportant work; they must not be used where discoloration would be objectionable.

(c) Shores supporting successive lifts shall be placed directly over those below, or so designed that the load will be transmitted directly to them.

(d) Forms shall be set to line and grade and so constructed and fastened as to produce true lines. Special care shall be taken to prevent bulging.

(e) Forms shall be lined with plywood for main-entrance surroundings.

Concrete Footing Lumber

This is calculated by:

(a) the amount of lumber touching wet concrete plus an allowance for bracing;

(b) adding for the amount and type of fastening devices;

(c) adding for carpenter labor in making and placing the forms, and also adding half the carpenter's time for carpenter's helper's time.

Assume that forms are to be made and placed for the first three numbered footings on p. 164 and that the material to be used is 2 × 12 dimension stock. Take the following steps:

Step 1:

(a) Find the area of lumber touching wet concrete;

(b) Surface area equals 12 sq ft;

(c) Add 10 percent for waste: 12 plus 1.2 = 13.2 sq ft;

(d) Multiply by the number of this type of footing forms:

$$3 \times 13.2 = 39.6 \text{ or, } 40 \text{ sq ft} = 80 \text{ bd ft of } 2 \times 12;$$

(e) Allow for half as much more material for stakes and braces: 20 sq ft = 40 bd ft of 2 × 4;

(f) Convert to a lumber order of:

$$4/2 \times 12 - 10'\text{-}0'' = 80$$
$$6/2 \times 4 - 10'\text{-}0'' = \underline{40}$$
$$120$$

Step 2: Allow for nails, wire, snap ties, or strap iron as required. (See Chap. 7, p. 158, for nailing schedule.)

Step 3:

(a) Allow for a carpenter making and setting these forms at a rate of 2 hr per form with about 1 hr helper time per form, including placing and removing.

(b) Carpenter will make, place, and remove about 120

bd ft of footing forms in 6 hr at a rate of 20 bd ft per hr or 100 bd ft in 6 hr.

(c) The carpenter will need one-half as much helper's time as his own time. Helper time would be 3 hr per 100 bd ft.

Step 4: On a large job where a temporary carpenter's machine shop area is set up, allow for the forms to be used four times. The cost of setting up and removing will remain constant. Apply a factor of 0.33 to the times given in Step 3(b) and (c).

Step 5: Make a code-numbered index card for footing forms as follows:

Code no. **500**	F
Footing forms 2″ stock Carpenter's time making, setting, and removing 5 hrs per 100 bd ft Helper's time 2½ hrs per 100 bd ft Nails—3½″ common 1.6 lbs per 100 bd ft Allow for using forms 4 times on a large job with factor 0.33	

Concrete for column footings is calculated by the surface area times the depth. Watch for the psi requirements specified. From an estimating point of view, although the amount of concrete required for each footing is very easy to estimate, there will be a very great difference in the cost of placing from job to job. Some jobs will be surface footings where the ready-mixed concrete truck can pull alongside and spout the concrete into place; on other jobs, the concrete may have to be wheeled a considerable distance; while on still other jobs the concrete may have to be bucketed to a great depth. The estimator must visualize all these variables. *Estimating is not an exact science; it requires men with wide building experience and sound judgment.*

It is common practice when estimating concrete to take off the concrete, the concrete forms, and the reinforcing for each item at the same time. This will lessen the risk of forgetting any one item. Your general estimate sheet should be headed in consecutive order:

Concrete footings	Concrete columns
Concrete footing forms	Concrete column forms
Concrete footing reinforcing	Concrete column reinforcing
and so on.	

Concrete column quantities are reckoned by the plan area of the column times the height. Watch very carefully for the psi of the concrete and make no allowance for the displacement of concrete by the reinforcing rods.

Concrete column reinforcing on large jobs is a subtrade. On small jobs the general contractor will estimate and complete the work with his own men. Read the drawings and specifications and allow for laps as shown on the tables.

As a rule of thumb, concrete beams on small jobs will require about 1 percent of the section area of beam for reinforcing. You may use this as a rough check that you have not grossly over- or underestimated such reinforcing. Architectural beams may require a great deal more than 1 percent end-section area.

15.11 CONCRETE REINFORCING-STEEL SPECIFICATIONS

All reinforcing steel may be read from the drawing schedule and specifications. List all bars by the lin ft and convert to weight per lin ft and cost, remembering that there is a different cost per lin ft according to the number of the bar. On large jobs, reinforcing for concrete is a subtrade.

Fabricated Bars

Grades of Bars—Reinforcing bars are furnished in several grades which vary in yield strength, ultimate tensile strength, percentage of elongation, bend test requirements and chemical composition. The particular grade is a very important factor in the design of concrete members. The engineer will state in his specifications or on his drawings the specific grades he wants furnished for the various parts of the reinforced concrete structure.

To obtain uniformity throughout the United States, the American Society for Testing and Materials (ASTM) has prepared standard specifications for these steels. It will be helpful to the iron-worker and inspector to know what these standards are since these grades will appear on bar bundle tags, in color coding, in rolled on marking on the bars or on bills of material. They are:

(a) A615 Standard Specification for Deformed and Plain Billet-Steel Bars for Concrete Reinforcement
  (1) Grade 40 (40,000 psi minimum yield strength).
  (2) Grade 60 (60,000 psi minimum yield strength).
(b) A616 Standard Specification for Rail-Steel Deformed Bars for Concrete Reinforcement
  (1) Grade 50.
  (2) Grade 60.
(c) A617 Standard Specification for Axle-Steel Deformed Bars for Concrete Reinforcement
  (1) Grade 40.
  (2) Grade 60.
(d) A706 Standard Specification for Low-Alloy Steel Deformed Bars for Concrete Reinforcement
  (1) Grade 60.

Reinforcing Bars

The standard type is the *deformed bar.* Each mill which rolls deformed bars may have a different pattern of deformation on the bar but all are rolled to conform to ASTM Specifications.

The following is an extract from a simple specification for a small warehouse:

(a) All reinforcing steel shall be New Billett Intermediate Grade Deformed Bars.
(b) Before being placed in position, metal reinforcing shall be thoroughly cleaned of loose mill and loose rust scale and of other coatings of any character that will destroy or reduce the bond. Reinforcement appreciably reduced in section shall be rejected. When there is any delay in depositing concrete, the reinforcement shall be reinspected and, where necessary, recleaned.
(c) Reinforcement shall be formed carefully to the dimensions indicated in the drawings or called for in the specifications. The diameter of bends shall not be less than 8 times the smallest diameter of the reinforcement bar.
(d) Reinforcing bars shall not be bent or straightened in such a manner that will injure the steel. Bars with kinks or sharp bends shall not be used.
(e) Reinforcing bars shall be positioned accurately and secured against displacement by using annealed iron wire of not less than no. 18 gage or suitable clips at intersections and shall be supported by metal chairs, spacers, hangers, and so on.
  Parallel bars shall not be placed closer in the clear than $\frac{1}{2}$ times the diameter of the bar.
(f) Suitable devices shall be used to hold the reinforcement in position both horizontally and vertically. These devices must be sufficiently rigid to prevent displacement of the reinforcement during the placing of the concrete. All such devices shall be submitted to and receive approval of the architects before using.

15.12 CONCRETE REINFORCED COLUMNS

The plan of columns may be round, square, rectangular, octagonal, or architectural. Round columns are formed with impregnated fiber tubes manufactured in varying diameters. For below-ground work the tubes are left in place, but above-ground a finer textured finish of fiber is used, which is later peeled off, leaving a finely finished concrete column. Architectural columns require special consideration and would have to be priced by the superintendent of the carpentry division or by the mill.

Square and rectangular columns are made up on the job. The dimensions are taken from the schedule of columns shown on one of the drawings. An example is as follows:

Schedule of Reinforced Concrete Columns		
Column number	Column size	Column reinforcing
A-1 A-2 A-3	$12'' \times 12''$	Reinforced no. 6 ϕ rods vertical tied $\frac{1}{4}'' \phi$ $12''$ cc
B-1 C-1	$16'' \times 16''$	— ditto —

Column forms are reckoned by:

(a) the amount of lumber touching wet concrete (the sheathing);

(b) the amount of lumber required for bracing the sheathing;

(c) the number of fastening devices.

There are two methods of framing square or rectangular columns (see Fig. 15.6).

The first method is by the use of square-edged, dressed dimension lumber—say, 1 × 6, 8, or 10 with 2 × 6 horizontal cleats, secured against deflection under load with steel strapping, wood, or metal yokes. With the first method the lumber may usually be used only once.

The second method uses plywood sheathing and vertical 2 × 4 cleats with steel strapping, wood, or metal yokes. Plastic-coated plywood sheathing will give excellent results and may be used many times. The extra initial cost will be more than offset by the number of times that it may be used. The vertical cleats do not come in direct contact with the wet concrete and may be used again for a similar purpose or in some other building operation.

When estimating the sheathing for columns, take the perimeter of the outside of the column sheathing times the height of the column. Thus a $12'' \times 12''$ column with $1''$ sheathing will be reckoned as $14'' \times 4 = 4'\text{-}4''$ perimeter $\times 10'\text{-}0''$ height = about, 44 sq ft. Convert the material to be used into either dimension lumber or plywood.

When erecting square or rectangular columns, it is usual (unless the columns are very short) to secure three sides around the base of the area of the column to be erected; then tie the reinforcing and finally secure the

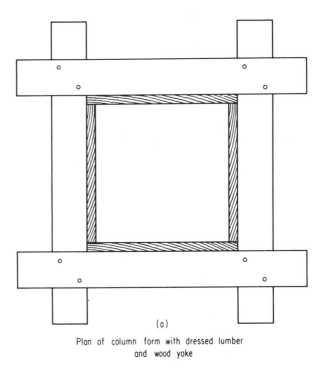

(a)

Plan of column form with dressed lumber and wood yoke

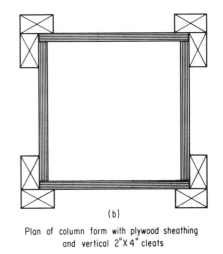

(b)

Plan of column form with plywood sheathing and vertical 2"X 4" cleats

Fig. 15.6

open side of the column and complete with steel strapping or metal yokes. Allow for either rental of yokes or for depreciation of own equipment.

Concrete columns are usually filled with wet concrete within thirty to forty minutes. This exerts a great pressure (according to the height of the column and type of mix) on the surface of the forms.

The weight of wet concrete is about 150 lbs per cu ft. It is more than twice the weight of water per cu ft. The amount of pressure exerted is 150 lb for every foot of depth. Thus a $12'' \times 12''$ column $10'\text{-}0''$ high filled with wet concrete will exert a pressure of $150 \times 10 =$

1,500 lb at its foot. At 9'-0" the pressure will be 1,350 lb decreasing by an increment of 150 lb per ft to the top of the column. This pressure will be momentarily increased by the impact of wet concrete during placing, and it will be still further increased up to 10 percent by vibrating the wet concrete. All columns and formwork must be constructed strong enough to support the load of wet concrete without deflection. The spacing of the yokes will require careful attention. They must be closer together at the bottom where the pressure is the greatest and spaced progressively wider apart toward the top.

The following is a guide in estimating the cost of column formwork:

(a) Take the number of sq ft of sheathing lumber or plywood touching wet concrete (assume that some column sheating requires 400 sq ft of sheathing).

(b) Allow for 1½ times as many bd ft of lumber for yokes, cleats, and bracing as sq ft of sheathing (in this case 600 bd ft of 2 × 4 would be needed).

(c) Allow 35 hr of carpenter's time and 30 hr of carpenter's helper's time to make, erect, and remove 1,000 bd ft of column formwork.

(d) Allow for rentals of metal yokes or depreciation of your own equipment.

15.13 FORM HARDWARE

The following material has been excerpted from literature supplied by the "Council of Forest Industries of British Columbia" and is reproduced here with the permission of the copyright holder.

Contractors should obtain their hardware only from recognized suppliers who can furnish proof of the strength of their accessories by laboratory or field tests. While different types of hardware have been developed, all standard types are suitable for use with plywood. Of particular interest as far as plywood is concerned are various kinds of high tensile steel ties or bolts, some of which have been specifically designed for plywood.

Snap-ties made from high tensile wire are also popular as an inexpensive method of form support. These ties are designed to prevent the forms from separating and achieve their ultimate strength on a gradual increase in pressure. Stripping the forms should be done only after concrete has set. With forms removed, a tie breaker is applied to exposed end of snap-tie and bent at right angles as close to the wall surface as possible. The unit is then turned with a clockwise circular motion until the snap-tie breaks at break-off point.

This type of tie, designed to be used where breakback is not required, consists of a straight unthreaded pencil rod with buttons. Clamps are slipped over the rod and bear against the walers.

A separate spreader must be used.

A bar tie may be used with wedges through the slots in the end of the tie.

For most commercial concrete finishes a rod or bar tie with integral spreaders is customary. Such ties generally have a notch or similar reduction in section which allows the tie to be broken back a set distance from the wall surface.

Rod ties are shaped to prevent rotation in the concrete when the rod is snapped off by twisting.

The bar tie shown on the left has a breakback notch and spurs which act as a spreader.

Twisted galvanized wire tie with integral spreader and breakback features.

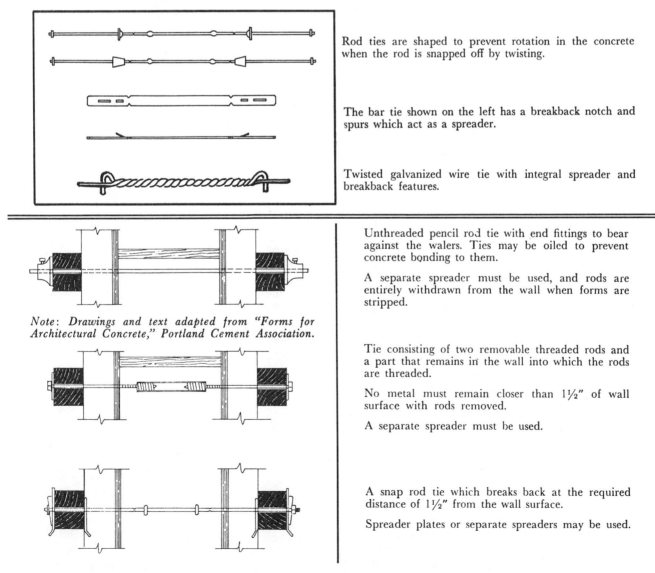

Unthreaded pencil rod tie with end fittings to bear against the walers. Ties may be oiled to prevent concrete bonding to them.

A separate spreader must be used, and rods are entirely withdrawn from the wall when forms are stripped.

Note: Drawings and text adapted from "Forms for Architectural Concrete," Portland Cement Association.

Tie consisting of two removable threaded rods and a part that remains in the wall into which the rods are threaded.

No metal must remain closer than $1\frac{1}{2}''$ of wall surface with rods removed.

A separate spreader must be used.

A snap rod tie which breaks back at the required distance of $1\frac{1}{2}''$ from the wall surface.

Spreader plates or separate spreaders may be used.

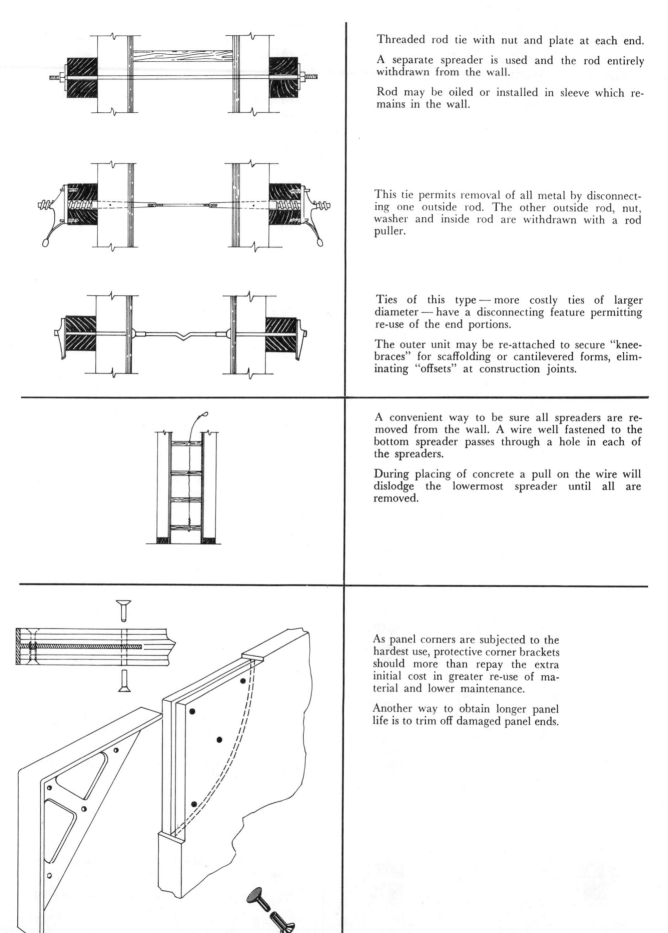

Threaded rod tie with nut and plate at each end.

A separate spreader is used and the rod entirely withdrawn from the wall.

Rod may be oiled or installed in sleeve which remains in the wall.

This tie permits removal of all metal by disconnecting one outside rod. The other outside rod, nut, washer and inside rod are withdrawn with a rod puller.

Ties of this type — more costly ties of larger diameter — have a disconnecting feature permitting re-use of the end portions.

The outer unit may be re-attached to secure "kneebraces" for scaffolding or cantilevered forms, eliminating "offsets" at construction joints.

A convenient way to be sure all spreaders are removed from the wall. A wire well fastened to the bottom spreader passes through a hole in each of the spreaders.

During placing of concrete a pull on the wire will dislodge the lowermost spreader until all are removed.

As panel corners are subjected to the hardest use, protective corner brackets should more than repay the extra initial cost in greater re-use of material and lower maintenance.

Another way to obtain longer panel life is to trim off damaged panel ends.

For walls 4' high: panels should be stacked horizontally two high, using top alignment waler and braces every 8'.

Use three rows of ties for a 3' wall and four rows for more than 3'.

Corners should be plumbed and braced. Top waler is placed after ties and bars have been placed.

This is another arrangement for 4' high walls using full 4' x 8' panels.

Other details are identical with those in the previous drawing.

This system is for walls 2' high or less, using 1500 pound ties and 2' x 8' panels.

A 2" x 4" toe plate should be nailed to footing within 48 hours of footing pour. This is used to align the outside form only.

Note alternate location of top ties. Plug these notches if ties are used over top of form. Vertical bars project above top of form.

2" x 4" used at vertical panel joints and braced.

Alternate methods of aligning wall forms.

Flared 2" metal channel section.

2" x 4" fastened with one nail per panel.

For 8' high walls, panels are applied vertically with horizontal steel bars.

Bar joints should be staggered.

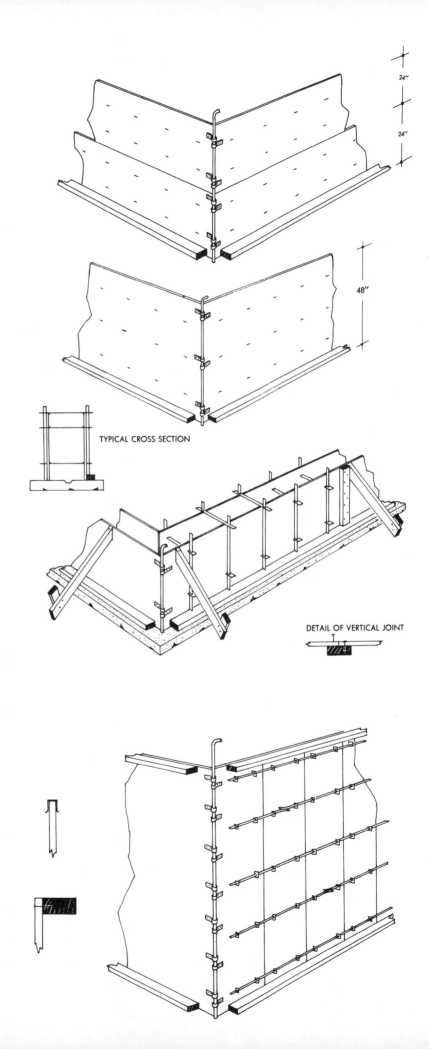

TYPICAL CROSS SECTION

DETAIL OF VERTICAL JOINT

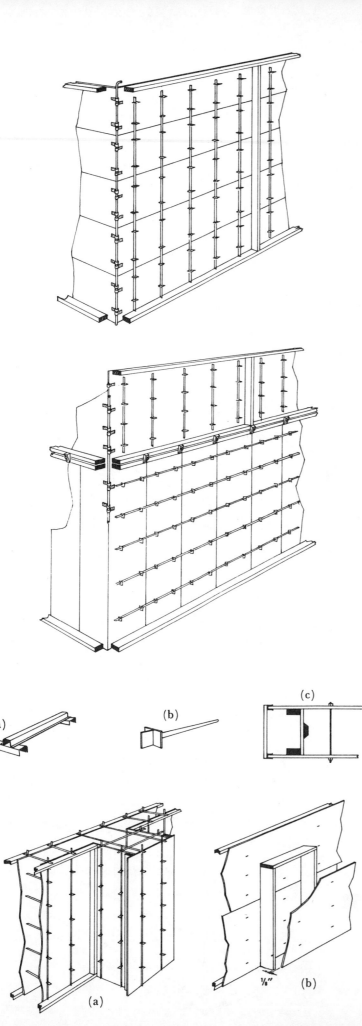

For wall heights up to 16': this system can have all horizontal panels or a combination of horizontal and vertical panels.

When panels are used horizontally, nail 2" x 4" studs at all vertical panel joints and braces.

When panels are used in combination horizontally and vertically use a double waler with standard snap ties at the break between the horizontal and the vertical.

Rod hole notched or drilled in lower form only.

(a) Channel top tie permits accurate spacing and alignment at top panel joints. Re-usable.

(b) Cleanout tool for cleaning tie slots of excess concrete after form removal.

(c) Suggested detail for wall bulkheads.

FORMING PILASTERS:

(a) For pilasters deeper than 10", ties must be used at right angles to the projections as illustrated.

(b) Leave-outs (or bucks) should be at least $\frac{1}{8}$" narrower than finished wall dimensions to facilitate threading bars through ties.

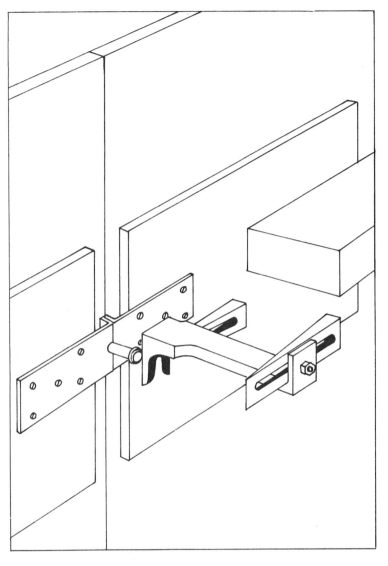

OTHER TYPES OF UNFRAMED PLYWOOD FORM PANELS

This type of panel has a combination clamp which draws panels together into a tight lock, secures the snap tie and lines up panel faces. It also provides shelf and wedge for walers.

The primary material is 1⅛″ Concrete Form Grade fir plywood to which are glued and nailed ½″ fir plywood plates.

Another patented form panel system using fir plywood is shown below.

Steel clamps are bolted to the plywood panels.

Panels contain a special slot which holds the snap tie.

Manufacturers say assembly process is especially applicable to stepped-up footings, pilasters, columns and beams.

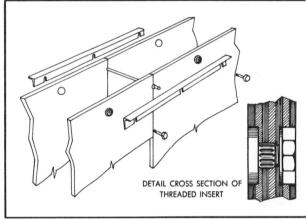

DETAIL CROSS SECTION OF
THREADED INSERT

FRAMED FORMS

These forms consist of a plywood face from ½″ to ¾″ thick with a wood or metal frame. Plywood dimensions may be 4′ x 8′ or 2′ x 8′ with miscellaneous filler pieces. A great variety of hardware is available. Ties pass between or through the panels depending on panel size. Framed forms are widely used for all straight wall foundation work, residential and commercial; all above ground concrete work, including walls, columns, slabs, spandrel beams and joists; heavy industrial concrete work; and architectural concrete. They assist in the production of the highest quality wall obtainable, and by the nature of their construction withstand rougher use than unframed forms. The framed form system lends itself to gang forming. Fewer but stronger ties are required.

Framed forms 32″ x 8′ are sometimes used to effect economies under certain combinations of pour, tie and form conditions and/or module considerations. Such panels placed horizontally result in vertical tie spacing of 32″. These 32″ panels can be made from 2 sheets of plywood (the 2 fall-off strips are combined to make up a third form).

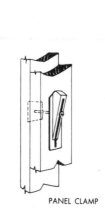

PANEL CLAMP

A bolt and wedge device to ensure a tight fit between pre-framed panel sections.

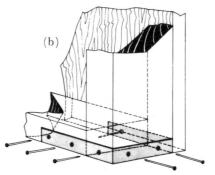

(a)   (b)

(a) Suggested nailing detail. Toe nail with one nail. Two nails through side frame into 2″ x 4″ member.

(b) Panel corner detail showing application of steel strap.

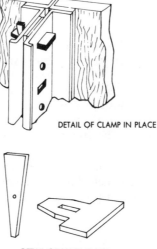

DETAIL OF CLAMP IN PLACE

DETAIL OF PANEL CLAMP

A typical 2′ x 8′ metal framed panel.

One of the advantages of these patented systems is that they offer extra protection to the edges of the plywood. The frames themselves have a long life and the plywood can be replaced or reversed at minimum cost. Initial cost of metal framed forms is high and they are rented in many instances.

COLUMNS

Several methods of forming square and rectangular columns from plywood are illustrated below. Column sizes should, wherever possible, be selected which will permit use of standard panels or minimize cutting waste.

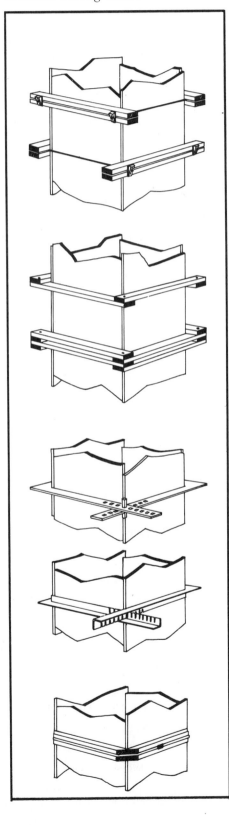

SQUARE COLUMN FORMS

Waler and tie rod.

As tie rod does not adequately brace the plywood this method is limited to columns having one side not wider than approximately 12″.

For square or nearly square columns this is an alternate arrangement.

Single timber walers bolted or nailed at corners.

Strength of single waler and corner joint limits this method to a maximum face width of 16″ for 2″ x 4″ and 20″ for 2″ x 6″.

Double timber waler bolted or nailed at corners.

Steel column clamps. Flat bar type which may be rented or purchased for any column size and serves to brace plywood as well.

Notched angle type. May be rented or purchased for any column size.

Steel strapping.

Lumber bracing is always required because the strap alone will not brace plywood form panels.

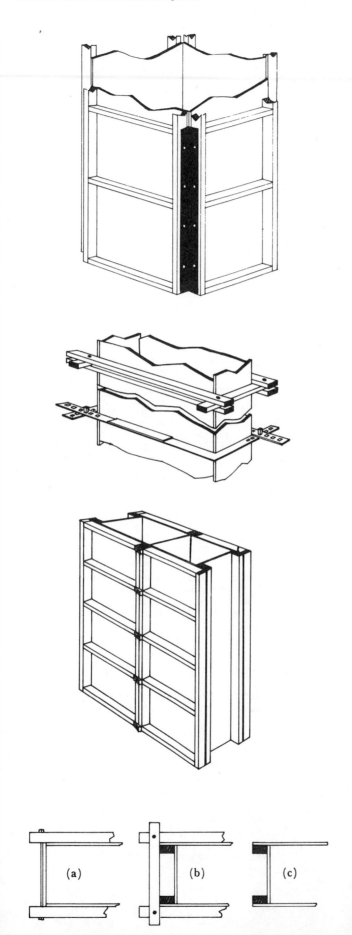

Standard wall panels. Corners are tied by angle irons.

RECTANGULAR COLUMN FORMS

Double walers can be of 2″ x 4″, 2″ x 6″, 4″ x 4″ or 2″ x 8″ lumber.

Special steel column clamps are readily available for column widths up to 6 feet.

Center section reinforced.

These standard wall panels use snap-back ties, plywood end plates and nailed vertical studs.

END PLATE DETAILS

(a) Steel bar tie.

(b) Bolted lumber cross bar.

(c) Nailed vertical studs.

Tie rod can be substituted for timber if face width is less than 12″.

Note "A": Multi-choice of accessories (stud rods, nut washers, handle washers, cone nuts, adapter cones etc.) See section 2, page 5 & 6 for illustrations.

Note "B": Multi-choice of accessories (coil bolts, flat washers, coil cones, spreader cones etc.) See section 2, page 4 for illustrations. 1/2"φ & 3/4"φ cont. th'd coil rod substituted for 1/2"φ & 3/4"φ coil bolts for certain applications. Loads shown above apply.

The following tables have been developed to act as a guide in selecting the proper forming system to use on your specific project. Your selection should take into consideration the following conditions:

(I) Form Lumber
(A) The rate of pour in vertical feet per hour.
(B) The setting temperature of the concrete.
(C) The use of Admixtures (Retardments).
(D) The number of reuses of the forms.
(E) The erection of forms—by hand or crane.
(F) The possible other uses of the material.

(II) Form Ties
(A) Form tie selected should develop the full allowable working load of the forms chosen.
(B) Specifications regarding depth of break-back can affect your decision.

* * *

The following (4) factors generally will govern the form pressures:

(1) **The rate of pour** is determined by the thickness of the wall and the length of wall to be poured, plus your placement capacity of the concrete.

(2) **The setting temperature** is very important; thus, when a design is selected—care must be taken that the temperature is maintained. For average weather conditions—we have for our design tables used:
50° F. for winter temperature.
70° F. for summer temperature.

(3) **The use of Admixtures (Retardments)** have a temperature lowering affect and thus should be taken into consideration. Consult your concrete supplier for further information.

(4) **Internal vibration** only has been taken into consideration in our design tables with depth of vibration limited to 4 feet below top of concrete surface.

(5) **Limiting Design Values**
(A) **Form Lumber:**—No. 1 Southern Pine, Construction Grade Douglas Fir, or Equal.
(B) **Extreme Fibre Stress** f = 1,875 P.S.I. Based on above lumber plus 25% increase due to short term loading.
(C) Deflection $D = \dfrac{l}{270}$

For forming problems not covered by our design tables—either consult our Engineering Department, or refer to the table below to determine your form pressure.

MAXIMUM LATERAL PRESSURE FOR DESIGN OF WALL FORMS

Rate of Pour Feet per Hr.	Maximum Lateral Pressure psf					
	90°F	80°F	70°F	60°F	50°F	40°F
1	250	262	278	300	330	375
2	350	375	407	450	510	600
3	450	488	536	600	690	825
4	550	600	664	750	870	1050
5	650	712	793	900	1050	1275
6	750	825	921	1050	1230	1500
7	850	938	1050	1200	1410	1725
8	881	973	1090	1246	1466	1795
9	912	1008	1130	1293	1522	1865
10	943	1043	1170	1340	1578	1935

Notes: (1) Table above based on ACI Committee 622 pressure formulas.
(2) Do not use design pressures in excess of 2000 psf, or 150 x height of fresh concrete in forms, whichever is less.
(3) Slump of concrete not to exceed 4".

General Notes:
(1) Reprints of this table are available upon request.
(2) For complete data on form designing, etc., obtain a copy of Handbook:
"Formwork for Concrete"
by: M. K. Hurd & A.C.I. Committee 347
Publisher: American Concrete Institute
Post Office Box 4754, Redford Station
Detroit 19, Michigan

PRACTICAL LUMBER AND FORM TIE TABLE

TIE LOADING DATA		UNIVERSAL FORM CLAMP CO. TIEING MATERIAL DATA						PRACTICAL FORM DESIGN DATA					MAXIMUM ALLOWABLE RATE OF POUR FT. PER HOUR	
		FORM TIES			TIEING ACCESSORIES				STUDS		WALES			
SAFE WORKING LOAD	ULTIMATE LOAD	TYPE	SPACING HORIZ.	SPACING VERT.	ITEM	CAT. NO.	CATALOG REFERENCE PAGE NO.	SHEATHING	SIZE	SPACING	SIZE	DB'L SPACING	50 F.	70 F.
3,000	4,500	3M Snaptie	2'0"	2'0"	Snap Tie Clamp	#3815	Sec. 2 Page 3	3/4" Plywood	2x4	12" O.C.	2x4	2'0"	3'4"	4'6"
3,000	4,000	Twistye	2'0"	2'0"	Twistye Clamp	#360	Sec. 2 Page 3	3/4" Plywood	2x4	12" O.C.	2x4	2'0"	3'4"	4'6"
3,000	*5,200	3/8"φ Mild Steel Rod	2'0"	2'0"	Form Clamp	#250	Sec. 2 Page 2	3/4" Plywood	2x4	12" O.C.	2x4	2'0"	3'4"	4'6"
5,000	6,500	5M Snaptie	2'0"	2'0"	Snap Tie Clamp	#3815	Sec. 2 Page 3	3/4" Plywood	2x4	12" O.C.	2x4	2'0"	6'0"	11'6"
5,000	6,600	3/8"φ C.D. Rod	2'0"	2'0"	See Note "A"			3/4" Plywood	2x4	12" O.C.	2x4	2'0"	6'0"	11'6"
5,000	*6,600	1/2"φ Mild Steel Rod	2'0"	2'0"	Form Clamp	#350	Sec. 2 Page 2	3/4" Plywood	2x4	12" O.C.	2x4	2'0"	6'0"	11'6"
6,000	9,000	1/2" x 6M Coil Tie	2'6"	2'6"	See Note "B"			3/4" Plywood	2x6	12" O.C.	2x6	2'6"	4'6"	6'3"
9,000	12,000	1/2"φ C.D. Rod	3'0"	3'0"	See Note "A"			3/4" Plywood	2x6	12" O.C.	2x6	3'0"	4'0"	5'9"
9,000	13,000	1/2" x 9M Coil Tie	3'0"	3'0"	See Note "B"			3/4" Plywood	2x6	12" O.C.	2x6	3'0"	4'0"	5'9"
9,000	13,500	3/4" x 9M Coil Tie	3'0"	3'0"	See Note "B"			3/4" Plywood	2x6	12" O.C.	2x6	3'0"	4'0"	5'9"
12,000	17,000	3/4" x 12M Coil Tie	3'0"	4'0"	See Note "B"			3/4" Plywood	3x6	12" O.C.	3x6	4'0"	4'8"	6'6"
13,000	17,000	1/2"φ HI-Ten Rod	3'0"	4'0"	See Note "A"			3/4" Plywood	3x6	12" O.C.	3x6	4'0"	5'0"	7'3"
14,000	18,000	5/8"φ C.D. Rod	3'0"	4'0"	See Note "A"			3/4" Plywood	3x6	12" O.C.	4x6	4'0"	5'6"	10'0"
20,000	25,000	5/8"φ HI-Ten Rod			See Note "A"									
20,000	25,000	3/4"φ C.D. Rod			See Note "A"									
25,000	35,000	3/4"φ HI-Ten Rod			See Note "A"									

3/4"φ C.D., 5/8"φ & 3/4"φ HI-Ten rods normally used in conjunction with steel forms. Consult Universal Form Clamp Co. Engineering Dept. for spacing & form design

*Indicates Yield Point of Rod

TABLE 5 Properties of Sections

Nominal Size (In.)	Actual Size (In.) Unseasoned	Actual Size (In.) Dry	Area (In.²)	Axis X-X S (In.³)	Axis X-X I (In.⁴)	Axis Y-Y S (In.³)	Axis Y-Y I (In.⁴)	Board Measure Per Lineal Foot	Approx. Weight Per Lineal Foot (Lbs.)
				At Dry Dimensions					
2 x 2	$1^{9}/_{16}$ x $1^{9}/_{16}$	$1^{1}/_{2}$ x $1^{1}/_{2}$	2.25	0.56	0.51	0.56	0.51	$^{1}/_{3}$	0.6
2 x 3	$2^{9}/_{16}$	$2^{1}/_{2}$	3.75	1.56	1.95	0.94	0.70	$^{1}/_{2}$	0.9
2 x 4	$3^{9}/_{16}$	$3^{1}/_{2}$	5.25	3.06	5.36	1.31	0.98	$^{2}/_{3}$	1.3
5	$4^{5}/_{8}$	$4^{1}/_{2}$	6.75	5.06	11.39	1.69	1.28	$^{5}/_{6}$	1.6
6	$5^{5}/_{8}$	$5^{1}/_{2}$	8.25	7.56	20.79	2.06	1.55	1	2.0
8	$7^{1}/_{2}$	$7^{1}/_{4}$	10.87	13.13	47.60	2.72	2.04	$1^{1}/_{3}$	2.6
10	$9^{1}/_{2}$	$9^{1}/_{4}$	13.87	21.38	98.89	3.47	2.60	$1^{2}/_{3}$	3.4
12	$11^{1}/_{2}$	$11^{1}/_{4}$	16.87	31.63	177.92	4.22	3.17	2	4.1
14	$13^{1}/_{2}$	$13^{1}/_{4}$	19.87	43.88	290.71	4.97	3.73	$2^{1}/_{3}$	4.8
3 x 4	$2^{9}/_{16}$ x $3^{9}/_{16}$	$2^{1}/_{2}$ x $3^{1}/_{2}$	8.75	5.11	8.95	3.65	4.56	1	2.4
5	$4^{5}/_{8}$	$4^{1}/_{2}$	11.25	8.44	18.99	4.73	5.85	$1^{1}/_{4}$	3.2
6	$5^{5}/_{8}$	$5^{1}/_{2}$	13.75	12.61	34.68	5.73	7.16	$1^{1}/_{2}$	3.8
8	$7^{1}/_{2}$	$7^{1}/_{4}$	18.12	21.90	79.39	7.55	9.44	2	5.0
10	$9^{1}/_{2}$	$9^{1}/_{4}$	23.12	35.64	164.8	9.63	12.04	$2^{1}/_{2}$	6.4
12	$11^{1}/_{2}$	$11^{1}/_{4}$	28.12	52.73	296.6	11.72	14.65	3	7.8
14	$13^{1}/_{2}$	$13^{1}/_{4}$	33.12	73.14	484.6	13.80	17.25	$3^{1}/_{2}$	9.2
4 x 4	$3^{9}/_{16}$ x $3^{9}/_{16}$	$3^{1}/_{2}$ x $3^{1}/_{2}$	12.25	7.15	12.52	7.15	12.51	$1^{1}/_{3}$	3.4
5	$4^{5}/_{8}$	$4^{1}/_{2}$	15.75	11.81	26.57	9.14	16.07	$1^{2}/_{3}$	4.4
6	$5^{5}/_{8}$	$5^{1}/_{2}$	19.25	17.65	48.54	11.23	19.65	2	5.4
8	$7^{1}/_{2}$	$7^{1}/_{4}$	25.37	30.66	111.2	14.80	25.90	$2^{2}/_{3}$	7.1
10	$9^{1}/_{2}$	$9^{1}/_{4}$	32.37	49.90	230.8	18.88	33.04	$3^{1}/_{3}$	9.0
12	$11^{1}/_{2}$	$11^{1}/_{4}$	39.37	73.82	415.2	22.96	40.18	4	10.9
14	$13^{1}/_{2}$	$13^{1}/_{4}$	46.37	102.4	678.4	27.05	47.34	$4^{2}/_{3}$	12.9
16	$15^{1}/_{2}$	$15^{1}/_{4}$	53.37	135.7	1,034.	31.13	54.48	$5^{1}/_{3}$	14.8
6 x 6	$5^{1}/_{2}$ x $5^{1}/_{2}$		30.25	27.73	76.26	27.73	76.26	3	8.3
8	$7^{1}/_{2}$		41.25	51.56	193.4	37.81	104.0	4	11.4
10	$9^{1}/_{2}$		52.25	82.73	393.0	47.90	131.7	5	14.4
12	$11^{1}/_{2}$		63.25	121.2	697.1	57.98	159.4	6	17.4
14	$13^{1}/_{2}$		74.25	167.1	1,127.7	68.06	187.2	7	20.3
16	$15^{1}/_{2}$		85.25	220.2	1,706.8	78.15	214.9	8	23.5
18	$17^{1}/_{2}$		96.25	280.7	2,456.4	88.23	242.6	9	26.5
8 x 8	$7^{1}/_{2}$ x $7^{1}/_{2}$		56.25	70.31	263.7	70.31	263.7	$5^{1}/_{3}$	15.5
10	$9^{1}/_{2}$		71.25	112.8	535.9	89.06	334.0	$6^{2}/_{3}$	19.6
12	$11^{1}/_{2}$	See Note (4)	86.25	165.3	950.5	107.8	404.3	8	23.8
14	$13^{1}/_{2}$		101.3	227.8	1,537.7	126.6	474.6	$9^{1}/_{3}$	27.7
16	$15^{1}/_{2}$		116.3	300.3	2,327.4	145.3	544.9	$10^{2}/_{3}$	32.0
18	$17^{1}/_{2}$		131.3	382.8	3,349.6	164.1	615.2	12	36.0
10 x 10	$9^{1}/_{2}$ x $9^{1}/_{2}$		90.25	142.9	678.8	142.9	678.8	$8^{1}/_{3}$	24.7
12	$11^{1}/_{2}$		109.3	209.4	1,204.0	173.0	821.7	10	30.0
14	$13^{1}/_{2}$		128.3	288.6	1,947.8	203.1	964.5	$11^{2}/_{3}$	35.5
16	$15^{1}/_{2}$		147.3	380.4	2,948.1	233.1	1,107.4	$13^{1}/_{3}$	40.5
18	$17^{1}/_{2}$		166.3	484.9	4,242.8	263.2	1,250.3	15	45.6
12 x 12	$11^{1}/_{2}$ x $11^{1}/_{2}$		132.3	253.5	1,457.5	253.5	1,457.5	12	36.5
14	$13^{1}/_{2}$		155.3	349.3	2,357.9	297.6	1,711.0	14	42.6
16	$15^{1}/_{2}$		178.3	460.5	3,568.7	341.6	1,964.5	16	49.1

Notes: (1) Weights for 2" thick material based on 20% moisture content and 35 pounds per cubic foot; for over 2" thick, 34% moisture content and 40 pounds per cubic foot.
(2) Section properties are based on dry dimensions.
(3) Board measure is based on nominal dimensions.
(4) Timbers (5 inches or more in least nominal dimension) are always dressed unseasoned, $^{1}/_{2}$ inch under nominal size.
(5) Metric section properties for lumber are published in CWC Datafile "Metric Section Properties".

15.14 CONCRETE-FORM TIE LOADING AND PRACTICAL FORM-DESIGN DATA

This important section deals with recommended (specified) tie-space loading and data on form design. A standard table is quoted on sheathing, the size and spacing of studs and wales, and the rate of pour in depth per ft for given concrete temperatures. Specialized companies design, fabricate, and sell or rent forms and accessories for any given engineering requirement. The estimator must consider whether it is cheaper to own or rent concrete forms.

15.15 CONCRETE WATER TANK ESTIMATE

The following exercise has been designed to help you think through an estimating problem from its inception to completion.

Problem

Make a neat, methodical, and complete cost estimate of the concrete water tank described in the following drawings and specifications (see Fig. 15.7).

Specifications

Excavations: Assume the lot to be level. The top of the tank is to finish 1'-0" above the grade. Strip and return 0'-8" of topsoil. Backfill and scatter the balance of spoil.

Outside Dimensions

Walls: 16'-0" × 12'-0" × 8'-0" deep × 0'-8" thick.

Bottom: 0'-10" thick.

Reinforcing

Use No. 8 rods 8" OC laterally and no. 3 rods 1'-0" OC vertically in the walls and No. 8 rods 1'-0" each way in the bottom.

Concrete

Volume mix 1:1½:2½ and a 5½-gal paste. Place a sheet of polyethylene under the concrete floor.

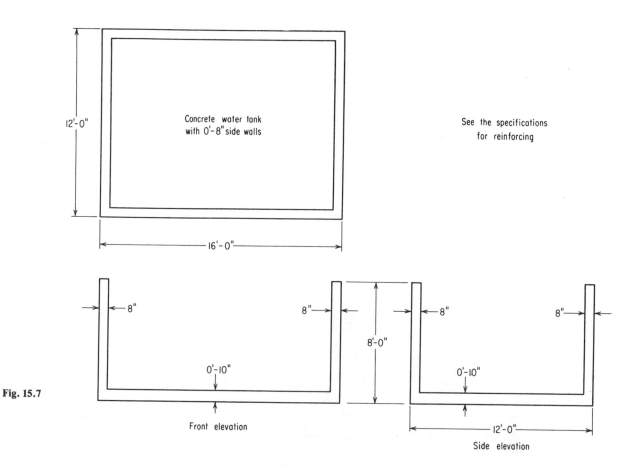

Fig. 15.7

References: Your A.I.A. file and library.

Stationery: Work-up sheets and general estimate sheets.

Prices: Use local prices for labor, materials, and rental of machinery. (If you use your own machinery, allow for depreciation.)

Method

1. Estimate the labor crew you will require.
2. Take off the excavations.
3. Estimate the volume of placed concrete required for the job.
4. Find the number of sacks of cement required.
5. Find the number of gal of water required to place the concrete.
6. Estimate the number of cu yd of sand.
7. Estimate the number of cu yd of gravel.
8. Find the weight, cost, and labor for reinforcing.
9. Find the cost of polyethylene required.
10. Find the amount of cement required for each batch of a 3-cu-ft delivery mix.
11. Find the amount of water required per batch.
12. Find the amount of sand required per batch.
13. Find the amount of gravel required per batch.
14. Estimate the number of batches required to do the job.
15. Estimate the time required to place the concrete allowing 3 min per batch.
16. Find the weight of the concrete tank when empty. *Allow for 1 cu ft of concrete weighing approximately 150 lb.*
17. Find how many gal of water the tank will contain when full. *One cu ft will contain 7½ American gal or 6¼ Canadian gal approximately.*
18. Find the weight of water that may be contained in the full tank. *The weight of 1 cu ft of water is 62¼ lb approximately.*
19. What is the combined weight of the concrete tank and the water when full?
20. Estimate an allowance of time for the crew to get started and time for the cleanup when finished.
21. Estimate the time required to strip and return the semipermanent forms to base.
22. Estimate the labor; fringe benefits of 5 percent and workmen's compensation of 2 percent all on the labor bill.
23. Allow overhead expenses of 6 percent.
24. Profit: allow 10 percent.

Remember that on some jobs you will have to haul water. Do not forget to add one-half as much more materials in the dry state to place a given amount of concrete.

After you have completed the work-up sheets, attach all the freehand calculations and/or the machine calculations to the work-up sheet and make out the general estimate sheet in the following order:

Excavation:

Topsoil
Mass
Backfill and scatter spoil

Concrete:

Sacks of cement
Sand
Gravel

Formwork:

Semipermanent per-sq-ft cost
Ramps
Runways
Fastening devices

Equipment:

Rentals
Depreciation of your own plant chargeable to this job
Gas, oil, and servicing

Labor:

Wages
Fringe benefits
Workmen's compensation

Overhead Expenses

Profit

Concrete Septic Tank Problem

You are to construct by volume mix a 600-gal septic tank for a country home sewerage-disposal system.

The design mix is 1 cement:2¼ sand:3 gravel by volume with a 5½-gal paste. The walls and floor are to be 0'-6" thick.

Method

Step 1: Check with the local authority for latest information. Allow 7½ gal per cu ft (in Canada allow 6¼ gal per cu ft).

Step 2: Assume that the tank is to be 4'-0" deep (inside dimensions). From this information find a reasonable width and length for the tank. The loose concrete top will be 3″ deep and steel-mesh reinforced.

Step 3: Make a sketch of the tank and dimension it.

Step 4: Find the actual volume of concrete required, including the loose slab, by subtracting the inside volume of the tank from the outside volume.

Step 5: Remember to make an allowance for extra cement, sand, and gravel in the dry state.

Step 6: Estimate the amount of cement, sand, and gravel required to place the concrete.

Step 7: Using a 3-cu-ft-capacity mixer, estimate the amount of water, cement, sand, and gravel required for each mix.

Step 8: Estimate the number of batches required to place all the concrete.

Step 9: Estimate the time required to place all the concrete (allow 3 min per batch).

Step 10: Using local prices, estimate the net cost to complete the project.

Step 11: Remember that your crew has to get started, and they also require time to clean up aftewards.

Step 12: Estimate the cost of cement, sand, gravel, tie wires, forms, rental or depreciation of equipment, gas and oil, traveling time, crew plus fringe benefits, workmen's compensation, company truck expenses.

Step 13: Remember to allow for overhead and profit.

15.16 GRAVEL BASE UNDER CONCRETE FLOORS

Local authorities require in some areas that a 0″-5″ gravel base be placed under basement concrete floors.

Example

Estimate for 0′-5″ of gravel to be placed under the concrete floor of Fig. 8.5, p. 164.

Step 1: The area to be covered with gravel is equal to the area of the subfloor (1,980 sq ft), less the estimated area of displacement by the width of the concrete wall, plus the width inside footing offset, 0′-10″ + 0′-6″ = 1′-4″. This portion is shown shaded in Fig. 15.8.

Step 2: The estimated area of displacement of gravel is the perimeter × 1′-4″ = 210 × 1⅓ = 280 sq ft.

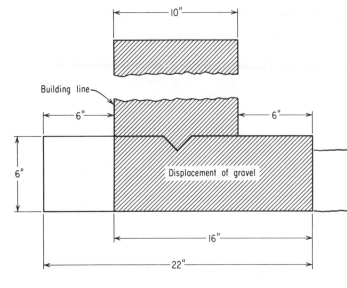

Fig. 15.8

Step 3: The area to be covered with gravel is 1,980 − 280 = 1,700 sq ft. Thus

$$\frac{1,700 \times 5 \times 1}{12 \times 27} = 26\tfrac{19}{27}$$

or 27 cu yd of gravel.

Problem

Using local prices and the same specifications as in the foregoing example, estimate the cost for delivery and placing of 0′-5″ of gravel under each concrete floor for Figs. 8.1 through 8.4, p. 000.

15.17 CONCRETE BASEMENT FLOORS

Concrete floors for small jobs are usually quoted at a price per sq ft placed and finished. The price varies with the depth of the concrete. You should obtain a price list from your local dealers. Larger jobs may be quoted per cu yd delivered and placed.

Example

Estimate the number of sq ft of placed and finished concrete required for Fig. 8.5, p. 164. Allow for a 0′-10″-thick basement wall.

Step1: Draw and dimension a sketch. Notice that the total overall area of the basic plan is 60′-0″ × 40′-0″ (outside measurements).

Step 2: Find the total basic rectangular plan area of the concrete floor. This includes the offsets and the recess but excludes the thickness of the concrete walls:

$$58'\text{-}4'' \times 38'\text{-}4'' = 58\tfrac{1}{3} \times 38\tfrac{1}{3} = 2{,}236\tfrac{1}{9} \text{ sq ft.}$$

Step 3: Find the area of the offsets and the recess:

Left-side offset	10×10	$= 100$
Bottom left offset	25×8	$= 200$
Right-side offset	$25\tfrac{2}{3} \times 5$	$= 128\tfrac{1}{3}$
		$428\tfrac{1}{3}$

Step 4: From the total rectangular plan area of the concrete floor, take the areas of the offsets and recess: $2{,}236\tfrac{1}{9} - 428\tfrac{1}{3} = 1{,}807\tfrac{7}{9}$, or approximately 1,810 sq ft.

Problem

Using your local price list, estimate the cost to place a concrete floor to a depth of $0'\text{-}3''$ for Figs. 8.1 through 8.4.

15.18 MESH: STEEL-WELDED REINFORCING

Fabricated reinforcing mesh is widely used in concrete floors, driveways, and roads. It is also used in other areas of building construction as temperature reinforcing. Some of the main things that an estimator should know about this type of reinforcing are as follows:

(a) It is fabricated in both square and rectangular mesh.

(b) The longitudinal gage of the wire may be of a heavier gage than the transverse wire.

(c) When there is any difference in the gages of the wire for any one type of mesh, the longitudinal wire is the heavier gage.

(d) The style of the mesh may be described as $4'' \times 4''$—9×12 (or 44-912), which means that the area of each mesh is $4'' \times 4''$, the longitudinal gage of the wire is no. 9, and the transverse wire is no. 12 (for a $4'' \times 8''$—9—12 or 48-912 welded steel mesh, each mesh is $4'' \times 8''$ and the gages of the wire are no. 9 and no. 12 respectively).

(e) The smaller the number of the gage, the thicker the wire.

QUESTIONS

1. What is the first consideration when estimating the amount of material required for concrete formwork?

2. State four important considerations in estimating the cost per square foot for each usage of forms for concrete work that are designed to be used 25 times.

3. (a) Concrete for foundations is estimated in units of _____ .

   (b) Concrete floors are estimated in units of _____ .
   (c) Concrete mortar is estimated in units of _____ .
   (d) Concrete blocks are estimated in units of _____ .
   (e) Prestressed concrete units are estimated in _____ .
   (f) Concrete reinforcing is estimated in units of _____ .
   (g) Concrete wall form ties are estimated in units of _____ .
   (h) Concrete equipment on the job is allowed for on the estimate by _____ .
   (i) Concrete ramps and runways are estimated by _____ .

4. Write out in full the meaning of first line of footing particulars on this schedule.

Schedule of Reinforced Concrete Footings		
Footing no.	Footing size	Footing reinforcing
A-1 A-2 A-4	$36'' \times 36'' \times 12''$	6 no. 4 ϕ rods EWH 6" cc
B-1 C-1	$48'' \times 48'' \times 12''$	no. 4 ϕ rods 6" cc EWH

5. Make plan drawings of two different types of wood forms for reinforced concrete columns.

6. A certain type of steel-welded reinforcing mesh for concrete has openings $4'' \times 8''$, the longitudinal wire is no. 8 gage, and the lateral wire is no. 12 gage. Write a commercial (abbreviated) way of describing this particular type of reinforcing mesh.

7. Give six considerations that an estimator must take into account in determining the types and numbers of form-ties required for any concrete estimate.

16
CONCRETE BLOCK AND BRICK CONSTRUCTION

This chapter is designed to give you an awareness of masonry block and brick estimating. The first part deals with concrete block construction and then develops, with some detail, types of bonds in brickwork and the effect of bonding in estimating.

As with all estimating there are so many variables in this area that it is impossible to be absolutely specific. Reasonable figures are given for average conditions. I have stated several times in the text and repeat again: There are no estimating figures comparable to your own. This is especially true with masonry construction.

Here are some of the things that can affect your estimate:

(a) the type of brick and the type of backing leaf to the face wall—that is, clay brick or large masonry units bonded with metal ties;

(b) the type of masonry or metal bond in which the units are laid up;

(c) the type of mortar and the kind of joints specified—it is easier for a mason to lay up units with lime mortar than with cement mortar;

(d) the thickness of the wall: the thicker the wall the less often the scaffold will have to be lifted; also the internal bricks will not require joints to be tooled by the mason, since apprentices can handle the inside bricks easily;

(e) the provision of adequate runways for buggies;

(f) the weather: hot weather requires more careful attention to wetting the bricks and caring for the mortar; cold, freezing weather will slow down the mason and his helper; also the bricks will have to be heated and so will the mortar—the work will have to be covered and kept heated;

(g) the higher the wall, the fewer number of bricks will be laid up per 8-hr day per unit gang of masons and helpers;

(h) the skill of the masons—when the industry is very busy, some less efficient masons may have to be employed;

(i) the efficiency of the mason foreman and the general foreman—for example, imagine having bricks unloaded too far from the masons;

(j) the most important single item that can make or break a job is the superintendent.

16.1 SIZES AND SHAPES OF CONCRETE MASONRY

The nominal sizes of concrete blocks are equal to the unit size plus one mortar joint in all three dimensions. As an example, a unit described as an 8 × 8 × 16 block has an actual manufactured size of $8'' \times 7\frac{3}{4}'' \times 15\frac{3}{4}''$ or $8'' \times 7\frac{5}{8}'' \times 15\frac{5}{8}''$; the $\frac{1}{4}''$ or $\frac{3}{8}''$ in each dimension is an allowance for the mortar (see Fig. 16.1).

When estimating masonry units, read the specifications very carefully for the type of block. There is a wide variety of size, texture, color, and structural qualities. There is a great emphasis on masonry unit construction at the present time. It is used extensively in warehouses, schools, hospitals, and public building construction, as structural members, curtain walls, and partition walls. Masonry is manufactured in both lightweight and standard heavyweight units. Use of this medium leads to a reduction in construction time and, consequently, to a saving in labor costs.

When estimating, take the face area of the unit concerned (unit plus mortar joints) and divide it into 100 sq ft of laid-up units. Thus a $7\frac{5}{8}'' \times 7\frac{5}{8}'' \times 15\frac{5}{8}''$ unit laid up will have a face area of 8 × 16 = 128 sq in. The number of such blocks contained in 100 sq ft of

wall area will be $100 \div \frac{128}{144} = 112.5$ blocks per 100 sq ft of wall area.

Some of the most difficult estimating problems arise in the estimating of time required to build some internal hospital walls. Such a wall may have to contain plumbers' pipes, heating engineers' lines, telephone conduits, electricians' conduits, bell wires, and TV outlets. In some walls more than 50 percent of all blocks will have to be cut and fitted.

On an average job with straight walls, a mason and his helper can lay up about 150 heavyweight blocks per day (8 × 8 × 16). Lightweight blocks will go faster—say, up to 180 per 8-hr day. In some hospital walls a mason would not lay up more than fifty to sixty—8 × 8 × 16 blocks in an 8-hr day.

Figure 16-1 was supplied by National Masonry Association and is reproduced here with the permission of the copyright holder.

Problem

Secure a price list from your nearest masonry unit block and tile manufacturer and make out an index card for every type of block offered, showing the number of blocks required to lay up 100 sq ft of wall area. Remember that index cards must be very carefully checked for accuracy. Once you have a good index-card system you can make a great saving in estimating time.

Remember that every unsuccessful bid is a loss to your company.

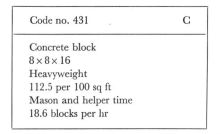

Code no. 431	C
Concrete block	
8 × 8 × 16	
Heavyweight	
112.5 per 100 sq ft	
Mason and helper time	
18.6 blocks per hr	

16.2 CONCRETE-BLOCK CONSTRUCTED WAREHOUSE

An extract of the specifications for a concrete-block constructed warehouse is as follows (see Fig. 16.2):

(a) Outside dimensions $20'-0'' \times 50'-0''$ and $11'-8''$ high, including the roof assembly (see Fig. 16.2).

(b) Allow for $7\frac{5}{8}'' \times 7\frac{5}{8}'' \times 15\frac{5}{8}''$ heavyweight concrete block construction. *(Note that 112.5 blocks displace 100 sq ft of built-up wall area and require $3\frac{1}{4}$ cu ft of mortar.)*

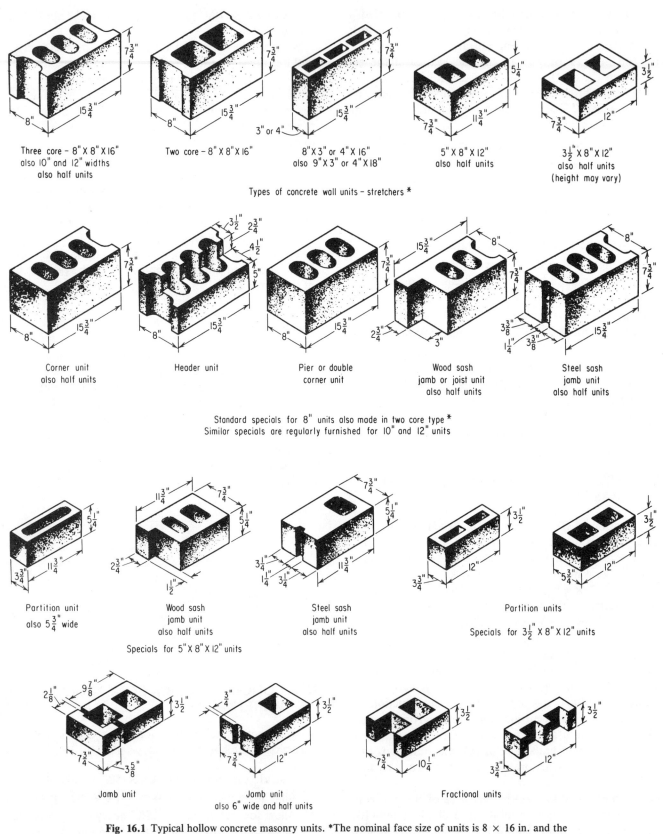

Fig. 16.1 Typical hollow concrete masonry units. *The nominal face size of units is 8 × 16 in. and the nominal thickness of units is 3, 4, 6, 8, 10, and 12 in. Units are generally manufactured to allow for ¼-, ⅜-, or ½-in. mortar joints. Designers should contact local sources of supply to determine dimensions of units available (Courtesy of national Concrete Masonry Association).

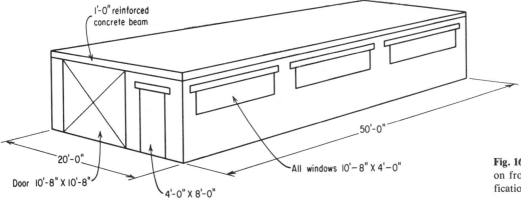

(c) Mortar is to be made of 1 part Portland cement with 20 percent by volume of lime and 3 parts of sand.

(d) Allow for two door openings $4'-0'' \times 8'-0''$ and $10'-8'' \times 10'-8''$, respectively.

(e) Allow for three windows each $10'-8'' \times 4'-0''$. *Note that doors and windows are always dimensioned length first.)*

(f) Allow for the small door and all windows to have precast reinforced lintels with end bearings of $0'-8''$.

(g) Allow for a 12″-deep continuous concrete beam over the top of the walls, reinforced with two no. 6 continuous reinforcing rods; concrete to be 2,500 psi, with a 5¾-gal paste and 1:1¼:2¼ mix.

Labor: Allow for a mason and helper to lay up 140 heavyweight concrete blocks per 8-hr day.

Mortar: To be made on the job.

Concrete: To be made on the job.

Scaffold: Allow for rentals.

Problems

1. What is the outside perimeter?

2. The walls are $0'-8''$ thick nominal. What is the inside perimeter?

3. What is the total area of all the openings?

4. What is the outside wall area, excluding the openings?

5. What is the inside wall area, excluding the openings?

6. What is the inside floor area?

7. How many concrete blocks will be required for the walls? (Disregard halves and take up to the next higher multiple of 10. Do not count for doubling the corners.)

8. Allowing for 3¼ cu ft of mortar for every 100 sq ft

of wall area, how much mortar to the next higher cu ft will be required to build the warehouse?

9. How many cu yd (to the next higher ⅓ cu yd) of concrete will be required for the continuous concrete beam?

10. What is the weight of the reinforcing rods for the beam? *(Note: Allow for a lap in the reinforcing rods equal to 16 times the diameter of the rod.)* See p. 000.

Estimate: You will need local prices, your reference library, and A.I.A. file to make this estimate.

Using local prices, make a complete labor and materials estimate for the following items:

(a) concrete blocks

(b) mortar

(c) precast lintels

(d) continuous concrete beam

(e) continuous concrete beam reinforcing

(f) scaffold rentals

Cost of Estimating:

(a) Allow for your time at $5 per hour. How much did working the problems for this warehouse cost?

(b) Would you like to pay an estimator the cost in (a), above?

(c) You must keep in mind that not only are neatness, accuracy, and method important, but the time factor is paramount.

16.3 BRICKWORK DEFINITIONS

The following definitions are given in explanation of some of the abbreviations and terms used in brickwork.

SCPI—Structural Clay Products Institute

ASTM—American Society for Testing Materials

CSA—Canadian Standards Association

SCR—Structural Clay Research

All these institutions publish excellent booklets, which you should have.

Course One horizontal row of masonry units that is laid up with a specified kind and thickness of mortar.

Stretcher A masonry unit laid up with its longest face showing on the face of the wall.

Header A brick laid up in a course with its end (header) showing on the face of a wall. The header is a bonding unit in walls of 0′-8″ thickness and over.

Metal Ties When brick walls are built up with stretcher courses only, they are bonded (bound) together with *metal ties.*

Face Brick A good quality brick that is laid up on the face of a wall where appearance is a dominant feature. It is laid up with cheaper backing bricks.

Backing Brick When the opposite side of a wall from the face brick is something other than a face brick, it is laid up with cheaper *backing bricks.* Larger masonry units may also be used as backing.

Bond When the appearance of masonry units are laid up so that no vertical joint is immediately above or below any other joint, the wall is in *bond.*

Wythe Each side of a cavity wall is a *wythe* (or *withe*). The opposite wythe of a face-brick wall is very often built up with larger masonry units. Each unit is 12″ or more square and may be obtained in varying thicknesses to meet the need. These units displace six or more backing bricks with only one operation of laying by the mason with a saving in labor costs. They are secured to the face bricks with metal ties.

Joints The visible finished mortar between horizontal and vertical masonry units are the *joints.* Specifications will state the kind of mortar, the thickness of horizontal and vertical joints, and the type of finished joints to be used throughout the work.

Cavity Walls The following is an extract from the Division of Research and National Research Council, Ottawa, Canada:

Cavity walls shall not be less than 0′-10″ thick, exclusive of parging or finish, and the cavity shall not be less than 2″ nor more than 3″ in width. The height of a cavity wall shall not be more than 25′-0″.

This latter definition shows that you must read your local and national building codes. Since cavity walls will require metal ties, they must be estimated. To find the number required, read (a) the specifications, (b) the local building code, and (c) the national building code.

16.4 BRICKWORK BONDING

There are a great number of variations of bonding for brick walls that are 0′-8″ and over in thickness. The architect or designer desires to combine beauty of design with sufficient strength for the required structure. In every type of masonry brick bond where facing bricks are used, there must be a compensating adjustment of face bricks to backing bricks. For a full treatment of bonding, secure a copy of *Bonds and Mortars in the Wall of Brick* from the Structural Clay Products Institute. Some of the most common types of bonds in brick walls, which every estimator should know, are as follows:

1. *Running bond* is a wall built with all stretcher courses and no vertical mortar joints immediately over or below any other vertical mortar joint. There are no masonry *brick bonding units* in this wall. The bonding agent is a metal tie. (See Fig. 16.3.)

   This type of wall is used for both cavity and brick veneer construction. *Let us for the purpose of this discussion assume that a running bond will require 100 percent face brick and 100 percent backing brick.*

2. *Stack bond* is a wall built with all stretcher courses and all vertical joints. (See Fig. 16.4.)

   This bond will require 100 percent face brick and 100 percent backing brick. The bonding agent is either a special header unit whose head equals in width its stretcher length, or steel reinforcing bedded into the horizontal mortar joints.

3. *Common bond* is similar to the running bond except that it may be specified to being bonded every fifth, sixth, or seventh course with a complete course of headers. (See Fig. 16.5.)

Fig. 16.3 Running bond.

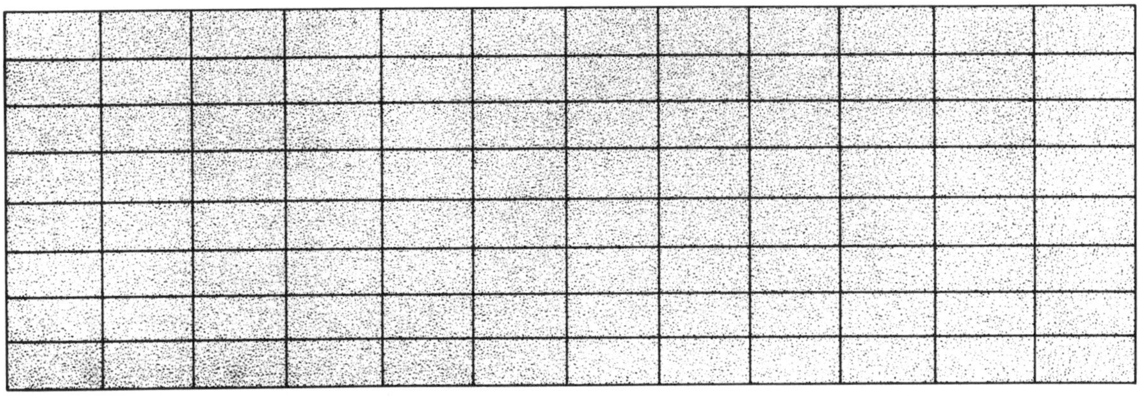

Fig. 16.4 Stack bond.

By inspection it may be seen that if expensive facing bricks are used with a cheaper backing brick, more facing bricks and less backing bricks will be required than would be necessary for a simple running bond.

Where a complete course of headers is specified for every fifth course in a 0'-8" wall, as many face bricks in five courses as can be laid up in a running bond of six courses will be required. A correction factor must be introduced. For the face bricks this would be a correction factor of ⅚ × 100 = 120 percent. For all common bonded brickwork having a complete fifth course of headers, add 20 percent extra to the estimated facing brick and deduct 20 percent from the estimated backing brick. *The correction factor is* ⅕. (a) Find the correction factor for facing brick in an 0'-8" common bond wall where every sixth course is a header. (b) Find the correction factor for facing brick in an 0'-8" common bond wall where every seventh course is a header. (c) How would you allow for facing and backing bricks in a 13" wall in (b)?

4. *English bond* has alternate courses of stretchers and headers. (See Fig. 16.6.)

This is a very strong brick wall bond, but it would

be very expensive if face bricks were used along with backing bricks. It would require as many face bricks to lay up two courses in the English bond as three courses in the running bond. Estimate for the running bond and apply the correction factor of + 50 percent to the facing brick estimate and − 50 percent of the backing brick. *Correction factor is* ½.

5. *English or Dutch bond* with full headers every sixth course and blind headers in intermediate courses. (See Fig. 16.7.)

The "design of the bond" is similar to the common bond. The only difference is that half-bricks (blind headers) are used to form the pattern of actual headers in every course except the sixth. As a consequence, the correction factor is the same as for the common bond, with actual headers every sixth course. *The correction factor is 16.7 or* ⅙.

6. *Flemish bond* has alternate stretchers and full headers in every course. This is a very popular and pleasing design. (See Fig. 16.8.)

By inspection it may be seen that where alternate stretchers and full headers are used in every course, it will require two face bricks to displace 1½ "visible" face bricks. The correction factor is

2" closer

Fig. 16.5 Common bond (two settings up). One setting up equals one header course and five stretcher courses.

Fig. 16.6 Old English bond. Alternating courses of stretchers and headers, stretchers perpendicularly aligned.

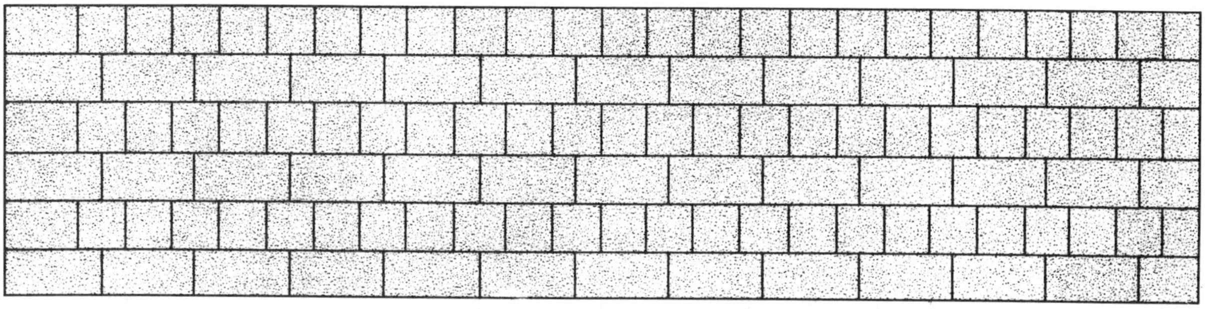

Fig. 16.7 English or Dutch bond with blind headers to form the pattern. Full headers every fifth, sixth, or seventh course.

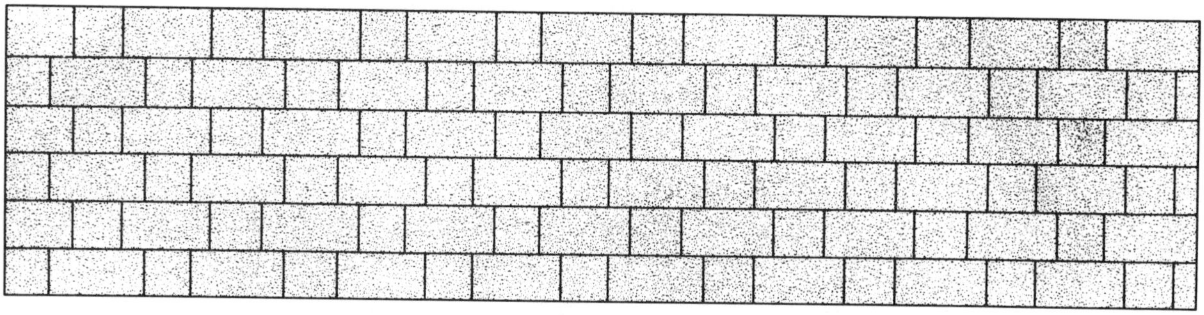

Fig. 16.8 Flemish bond.

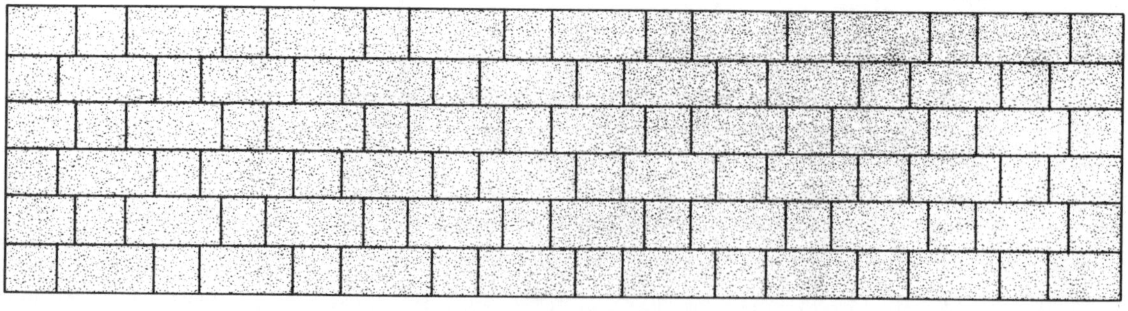

Fig. 16.9 Flemish bond with blind headers and full headers every sixth course.

$$\frac{2}{1\frac{1}{2}} \times 100 = 133\frac{1}{3}\%$$

Such a brickwork design requires $33\frac{1}{3}\%$ more face bricks and $33\frac{1}{3}\%$ less backing bricks than for a wall with a running bond; therefore, *the correction factor is* $\frac{1}{3}$.

7. *Flemish bond* has stretchers and headers every sixth course and intermediate courses with **blind headers** (halves). (See Fig. 16.9.)

   The first five courses will require the same number of face bricks as for the running bond. The sixth course will require 2 bricks to displace $1\frac{1}{2}$ visible bricks, or $\frac{2}{3} = 1\frac{1}{3}$ times as many face bricks in the sixth course for the Flemish bond than if the running bond was used throughout. The six complete courses will require

$$\frac{5 + 1\frac{1}{3}}{6} \times 100 = 105.6\%$$

The correction factor is $\frac{1}{18}$.

8. *Flemish cross bond* has alternating courses of stretchers and full headers. Note carefully that one course is in running bond and the next course is in Flemish bond. (See Fig. 16.10.)

   The correction factor will be: (a) 1 facing brick to displace 1 facing brick in the running bond for alternate courses; (b) $1\frac{1}{3}$ facing brick to displace 1 visible facing brick in the Flemish bond for alternate courses. Two courses will require

$$\frac{1 + 1\frac{1}{3}}{2} \times 100 = 116.7\%$$

The correction factor is $\frac{1}{6}$.

9. *Double header and stretcher every sixth course,* along with the other five courses, is the running bond. (See Fig. 16.11.)

   It will require 3 face bricks to displace 2 visible face bricks in the sixth course. The percentage factor, therefore, will be

$$\frac{5 + 1\frac{1}{2}}{6} \times 100 = 108.3$$

The correction factor is 8.3.

Find the correction factor for a double-header course in every fifth course with the other four courses in running bond.

16.5 ESTIMATING INFORMATION FOR CANADIAN CLAY BRICK

The following material has been excerpted from literature supplied by the Clay Brick Association of Canada and is reproduced here with permission.

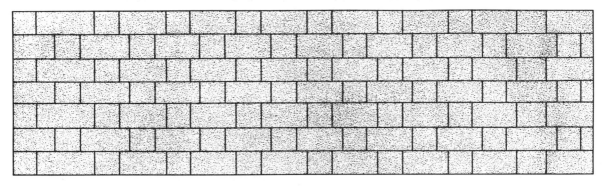

16.10 Flemish cross bond. Alternate course of stretcher and full headers.

Fig. 16.11 Double header and stretcher every sixth course.

16.6 ESTIMATING TABLES

The following estimating tables and technical notes for brick and tile construction are published by courtesy of the Medicine Hat Brick and Tile Company, Alberta, Canada. These tables are typical of many such publications available to you in your own state or province. Notice that it is published with the A.I.A. file no. 5/10 or 3 for your convenience in filing. Turn to the estimating tables and read "Bond Quantity Adjustments," p. 287.

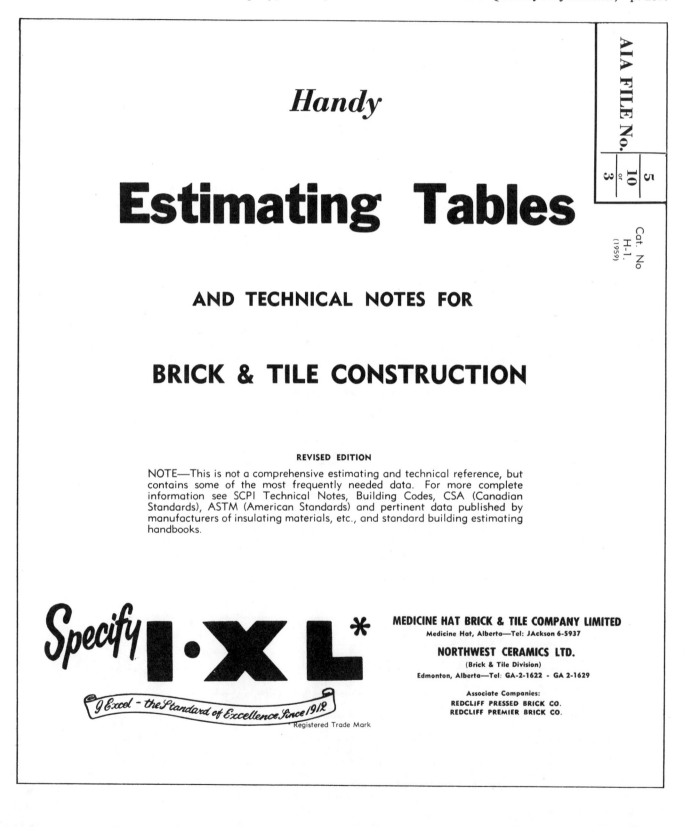

Handy

Estimating Tables

AND TECHNICAL NOTES FOR

BRICK & TILE CONSTRUCTION

AIA FILE No. 5 / 10 or 3

Cat. No
H-1.
(1959)

REVISED EDITION

NOTE—This is not a comprehensive estimating and technical reference, but contains some of the most frequently needed data. For more complete information see SCPI Technical Notes, Building Codes, CSA (Canadian Standards), ASTM (American Standards) and pertinent data published by manufacturers of insulating materials, etc., and standard building estimating handbooks.

Specify **I·XL***

I Excel - the Standard of Excellence Since 1912

Registered Trade Mark

MEDICINE HAT BRICK & TILE COMPANY LIMITED
Medicine Hat, Alberta—Tel: JAckson 6-5937

NORTHWEST CERAMICS LTD.
(Brick & Tile Division)
Edmonton, Alberta—Tel: GA-2-1622 - GA 2-1629

Associate Companies:
REDCLIFF PRESSED BRICK CO.
REDCLIFF PREMIER BRICK CO.

COURSING TABLE
FOR FACING BRICK & TILE

NUMBER OF COURSES	ROMAN BRICK 1½"x 3¾"x 11½"		SCR BRICK 2¼"x5½"x11½" NORMAN BRICK 2¼"x3¾"x11½" STANDARD BRICK 2¼"x 3¾"x 8"		TEXTURED FACING TILE #316 Series	#216 Series	NUMBER OF COURSES
	6C=11½"	8C=16"	6C=16"	4C=11"	3C=16"	2C=16"	
1	0'— 1⅞"	0'— 2"	0'— 2 2/3"	0'— 2¾"	0'— 5⅜"	0'—8"	1
2	0'— 3¾"	0'— 4"	0'— 5 1/3"	0'— 5½"	0'—10⅝"	1'—4"	2
3	0'— 5⅝"	0'— 6"	0'— 8 "	0'— 8¼"	1'— 4 "	2'—0"	3
4	0'— 7½"	0'— 8"	0'—10 2/3"	0'—11 "	1'— 9⅜"	2'—8"	4
5	0'— 9⅜"	0'—10"	1'— 1 1/3"	1'— 1¾"	2'— 2⅝"	3'—4"	5
6	0'—11¼"	1'— 0"	1'— 4 "	1'— 4½"	2'— 8 "	4'—0"	6
7	1'— 1⅛"	1'— 2"	1'— 6 2/3"	1'— 7¼"	3'— 1⅜"	4'—8"	7
8	1'— 3 "	1'— 4"	1'— 9 1/3"	1'—10 "	3'— 6⅝"	5'—4"	8
9	1'— 4⅞"	1'— 6"	2'— 0 "	2'— 0¾"	4'— 0 "	6'—0"	9
10	1'— 6¾"	1'— 8"	2'— 2 2/3"	2'— 3½"	4'— 5⅜"	6'—8'	10
11	1'— 8⅝"	1'—10"	2'— 5 1/3"	2'— 6¼"	4'—10⅝"	7'—4"	11
12	1'—10½"	2'— 0"	2'— 8 "	2'— 9 "	5'— 4 "	8'—0"	12
13	2'— 0⅜"	2'— 2"	2'—10 2/3"	2'—11¾"	5'— 9⅜"	8'—8"	13
14	2'— 2¼"	2'— 4"	3'— 1 1/3"	3'— 2½"	6'— 2⅝"	9'—4"	14
15	2'— 4⅛"	2'— 6"	3'— 4 "	3'— 5¼"	6'— 8 "	10'—0"	15
16	2'— 6 "	2'— 8"	3'— 6 2/3"	3'— 8 "	7'— 1⅜"	10'—8"	16
17	2'— 7⅞"	2'—10"	3'— 9 1/3"	3'—10¾"	7'— 6⅝"	11'—4"	17
18	2'— 9¾"	3'— 0"	4'— 0 "	4'— 1½"	8'— 0 "	12'—0"	18
19	2'—11⅝"	3'— 2"	4'— 2 2/3"	4'— 4¼".	8'— 5⅜"	12'—8"	19
20	3'— 1½"	3'— 4"	4'— 5 1/3"	4'— 7 "	8'—10⅝"	13'—4"	20
21	3'— 3⅜"	3'— 6"	4'— 8 "	4'— 9¾"	9'— 4 "	14'—0"	21
22	3'— 5¼"	3'— 8"	4'—10 2/3"	5'— 0½"	9'— 9⅜"	14'—8"	22
23	3'— 7⅛"	3'—10"	5'— 1 1/3"	5'— 3¼"	10'— 2⅝"	15'—4"	23
24	3'— 9 "	4'— 0"	5'— 4 "	5'— 6 "	10'— 8 "	16'—0"	24
25	3'—10⅞"	4'— 2"	5'— 6 2/3"	5'— 8¾"	11'— 1⅜"	16'—8"	25
26	4'— 0¾"	4'— 4"	5'— 9 1/3"	5'—11½"	11'— 6⅝"	17'—4"	26
27	4'— 2⅝"	4'— 6"	6'— 0 "	6'— 2¼"	12'— 0 "	18'—0"	27
28	4'— 4½"	4'— 8"	6'— 2 2/3"	6'— 5 "	12'— 5⅜"	18'—8"	28
29	4'— 6⅜"	4'—10"	6'— 5 1/3"	6'— 7¾"	12'—10⅝"	19'—4"	29
30	4'— 8¼"	5'— 0"	6'— 8 "	6'—10½"	13'— 4 "	20'—0"	30
31	4'—10⅛"	5'— 2"	6'—10 2/3"	7'— 1¼"	13'— 9⅜"	20'—8"	31
32	5'— 0 "	5'— 4"	7'— 1 1/3"	7'— 4 "	14'— 2⅝"	21'—4"	32
33	5'— 1⅞"	5'— 6"	7'— 4 "	7'— 6¾"	14'— 8 "	22'—0"	33
34	5'— 3¾"	5'— 8"	7'— 6 2/3"	7'— 9½"	15'— 1⅜"	22'—8"	34
35	5'— 5⅝"	5'—10"	7'— 9 1/3"	8'— 0¼"	15'— 6⅝"	23'—4"	35
36	5'— 7½"	6'— 0"	8'— 0 "	8'— 3 "	16'— 0 "	24'—0"	36
37	5'— 9⅜"	6'— 2"	8'— 2 2/3"	8'— 5¾"	16'— 5⅜"	24'—8"	37
38	5'—11¼"	6'— 4"	8'— 5 1/3"	8'— 8½"	16'—10⅝"	25'—4"	38
39	6'— 1⅛"	6'— 6"	8'— 8 "	8'—11¼"	17'— 4 "	26'—0"	39
40	6'— 3 "	6'— 8"	8'—10 2/3"	9'— 2 "	17'— 9⅜"	26'—8"	40
41	6'— 4⅞"	6'—10"	9'— 1 1/3"	9'— 4¾"	18'— 2⅝"	27'—4"	41
42	6'— 6¾"	7'— 0"	9'— 4 "	9'— 7½"	18'— 8 "	28'—0"	42
43	6'— 8⅝"	7'— 2"	9'— 6 2/3"	9'—10¼"	19'— 1⅜"	28'—8"	43
44	6'—10½"	7'— 4"	9'— 9 1/3"	10'— 1 "	19'— 6⅝"	29'—4"	44
45	7'— 0⅜"	7'— 6"	10'— 0 "	10'— 3¾"	20'— 0 "	30'—0"	45
46	7'— 2¼"	7'— 8"	10'— 2 2/3"	10'— 6½"	20'— 5⅜"	30'—8"	46
47	7'— 4⅛"	7'—10"	10'— 5 1/3"	10'— 9¼"	20'—10⅝"	31'—4"	47
48	7'— 6 "	8'— 0"	10'— 8 "	11'— 0 "	21'— 4 "	32'—0"	48
49	7'— 7⅞"	8'— 2"	10'—10 2/3"	11'— 2¾"	21'— 9⅜"	32'—8"	49
50	7'— 9¾"	8'— 4"	11'— 1 1/3"	11'— 5½"	22'— 2⅝"	33'—4"	50

BOND QUANTITY ADJUSTMENTS

To be applied to 4" wall quantities for estimating the required number of facing brick in various bonds.

(ADD TO FACING BRICK AND DEDUCT FROM BACKUP)

Type of Bond	Correction Factor	Type of Bond	Correction Factor
Common bond with full header every 5th course	20% or 1/5	Flemish bond, alternate stretchers and full headers every course	33.3% or 1/3
Common bond with full header every 6th course	16.7% or 1/6	Flemish bond, with stretchers and full headers every 6th course, intermediate courses with blind headers	5.6% or 1/18
Common bond with full header every 7th course	14.3% or 1/7		
English bond, alternate courses full headers	50% or 1/2	Flemish cross bond, alternate courses with stretchers and headers	16.7% or 1/6
English or Dutch bond with full headers every 6th course, and blind headers in intermediate courses	16.7% or 1/6	Double header and stretcher every 6th course	8.3% or 1/12
		Double header and stretcher every 5th course	10% or 1/10

ESTIMATING TABLES

QUANTITY OF BRICK OR FACING TILE AND MORTAR PER 100 SQ. FT. OF WALL

Type of Brick or Tile	Size (Inches)	No. of Brick or Tile		Cu. Ft. of Mortar (Inc. 10% added for waste)	
		3/8" joints	1/2" joints	3/8" joints	1/2" joints
American Standard Brick	2¼x3¾x8	654	617	6.1	7.7
Roman Brick	1½x3¾x11½	640	600	6.8	9.0
Norman Brick	2¼x3¾x11½	460	440	5.4	7.2
SCR Brick	2¼x5½x11½	–	440	–	10.1
316 Textured Facing Tile	4⅞x11½ face				
4" Wythe		–	225	–	2.8
6" Wythe		–	225	–	3.0
8" Wythe		–	225	–	3.2
10" Wythe		–	225	–	3.4
216 Textured Facing Tile	7½x11½ face				
4" Wythe		–	150	–	2.4
6" Wythe		–	150	–	2.6
8" Wythe		–	150	–	2.8
10" Wythe		–	150	–	3.0

QUANTITY OF PARTITION TILE AND MORTAR PER 100 SQ. FT.

Type of Tile (Inches)	Laid in	No. of Tile 1/2" joint	Cu. Ft. Mortar (Incl. 10% added for waste) 1/2" joint
3x12x12 Partition Tile	Horizontal Bed	100	1.8
4x12x12 Partition Tile	Horizontal Bed	100	2.0
5x12x12 Partition Tile	Horizontal Bed	100	2.7
6x12x12 Partition Tile	Horizontal Bed	100	3.0
8x12x12 Partition Tile	Horizontal Bed	100	3.7
10x12x12 Partition Tile	Horizontal Bed	100	4.2
12x12x12 Partition Tile	Horizontal Bed	100	4.9

NOTE: In estimating order quantity of Brick & Tile it is standard practice to increase quantities 5% for Brick and 2% for Tile. For additional estimating data see SCPI Technical Notes 1-12.

VOLUME FACTORS OF VARIOUS CONCRETE MIXES

KIND OF CONCRETE WORK	Mix by Volume Job Damp Materials			Work-ability or Con-sist-ency	Water Added at Mixer Per Bag, Gallons	A One Bag Batch Makes This Volume of Concrete Cu. Ft.	Total Water Per Bag, Gallons	Materials for One Cubic Yard of Concrete				Materials for 100 Square Feet 1" in Thickness		
	Cement Bags	Sand Cu. Ft.	Stone, Gravel Cu. Ft.					Cement Bags	Sand Cu. Ft.	Stone, Gravel Cu. Ft.	Water Added at Mixer Gallons	Cement Bags	Sand Cu. Ft.	Stone, Gravel Cu. Ft.
Footings, Heavy Foundations	1	3.75	5	stiff	6.4	6.2	8.00	4.3	16.3	21.7	27.6	1.34	5.02	6.71
Watertight Concrete for Cellar Walls and Walls Above Ground	1	2.5	3.5	medium	4.9	4.5	6.00	6.0	15.0	21.0	29.5	1.85	4.63	6.48
Driveways Floors Walks } One Course	1	2.5	3	stiff	4.4	4.1	5.50	6.5	16.3	19.5	28.7	2.03	5.09	6.08
Driveways Floors Walks } Two Course	1 / 1	Top 2 Base 2.5	0 / 4	stiff / stiff	3.6 / 4.9	2.14 / 4.8	6.00	12.6 / 5.7	25.2 / 14.2	22.8	45.3 / 27.8	3.89 / 1.75	7.78 / 4.38	7.01

AVERAGE WEIGHT OF SOLID BRICK WALLS

Brick assumed to weigh 4½ lbs. each — ½" mortar joints

Area in Sq. Ft.	4-inch Wall	8-inch Wall	12-inch Wall
1	36.782 lb.	78.808 lb.	115.414 lb.
10	368	788	1,154
20	736	1,576	2,308
30	1,103	2,364	3,462
40	1,471	3,152	4,617

TABLE FOR ESTIMATING FIREBRICK

For estimating on fire-brickwork, use the following figures: From 400 to 600 pounds of high temperature cement or fire clay are enough to lay one thousand nine-inch straight brick.
1 square foot 4½-inch wall requires 6 nine-inch straight brick.
1 square foot 9-inch wall requires 12 brick.
1 square foot 13½-inch wall requires 18 brick.
1 cubic foot of fire-brickwork requires 17 brick.
1 cubic foot of fire-brickwork weighs 125 to 140 pounds.
1000 brick (closely stacked) occupy 56 cubic feet.
1000 brick (loosely stacked) occupy 72 cubic feet.

NUMBER OF FACE AND COMMON BRICK IN PIERS

Laid with ½" Joints

Size	8" x 8"			8" x 12"			12" x 12"			12" x 16"			16" x 16"		
Feet High	Face	Common	Total	Face	Common	Total	Face	Common	Total	Face	Common	Total	Face	Common	Total
1	9	0	9	14	0	14	18	2	20	24	4	28	28	8	36
2	18	0	18	27	0	27	36	5	41	43	9	54	54	18	72
3	26	0	26	39	0	39	52	7	59	65	13	78	78	26	104
4	35	0	35	53	0	53	70	9	79	88	18	106	104	36	140
5	44	0	44	66	0	66	88	11	99	110	22	132	132	44	176
6	52	0	52	79	0	79	104	13	117	132	26	158	156	52	208
7	61	0	61	92	0	92	122	15	137	153	31	184	182	62	244
8	70	0	70	105	0	105	140	18	158	175	35	210	210	70	280
9	79	0	79	119	0	119	157	20	177	199	39	238	238	78	316
10	88	0	88	132	0	132	176	22	198	220	44	264	264	88	352

I·XL BRICK & TILE - *the standard*

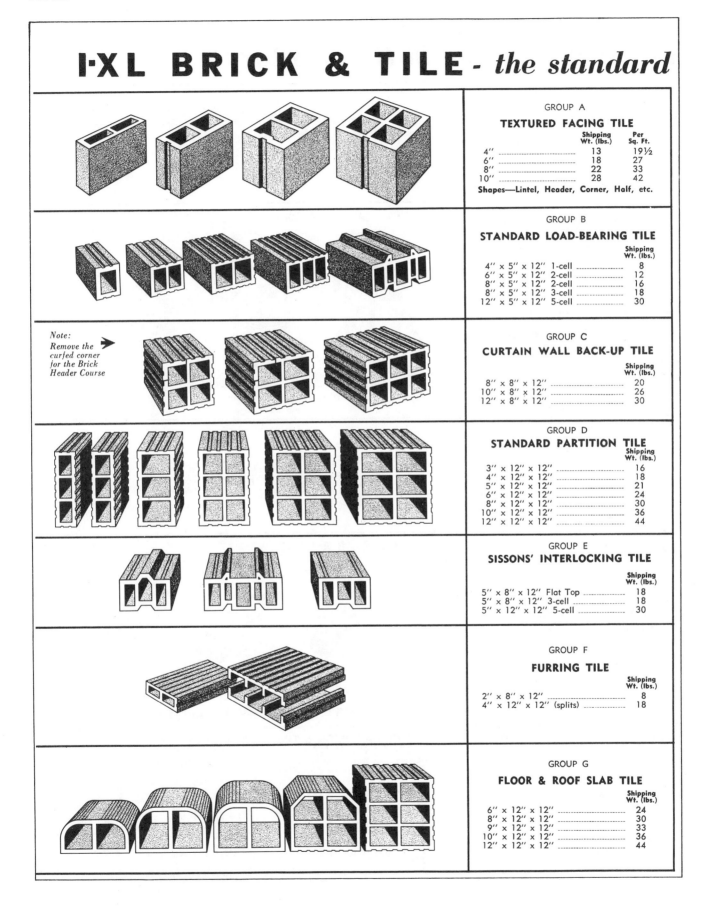

GROUP A

TEXTURED FACING TILE

	Shipping Wt. (lbs.)	Per Sq. Ft.
4″	13	19½
6″	18	27
8″	22	33
10″	28	42

Shapes—Lintel, Header, Corner, Half, etc.

GROUP B

STANDARD LOAD-BEARING TILE

	Shipping Wt. (lbs.)
4″ x 5″ x 12″ 1-cell	8
6″ x 5″ x 12″ 2-cell	12
8″ x 5″ x 12″ 2-cell	16
8″ x 5″ x 12″ 3-cell	18
12″ x 5″ x 12″ 5-cell	30

Note:
Remove the curved corner for the Brick Header Course ➤

GROUP C

CURTAIN WALL BACK-UP TILE

	Shipping Wt. (lbs.)
8″ x 8″ x 12″	20
10″ x 8″ x 12″	26
12″ x 8″ x 12″	30

GROUP D

STANDARD PARTITION TILE

	Shipping Wt. (lbs.)
3″ x 12″ x 12″	16
4″ x 12″ x 12″	18
5″ x 12″ x 12″	21
6″ x 12″ x 12″	24
8″ x 12″ x 12″	30
10″ x 12″ x 12″	36
12″ x 12″ x 12″	44

GROUP E

SISSONS' INTERLOCKING TILE

	Shipping Wt. (lbs.)
5″ x 8″ x 12″ Flat Top	18
5″ x 8″ x 12″ 3-cell	18
5″ x 12″ x 12″ 5-cell	30

GROUP F

FURRING TILE

	Shipping Wt. (lbs.)
2″ x 8″ x 12″	8
4″ x 12″ x 12″ (splits)	18

GROUP G

FLOOR & ROOF SLAB TILE

	Shipping Wt. (lbs.)
6″ x 12″ x 12″	24
8″ x 12″ x 12″	30
9″ x 12″ x 12″	33
10″ x 12″ x 12″	36
12″ x 12″ x 12″	44

of excellence for half a century

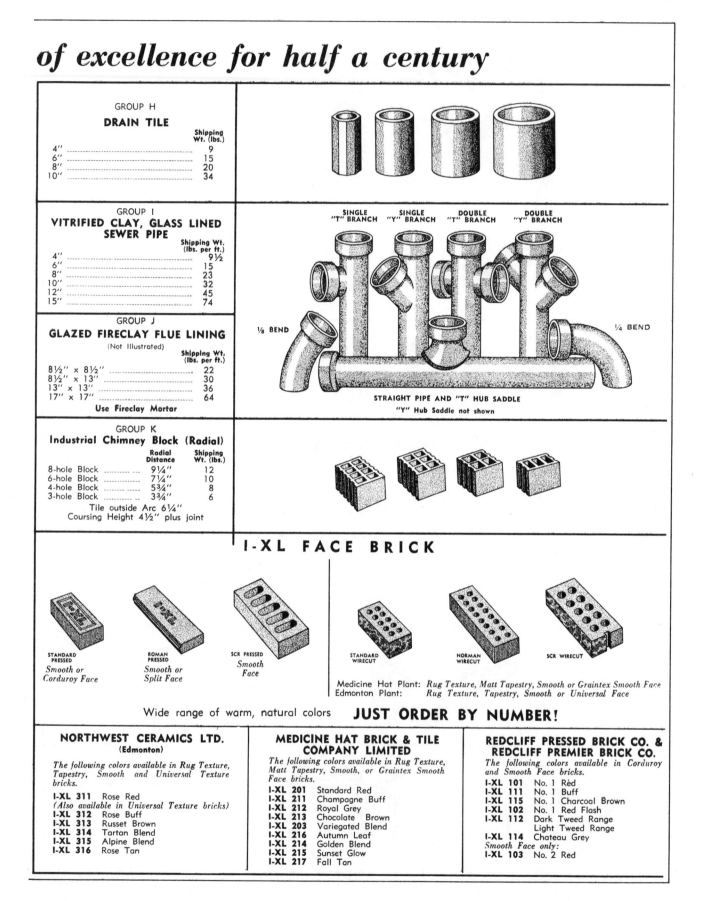

GROUP H
DRAIN TILE

	Shipping Wt. (lbs.)
4"	9
6"	15
8"	20
10"	34

GROUP I
VITRIFIED CLAY, GLASS LINED SEWER PIPE

	Shipping Wt. (lbs. per ft.)
4"	9 ½
6"	15
8"	23
10"	32
12"	45
15"	74

GROUP J
GLAZED FIRECLAY FLUE LINING
(Not Illustrated)

	Shipping Wt. (lbs. per ft.)
8½" x 8½"	22
8½" x 13"	30
13" x 13"	36
17" x 17"	64

Use Fireclay Mortar

GROUP K
Industrial Chimney Block (Radial)

	Radial Distance	Shipping Wt. (lbs.)
8-hole Block	9¼"	12
6-hole Block	7¼"	10
4-hole Block	5¾"	8
3-hole Block	3¾"	6

Tile outside Arc 6¼"
Coursing Height 4½" plus joint

SINGLE "T" BRANCH SINGLE "Y" BRANCH DOUBLE "T" BRANCH DOUBLE "Y" BRANCH

⅛ BEND   ¼ BEND

STRAIGHT PIPE AND "T" HUB SADDLE
"Y" Hub Saddle not shown

I-XL FACE BRICK

STANDARD PRESSED
Smooth or Corduroy Face

ROMAN PRESSED
Smooth or Split Face

SCR PRESSED
Smooth Face

STANDARD WIRECUT

NORMAN WIRECUT

SCR WIRECUT

Medicine Hat Plant: *Rug Texture, Matt Tapestry, Smooth or Graintex Smooth Face*
Edmonton Plant: *Rug Texture, Tapestry, Smooth or Universal Face*

Wide range of warm, natural colors   **JUST ORDER BY NUMBER!**

NORTHWEST CERAMICS LTD.
(Edmonton)

The following colors available in Rug Texture, Tapestry, Smooth and Universal Texture bricks.

I-XL 311	Rose Red
(Also available in Universal Texture bricks)

I-XL 312	Rose Buff
I-XL 313	Russet Brown
I-XL 314	Tartan Blend
I-XL 315	Alpine Blend
I-XL 316	Rose Tan

MEDICINE HAT BRICK & TILE COMPANY LIMITED

The following colors available in Rug Texture, Matt Tapestry, Smooth, or Graintex Smooth Face bricks.

I-XL 201	Standard Red
I-XL 211	Champagne Buff
I-XL 212	Royal Grey
I-XL 213	Chocolate Brown
I-XL 203	Variegated Blend
I-XL 216	Autumn Leaf
I-XL 214	Golden Blend
I-XL 215	Sunset Glow
I-XL 217	Fall Tan

REDCLIFF PRESSED BRICK CO. & REDCLIFF PREMIER BRICK CO.

The following colors available in Corduroy and Smooth Face bricks.

I-XL 101	No. 1 Red
I-XL 111	No. 1 Buff
I-XL 115	No. 1 Charcoal Brown
I-XL 102	No. 1 Red Flash
I-XL 112	Dark Tweed Range
	Light Tweed Range
I-XL 114	Chateau Grey

Smooth Face only:

I-XL 103	No. 2 Red

MORTAR DATA

MORTAR COLORS should consist of inorganic compounds. As a general rule they should not (with the exception of Carbon Black) be used in concentrations exceeding 10 to 15 per cent of the weight of the cement. The use of Carbon Black should be limited to 2 to 3 per cent of the weight of cement.

FOR BEST RESULTS mix with white cement, white lime and white silica sand. Mix color with the cementitious materials prior to addition of the sand. To keep color uniform it is generally preferable to lay up the masonry in regular grey mortar, then rake out joints and grout in the colored mortar continuously in cases where work interruptions can be expected.

BASIC RULES FOR COLORED MORTAR

1) When extending or adding to existing buildings, ensure that mortar color matches the mortar previously used. Overlooking this point (and it is too often overlooked!) is detrimental to the final appearance of the entire structure.

2) Because of the small amounts of color additive used in many cases (see table at right), errors in measurement can affect the color of the mortar. In addition to careful measurement, it is a good idea to mix colored mortar in sufficiently large batches, with a 1/3 carry-over for mixing into the next batch.

3) In larger buildings—or wherever desired—the masonry can be laid up in standard grey mortar—the joints then raked out and, when ready, the colored mortar grouted in and tooled to the required finish. Joints should first be dampened and, to avoid "crazing", the grout should be sufficiently dry to prevent residual shrinkage.

4) Care should be taken to prevent colored mortar from "sloughing over" and staining the face of the brickwork.

MORTAR COLORS

Mortar coloring compounds are mixtures of various metallic compounds and carbon black, white cement, white sand, etc.

The following is an incomplete but commonly available list of commercially available colors.

	Northern Pigment Company Limited	Sonobrite (Sonneborn Ltd.)	COMMENT
Green	M - 1	Strong green	½ bag color/bag cement—
Buff	M - 2	Yellow	Tint color only
Black	M - 3	Carbo-jet	1 bag color/bag cement—
Red	M - 4	Tile red	Standard color
Chocolate	M - 5	Chocolate brown	2 bags color/bag cement—
Cocoa	M - 6	Medium brown	Dark color

HOW TO MIX MORTAR COLOR

The full value of mortar color is only obtained by thorough mixing. Careless mixing not only wastes the color, but also may result in spots and streaks. Each batch should be mixed an equal length of time.

The color, sand and cement should be mixed together dry until the mass is free from spots and streaks of color when smoothed over with a trowel. This mixing may be done by passing the materials several times through a screen of ¼-inch mesh.

After adding water, the mixing must be continued until all spots and streaks of color disappear. It is essential to uniformity that the same amount of water be used in each batch.

MORTAR TYPES: Their uses and compressive strengths

ASTM Mortar Type		Approximate Proportions By Volume (See ASTM C270-57 for comprehensive details)			Average Compressive Strength (PSI)		Recommended Uses
Former Description	New Description	Portland Cement :	Hydrated Lime or Lime Putty :	Sand Aggregate	7 days	28 days	
A1	M	1 :	¼ :	3	1500	2500	High strength mortar suitable for general use, but especially recommended for Reinforced Brick Masonry and Plain Masonry below grade and in contact with earth.
A2	S	1 :	½ :	4½	1100	1800	Suitable for most reinforced Brick Masonry depending on strength required. Ideal for cavity wall to full allowable heights of 25 feet.
B	N	1 :	1 :	6	450	750	Medium strength. For general use including exposed, load bearing applications.
C	O	1 :	2 :	6	200	350	Low strength. For non-load bearing walls. Can also be used in load bearing applications where exposure is not severe and compressive stresses do not exceed 100 lbs. psi. Not recommended where high lateral strength is required.

QUANTITY OF MORTAR PER 1,000 BRICKS

Width of Joint (Inches)	Thickness of Wall, Inches			
	4		8, 12, 16 and over	
	Cu. Ft.	Cu. Yd.	Cu. Ft.	Cu. Yd.
¼	5.7	0.21	8.7	0.32
⅜	8.7	0.32	11.8	0.44
½	11.7	0.43	15.0	0.56
⅝	14.8	0.55	18.3	0.68
¾	17.9	0.66	21.7	0.80

HOW TO MAKE WATER REPELLENT MORTAR

Add 2% by weight of calcium stearate (a powder) to the cement, lime or mixture of the two used in the mortar. Mortar so treated has been found very effective in preventing the passage of moisture where only ordinary forces are met. Joints of coping, window sills and parapet walls all need special care. In repointing joints that have cracked, water repellent mortar is called for, as such joints are usually in highly exposed locations.

FOR CHEMICAL-RESISTANT MORTAR see SCPI Technical Notes, Vol. 4, No. 11.

MORTAR AGGREGATE should be **only** clean, well-graded, sharp sand—free from organic material, iron pyrites or any scum-producing ingredients. Mortar Aggregates should meet the minimum requirements of ASTM Spec. C144.

ADMIXTURES— Laboratory tests seem to indicate that water repellents and clay plasticizing agents increase the workability of mortars, and their resistance to weathering. The use of salt for lowering the freezing point and sugar for retarding set should be prohibited.

NOTES ON

SCR *insulated cavity wall*

CONSTRUCTION TECHNIQUE

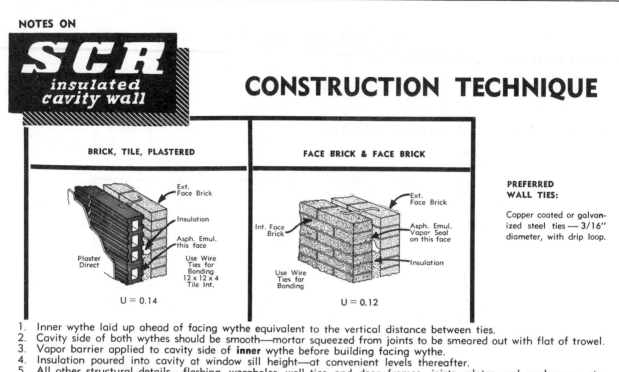

BRICK, TILE, PLASTERED	FACE BRICK & FACE BRICK	

BRICK, TILE, PLASTERED — Ext. Face Brick, Insulation, Asph. Emul. this face, Plaster Direct, Use Wire Ties for Bonding 12 x 12 x 4 Tile Int. — U = 0.14

FACE BRICK & FACE BRICK — Int. Face Brick, Ext. Face Brick, Asph. Emul. Vapor Seal on this face, Insulation, Use Wire Ties for Bonding — U = 0.12

PREFERRED WALL TIES:

Copper coated or galvanized steel ties — 3/16″ diameter, with drip loop.

1. Inner wythe laid up ahead of facing wythe equivalent to the vertical distance between ties.
2. Cavity side of both wythes should be smooth—mortar squeezed from joints to be smeared out with flat of trowel.
3. Vapor barrier applied to cavity side of **inner** wythe before building facing wythe.
4. Insulation poured into cavity at window sill height—at convenient levels thereafter.
5. All other structural details—flashing, weepholes, wall ties and door frames, joists, plates and anchorage, etc. follow conventional practice.

INSULATION REQUIREMENTS

When in place, the insulation must permit the cavity to continue to function as a barrier to moisture penetration and not permit moisture to be transmitted across the cavity.

It must be capable of supporting its own weight in the cavity without settling, wet or dry.

It should be inorganic or have comparable rot, termite and fire resistance properties, and have high insulating efficiency. (Fibreglas pour-type or Vermiculite pour-type or equivalent).

WATER EMULSION ASPHALT PAINT

Flintkote 71 or equivalent. If necessary, thin with **water** to brushing consistency. Apply with soft bristle brush to avoid scrubbing out mortar joints—keep brush in water when not in use.

THERMAL RATINGS FOR TYPICAL WALL CONSTRUCTIONS

(Approximate ratings shown)

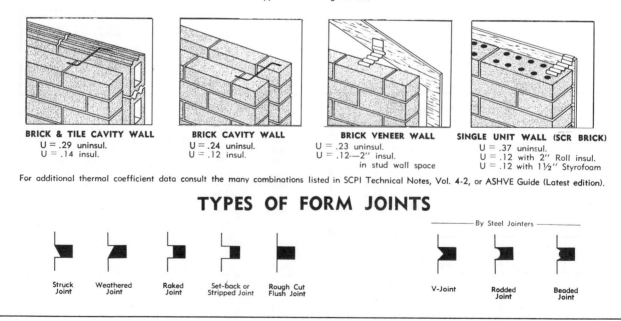

BRICK & TILE CAVITY WALL
U = .29 uninsul.
U = .14 insul.

BRICK CAVITY WALL
U = .24 uninsul.
U = .12 insul.

BRICK VENEER WALL
U = .23 uninsul.
U = .12—2″ insul.
in stud wall space

SINGLE UNIT WALL (SCR BRICK)
U = .37 uninsul.
U = .12 with 2″ Roll insul.
U = .12 with 1½″ Styrofoam

For additional thermal coefficient data consult the many combinations listed in SCPI Technical Notes, Vol. 4-2, or ASHVE Guide (Latest edition).

TYPES OF FORM JOINTS

——— By Steel Jointers ———

Struck Joint Weathered Joint Raked Joint Set-back or Stripped Joint Rough Cut Flush Joint

V-Joint Rodded Joint Beaded Joint

MISCELLANEOUS TECHNICAL DATA

NOTES ON FLASHING AND REMOVAL OF EFFLORESCENCE, SCUM OR STAINS

FLASHING—No flashing at all is better than poor flashing. The intention of flashing is to direct the travel of any moisture that might collect in the masonry. This travel is directed to weep holes, which should be provided at intervals of 18" to 2', permitting any water which accumulates on the flashing to drain away. To merely specify flashing and then trust to chance that it will be effective has proved disappointing in many installations. Flashing material should be selected that is not subject to corrosion, fracture or other loss of life expectancy—whether it is metallic, mastic or fibre. The vulnerable parts of a masonry structure where flashing is necessary and recommended are:

- At the grade line
- Sills and heads of openings
- Intersection of vertical and horizontal surfaces
- Spandrel beams
- Parapet walls
- Locations with high vapor differentials on either side.

For further information see SCPI Technical Notes, Vol. 1, Nos. 2 and 6

HOW TO REMOVE EFFLORESCENCE OR SCUM ON BRICK AND TILE

(See SCPI Technical Notes Vol. 7, Nos. 2 and 3)

Efflorescence cn brick or tile masonry walls is usually a light powder or crystallization caused by water soluble salts, deposited on the surface upon evaporation of the water. The salts (usually calcium sulphate, magnesium sulphate, sodium chloride, sodium sulphate or potassium sulphate are found in surface deposits of sand and gravel, etc. and in ground water. When in sufficient concentration they cause trouble, whether in the mortar, sand, mortar water, or even as ingredients of masonry units. In high quality clay products, surface clays are avoided because of this problem.

The presence of moisture in sufficient quantities to carry the salts to the surface is a second cause of efflorescence. Such moisture may be due to faulty construction, or too much water used in construction. In the latter case, a final cleaning will usually solve the problem. If it persists, try to find where the water is entering the wall, so the situation can be corrected.

Treatment: If the walls are water-tight, efflorescence will usually disappear after a few rains. If not, water and a mild detergent applied with a scrubbing brush will frequently do the job. If neither treatment works, wet the wall, then scrub it with water containing not more than one part muriatic (hydrochloric) acid to nine parts water. Immediately thereafter, thoroughly rinse the wall with plain water. It is vital that the wall be well rinsed with water, both before and after acid washing. Woodwork and exposed skin, etc., should be protected from the acid. Sometimes, it is desirable to give the surface a final washing with water containing approximately 5% household ammonia as a neutralizer. Before applying acid or basic reagents, make a small test of proposed cleaning procedure to see that immediate or eventual success is assured.

REMOVAL OF YELLOW AND GREENISH STAINS FROM BUFF BRICK AND TILE

Occasionally, a green or yellow stain will appear on buff or grey facing brick or tile. This may be a form of efflorescence resulting from vanadium or molybdenum compounds in the clay unit. If the surface is exposed to the weather, the stains will usually disappear within a few months. If the stain persists, chemical removal may be undertaken. Hydrochloric acid should **not** be used in attempting to remove efflorescence resulting from such compounds. The acid may react with the vanadium or molybdenum compounds, converting them to an insoluble brown stain that is practically impossible to remove, other than with abrasives.

A cleaning method that has been used successfully in many such cases is to wash the wall with a solution of caustic soda, such as one part "Drano" to ten parts water. Here again, the wall should be washed with clear water, both before and after the application of the caustic soda solution. Treat not more than 10 sq. ft. at a time. Precautions should be taken to protect the clothing and skin of the person using the solution.

There has also been used effectively a solution of 10% nitric acid, 10% hydrogen peroxide and 80% water. It is also reported that a 10% solution of potassium carbonate is effective in the removal of green and yellow stains in buff brick.

FIRE RESISTANCE RATINGS FOR I-XL MASONRY WALLS

	Nominal Wall Thickness (inches)	Wall Constructed of:	No. of Cells in Wall Thickness	Percent Solid %	PLAIN (No plaster)	PLASTERED (On at ½" of 1-3 Sanded Gypsum Plaster) One Side (Fire Side)	PLASTERED Two Sides
"A"—Tile commonly used for partitions.			"B"—Tile commonly used for load-bearing walls.			"C"—Brick-faced tile back-up walls.	
A	3	3x12x12	1	57.2	25m	45m	70m
A	4	4x12x12	1	47.2	25m	50m	75m
AB	4	4x 5x12	1	48.5	25m	50m	75m
AB	4 (TFT)	4x 5x12	1	61.7	30m	75m	90m
A	6	6x12x12	1	36.8	22m	56m	78m
AB	6 (TFT)	6x 5x12	2	63.0	106m	160m	210m
A	8	8x12x12	2	41.6	85m	112m	170m
AB	8	8x 5x12	2	47.7	114m	172m	230m
B	8	8x 5x12	3	50.7	2.8h	3.7h	4.6h
B	8 (TFT)	8x 5x12	2	52.0	2.2h	3.3h	4.4h
A	10	10x 8x12	3	39.3	2.1h	2.9h	3.7h
B	10 (TFT)	10x 5x12	3	56.0	3.5h	4.4h	5.8h
A	12	12x12x12	3	38.5	2.3h	3.3h	3.8h
A	12	12x 8x12	3	41.6	2.6h	3.7h	4.3h
B	12	12x 5x12	5	53.3	5.0h	6.0h	7.0h
C	8 Brick & Tile	AT LEAST 40% solid tile in backup			3.5h	4.0h	
C	12 " "	40% solid tile in backup			6.0h	7.0h	
C	12 " "	50% solid tile in backup			7.0h	8.0h	
AC	4	Single Brick		75% min.	1.3h	1.75h	2.5h
C	8	2 Bricks Thick			5.0h	6.0h	7.0h
C	12	3 Bricks Thick			10.0h	10.0h	12.0h

NOTE—These figures are for walls with no framed-in combustible, or only incombustible framed-in members in wall. To obtain equivalent fire ratings on the basis of pounds of combustibles per sq. ft. of floor area, the following conversions can be used: 1 hour—10 pounds; 1½ hours—15 pounds; 2 hours—20 pounds, etc.

16.7 BRICKWORD ESTIMATING

There are two ways to estimate brickwork: (a) the square-foot method, and (b) the cubic-foot method.

The square-foot method is very direct and simple providing the walls are not badly cut up by different thicknesses of walls, piers, buttresses, and so on.

Example

Step 1: See the specimen general estimate sheet, p. 295.

Step 2: Complete the name of the building, location, and so on above the double lines on every sheet applicable to each specific job estimate. *Be sure to write in the sheet numbers. This is very important to insure that no sheets are lost.*

Step 3: Under "Description of Work," identify the type of brick. In this case, it is standard pressed corduroy face bricks in Flemish bond and ½″ struck joint. (See Fig. 16.12.)

Step 4: In the "Dimension" column, write the overall wall dimensions, 129′-0″ × 27′-0″.

Step 5: In the first "Extension" column, write the calculated sq ft, 3,483, and underline. Do not write sq ft after the number, since all construction men know this. *Do not write any word, figure, sign, or symbol on the general estimate sheets unless it is indispensable.*

Step 6: In the first "Extension" column, deduct "Outs." *This is the most critical part of brickwork estimating.* "Outs" are the deductions made from the overall wall area for such items as glass block, decorative ornaments, exposed concrete columns and beams, concrete stone lintels, and window sills, or any other item that displaces brickwork. See windows and lintels on the specimen general estimate sheet.

Step 7: Where windows are shown as "Outs" on the specimen sheet, the number of equal-size windows are shown in the "No. of pieces" column. Notice that two different sizes of windows are shown and that the aggregate total of windows in the west wall is 18. *When estimating brickwork, be quite sure to count the total number of window openings on the elevation drawings and then check that you have indeed accounted for the same number of openings or displacements on your general estimate sheets.*

Step 8: Again examine the general estimate sheet: "West Wall," extension no. 1, "3,483" is underlined. This underlining indicates that this figure is not to be taken into any further account in this column, but it is carried as a live figure to extension column 2. The "Outs" are all entered and cast up in column 1 and carried as an underlined total of 1,760 to column 2. Total 1,760 is now deducted from the live figure of 3,483 in column 2 and is carried to "Total Estimated Quantity" column as 1,723.

Step 9: Each wall is taken off in this manner and the total sq ft of each wall is then carried to a "Summary of Face Bricks," as shown, for a final total of 6,460 where the indication sq ft is written for the first time.

Step 10: Consult any brickwork reference from your A.I.A. file or reference library for the number of bricks required per 100 sq ft of wall (for the specified bond), which in this case is 617 for a common bond. Add 33⅓ percent for correction factor for Flemish bond (see p. 287).

Step 11: In this case the correction factor for face bricks in Flemish bond amounted to 13,286 bricks. Assuming that the wall is 0′-8″ thick and that the backing bricks are the same size as the face bricks, the same correction factor would have to be applied to the backing bricks. In this case the backing bricks would be 39,858 less 13,286, making 26,572 backing bricks:

Face bricks	39,858	Backing bricks	39,858
Add correction		Deduct correction	
factor	13,286	factor	13,286
	53,144		26,572

Step 12: Add for waste. The waste on brick will vary with the skill of the masons and their helpers; a usual wastage factor is 5 percent. It should be the policy of any company to employ a loyal, skilful staff who enjoy the confidence of the administrative staff, so that errors and waste are always at a minimum.

16.8 BRICKWORK ESTIMATING: CUBIC-FOOT METHOD

When estimating brick-built buildings that are cut up with piers, offsets, and several thicknesses of walls, it is advantageous to estimate by the cu ft of brick masonry contained in the building and apply a factor of the

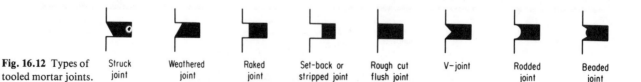

Fig. 16.12 Types of tooled mortar joints.   Struck joint   Weathered joint   Raked joint   Set-back or stripped joint   Rough cut flush joint   V-joint   Rodded joint   Beaded joint

ESTIMATE No. _47_

BUILDING _Rosemont School_

SHEET No. _23 of 39_

LOCATION _Mansfield_

ESTIMATOR _J. Smith_

ARCHITECTS _Bromley & Olsen_

CHECKER _H. D._

GENERAL ESTIMATE

SUBJECT _Schools_

DATE _17 June_

DESCRIPTION OF WORK	No. Pieces	DIMENSIONS			Extensions	Extensions	Total Estimated Quantity	Unit Price Mat'l	Total Estimated Material Cost	Unit Price Labor	Total Estimated Labor Cost
Brickwork:											
Standard pressed corduroy face bricks in Flemish bond, alternate stretchers and full headers every course; layed up with ½" struck joints											
West Wall:		129'-0"x27'-0"			3483	3483					
Outs:											
Windows	12/	6'-6"x8'-0"			624						
	6/	13'-0"x8'-0"			624						
Lintels	2/	118'-0"x1'-6"			354						
	3/	118'-0"x0'-8"			158	1760	1723				
East Wall:											
Outs and etc.											
Summary of face Bricks:											
West Wall:					1723						
North Wall:					1047						
East Wall:					2510						
South Wall:					1180	6460 sq. ft.					
		6460 @ 617 per 100 sq. ft.				39858	bricks				
Flemish bond add 33 1/3% correction factor						13286					
						53144				70/m	$ 3720
		add 3% for waste				1594					
						54738		120/m	$ 6578 —		

number of bricks contained in 1 cu ft. You may very quickly reckon how many bricks will displace 1 cu ft by taking the size of one brick plus the horizontal thickness of one mortar joint, plus the width of one brick, plus the width of one mortar joint, and dividing the capacity of one laid-up brick into 1,782 cu in. or 1 cu ft. For example, assume that a brick building is to be erected standard-sized bricks $8'' \times 2\frac{1}{4}'' \times 3\frac{3}{4}''$. Assume that the horizontal mortar joints are $\frac{1}{2}''$ and the vertical mortar joints are $\frac{1}{4}''$. The volume of one brick laid up in mortar is now

$$8\frac{1}{2}'' \times 2\frac{3}{4}'' \times 4'' = 90\frac{3}{4} \text{ cu in.}$$
$$1 \text{ cu ft.} = 1,728 \text{ cu in.}$$
$$1,728 \div 90\frac{3}{4} = 19 \text{ bricks (approx.)}$$

For this particular specification, estimate the number of cu ft of brickwork and multiply by 19. Carried to three decimal places it becomes 19.041.

When estimating by this method be sure to take off the quantities wall by wall and check (✔) off each wall on the drawing as it is dealt with. Deduct for "Outs" in the same manner as for the sq-ft method.

Care must be taken where the walls are built with facing bricks and backing bricks. The following method is suggested:

(a) Estimate all the bricks required by the cu-ft method.

(b) Estimate all the face bricks required by the *sq-ft method* and deduct (b) from (a).

(c) Add for waste in each case. The waste would be less for backing bricks than for facing bricks.

Problem

(a) Find the factor to apply in the cu-ft method of estimating common brick size $8'' \times 2\frac{1}{4}'' \times 3\frac{3}{4}''$. The mortar joints are $\frac{5}{8}''$ horizontal and $\frac{3}{8}''$ vertical. Take this factor to three decimal places.

(b) The size of a certain modular brick is $7\frac{1}{2}'' \times 3\frac{1}{2}'' \times 2\frac{1}{8}''$. Assuming that three courses will lay up $8''$ in height and that each brick with its modular bricks would lay up 1 cu ft of brick wall?

Note that not all modular bricks are exactly the same size, but they are all designed to lay up three courses in $8''$ of height, with the mortar joint $8''$ in length. The difference in the manufacture of modular sizes allows for a difference in thickness of mortar joints.

In all cases you must read the specifications to find the type and size of bricks and the type and thickness of mortar joints. With this information you can very quickly find your own adjusting factor of number of bricks per cu ft. If you do not have a card-index system, you should consider starting one and work out the factor to three decimal places for all sizes of bricks with different sizes of mortar joints.

Problem: Brick Warehouse Estimate

Examine the dimensioned drawing of the warehouse in Fig. 16.2, and, following the specifications and using the work-up sheets and the general estimate sheets, make a complete cost estimate of the brickwork required to complete the job. Allow for a waste of only 3 percent for bricks.

Specifications:

(a) Allow for all walls to be $0'-8''$ thick with common clay face bricks $8'' \times 2\frac{1}{4}'' \times 3\frac{3}{4}''$ and clay backing bricks of the same size.

(b) Allow for the small door and all windows to have precast lintels with bearings of $0'-8''$.

(c) There shall be five stretcher courses and one header course in all walls (common bond).

(d) There shall be 48 courses and 48 mortar joints above a concrete foundation in all walls.

(e) Struck mortar joints shall be approximately $\frac{1}{2}''$ bed and $\frac{1}{4}''$ vertical joints in all walls.

Price List:

Face bricks	$98 per M
Backing bricks	$88 per M
Mortar	$14 per cu yd
Allow	6 percent for fringe benefits
Allow	6 percent overhead expenses
Allow	10 percent for profit

Labor Rates: Allow for a mason and his helper to lay up about 825 of this type of brick per 8-hr day, at a rate of 9.7 hr per 1,000.

Rentals: Allow for rentals of tubular scaffold at the rate of $8 per day.

Use the handy estimating tables shown in this chapter for the number of bricks per sq ft of wall.

16.9 REFERENCE TABLES FOR ESTIMATING BRICK, CONCRETE BLOCK, AND MORTAR

The tables on p. 297 should be used when making estimating extensions for masonry units and mortar. Notice that no allowance has been made for waste. It is recommended that since an allowance for waste is always made on brickwork, and as the mortar is estimated against the total number of bricks required, no further allowance should be necessary.

For concrete block, make an allowance of 5 percent for waste on mortar. An untidy and careless mason can be very wasteful with mortar.

TABLE 1 BRICK AND MORTAR FOR 100 SQUARE FEET OF SOLID BRICK WALL (Quantities are net—allow for waste)

JOINT THICKNESS	1/4"	3/8"	1/2".	5/8"
4-INCH WALL				
Bricks—No.	698	**655**	616	581
Mortar—Cu Ft	4.0	**5.7**	7.2	8.6
Cu Yd	.148	**.211**	.267	.318
*Wt—1 Sq Ft Wall—Lb	33.6	**34.0**	34.2	34.6
8-INCH WALL				
Bricks—No.	1396	**1310**	1232	1161
Mortar—Cu Ft	12.2	**15.5**	18.6	21.3
Cu Yd	.452	**.574**	.689	.788
*Wt—1 Sq Ft Wall—Lb	72.6	**73.2**	73.9	74.3
12-INCH WALL				
Bricks—No.	2095	**1965**	1848	1742
Mortar—Cu Ft	20.3	**25.3**	29.9	34.0
Cu Yd	.752	**.937**	1.11	1.26
*Wt—1 Sq Ft Wall—Lb	111.4	**112.4**	113.4	114.1
16-INCH WALL				
Bricks—No.	2793	**2620**	2464	2322
Mortar—Cu Ft	28.5	**35.1**	41.4	46.7
Cu Yd	1.06	**1.30**	1.53	1.73
*Wt—1 Sq Ft Wall—lb	150.1	**151.0**	152.9	153.9

*Approximate wt wall = No. Bricks × 4.1 lb + Cu Ft Mortar × 125 lb.

Common brick 2¼" × 3¾" × 8". All joints filled.

TABLE 2 BLOCK AND MORTAR FOR 100 SQUARE FEET OF BLOCK WALL (Quantities are net—allow for waste)

MATERIALS	CONCRETE BLOCK Height × Length		
	7⅝" × 15⅝"	5" × 11¾"	3⅝" × 15⅝"
4-INCH WALL			
Blocks—No.	112.5	220	225
Mortar—Cu Ft	2.3	3.6	3.9
Cu Yd	.085	.133	.144
*Wt 1 Sq Ft Wall—Heavy Agt	28.5	30.0	30.5
—Lbs —Light Agt	20.5	21.5	21.5
6-INCH WALL			
Blocks—No.	112.5	220	225
Mortar—Cu Ft	2.3	3.6	3.9
Cu Yd	.085	.133	.144
*Wt 1 Sq Ft Wall—Heavy Agt	43.5	45.0	45.5
—Lbs —Light Agt	29.5	30.5	30.5
8-INCH WALL			
Blocks—No.	112.5	220	225
Mortar—Cu Ft	2.3	3.6	3.9
Cu Yd	.085	.133	.144
*Wt 1 Sq Ft Wall—Heavy Agt	55.0	56.5	57.0
—Lbs —Light Agt	36.0	37.0	37.0
12-INCH WALL			
Blocks—No.	112.5	—	—
Mortar—Cu Ft	2.3	—	—
Cu Yd	.085	—	—
*Wt 1 Sq Ft Wall—Heavy Agt	79.5	—	—
—Lbs —Light Agt	49.0	—	—

Data based on ⅜-inch mortar joint, with face-shell bedding.

*Approx wt of wall = actual wt single unit × no. units + 125 × cu ft mortar.

Actual wt of wall with heavy agt block usually within ± 7% table value.

Actual wt of wall with light agt block usually within ± 17% table value.

TABLE 3 1:3 MORTAR—QUANTITIES FOR 1000 BRICK—SIZE 2¼ × 3¾ × 8 in. (Quantities are net—allow for waste)

DRY MORTAR SAND Joint	Mortar Cu Ft	COARSE Cement Bags	Sand Cu Yd	MEDIUM Cement Bags	Sand Cu Yd	Cement Bags	Sand Cu Yd	FINE Cement Bags	Sand Cu Yd
4-INCH WALL									
¼"	5.7	1.9	.21	1.8	.20	1.7	.19	1.6	.18
⅜"	**9.0**	**3.0**	**.33**	**2.8**	**.32**	**2.7**	**.30**	**2.6**	**.28**
½"	11.7	3.9	.43	3.7	.41	3.5	.39	3.3	.37
⅝"	14.8	4.9	.55	4.7	.52	4.4	.49	4.2	.47
8-INCH WALL									
¼"	8.7	2.9	.32	2.7	.31	2.6	.29	2.5	.27
⅜"	**11.7**	**3.9**	**.43**	**3.7**	**.41**	**3.5**	**.39**	**3.3**	**.37**
½"	15.1	5.0	.56	4.8	.53	4.5	.50	4.3	.48
⅝"	18.3	6.1	.68	5.8	.64	5.5	.61	5.2	.58
12-INCH WALL									
¼"	9.7	3.2	.36	3.1	.34	2.9	.32	2.8	.31
⅜"	**13.0**	**4.3**	**.48**	**4.1**	**.46**	**3.9**	**.43**	**3.7**	**.41**
½"	16.3	5.4	.60	5.1	.57	4.9	.54	4.7	.51
⅝"	19.6	6.5	.73	6.2	.69	5.9	.65	5.6	.62
16-INCH WALL									
¼"	10.2	3.4	.38	3.2	.36	3.1	.34	2.9	.32
⅜"	**13.4**	**4.5**	**.50**	**4.2**	**.47**	**4.0**	**.45**	**3.8**	**.42**
½"	16.8	5.6	.62	5.3	.59	5.0	.56	4.8	.53
⅝"	20.1	6.7	.74	6.3	.71	6.0	.67	5.7	.63
Unit H × L	MORTAR QUANTITIES FOR 100 BLOCK								
8 × 16	2.1	.70	.078	.66	.074	.63	.070	.60	.066
5 × 12	1.5	.50	.056	.47	.053	.45	.050	.43	.047
4 × 16	1.6	.53	.059	.50	.056	.48	.053	.46	.050

Brick data based on common brick, all walls solid with full bed and head joints. Blocks data based on ⅜-in. mortar joints.

(Courtesy of Canada Masonry Cement).

Observe in Table 3, page 297 that the coarser the sand, the more cement required per cu ft. Why is ths?

16.10 ESTIMATING THE NUMBER OF BRICKS IN CHIMNEYS

The usual method for finding the number of bricks required in a chimney is to find the number of bricks per ft in the height of chimney and to multiply by the total height taken in ft.

In order to find the number of bricks required per ft in height, we must first know how many bricks are required for *one course* in height of the chimney and then how many courses there are in 1 ft in height.

Fig. 16.13 shows various sizes of chimneys and flues and how the bricks are bonded in each case. For each different size of flue and chimney, two layers of courses of brick are shown. This is done in order to show how the courses will bond. In some cases, one course will require more bricks than the other, as, for example, chimney (f). This will require 12 bricks for one course and 13 bricks for the other course. This makes an average of 12½ bricks per course.

From Fig. 16.14 we can see how many courses make up 1 ft in height. We can also figure out how many courses make up 1 ft in height. The standard brick is 2¼″ thick and the average joint is ½″ thick. This makes 2¾″ in height for every course. In 1 ft or 12″ in height, there are 12 ÷ 2.75 = 4.36 courses.

Chimney (a) in Fig. 16.13 is shown to have 6 bricks to each course, which would require 6 × 4.36 = 26.16— say, 27 bricks per ft in height. Chimney (f), which we found above to have an average of 12½ bricks per course, would require 12½ × 4.36 = 54.5 bricks per ft—say, 55 bricks per ft in height. To find the total bricks in a chimney, we must multiply the bricks per ft in height by the total height. Thus a chimney of the type (a), with 27 bricks per ft in height, if built 30-ft high would require 30 × 27 = 810 bricks.

Chimneys of types (m), (n), and (o), which are built as part of a wall, do not require as many extra bricks. Type (m) requires 4 extra bricks for each course. Type (n) requires only 2 extra bricks—that is, 2 more bricks for each course than would be required if the wall were built straight without the chimney.

How to Estimate Fireplaces

Fireplaces for chimneys of irregular shape that cannot be figured by lin ft in height are often figured by the cu-ft method. By this method the total cu ft of brickwork required is figured and the openings for the flues or ash pits are deducted.

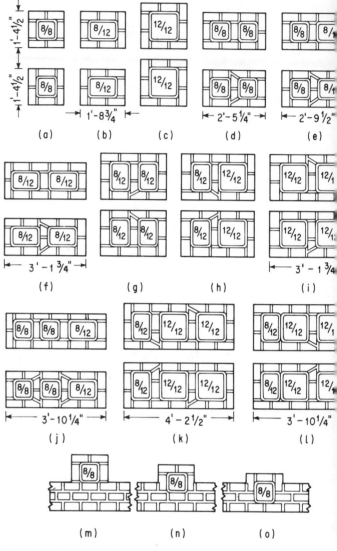

Fig. 16.13 Different sizes of chimneys and flues with their brick bonds.

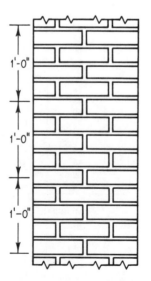

Fig. 16.14 Showing number of courses in 1 foot of height.

QUESTIONS

1. List ten items that can affect an estimate in both concrete block and brick construction.

2. How many concrete blocks, each $8'' \times 8'' \times 16''$ (when laid up), would be required for 750 sq ft of wall surface area?

3. Define lightweight and standard heavyweight concrete blocks.

4. Define the following brickwork terms:
   (a) course
   (b) stretcher
   (c) header
   (d) metal ties
   (e) face bricks
   (f) backing bricks
   (g) bond
   (h) wythe
   (i) joints
   (j) cavity wall
   (k) running bond
   (l) stack bond
   (m) common bond
   (n) English bond
   (o) Flemish bond

5. Complete the following diagram for the type of tooled brick mortar joint stated in each frame. The first one has been completed as a guide.

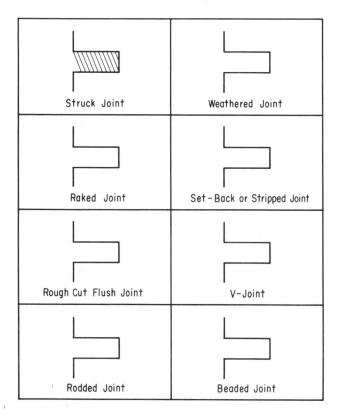

6. Which of the tooled joints in Question 5 would be most demanding on the mason's time?

17

FRAME CONSTRUCTION OF WALLS: EXTERIOR, INTERIOR, AND CLADDING

This chapter covers methods of estimating the quantities of rough carpentry frame wall construction and the cladding of such walls. Cladding comprises everything or anything that may be incorporated into frame walls (except service lines) from the outside to the finish on the inside; it also embraces interior wall cladding. See Fig. 9.1, p. 170, for a typical wall section.

Different methods of estimating wall framing are available to the builder according to the type of construction—namely, a single unit, multiple units, or prefabricated engineered units. *Remember there are no estimating figures, tables, or methods comparable to your own. They should be built up over a period of years against the background of local conditions.*

17.1 ROUGH CARPENTRY STUD WALL FRAMING

A Typical Specification for an Apartment Block

All outside wall and interior partitions shall be as noted on the drawings and as hereinafter specified.

All outside frame walls shall have 2 × 4 studs at 16″ OC with one bottom plate and two top plates.

All corridor partitions, all partitions between stairwells and suites, and all partitions between suites shall be staggered-stud partitions.

Staggered partitions shall be two rows of 2 × 4 at 16″ OC with a single bottom plate and two top plates of 2 × 6 material.

All staggered stud partitions shall be insulated with 2″ paper-backed fiberglass or other acceptable insulation approved by the architect.

All interior partition frame walls shall be provided with girths (girts) in their middle section.

All sides of openings such as doors, windows, and so on shall have double studs at the sides and two 2 × 8 on edge at the head. In addition to this, all openings in bearing partitions and openings over 3′-0″ shall be trussed.

Provide all girths, headers, fire stops, ribands, grounds, and nailing plates in all wall partitions and floor assemblies as necessary for all supports and backing for other trades to secure their units.

Note how frequently the word "all" appears in specifications. It is an embracing word and leaves no doubt in the mind of the estimator of the intentions of the architect. The word "shall" implies a mandatory feature.

17.2 STAGGERED-STUD WALL

This is one type of sound- and fireproofing construction and is acceptable in some areas (see Fig. 17.1).

17.3 ESTIMATING WOOD WALL FRAMING

The house design (Fig. 17-2) on pages 302-303 has been excerpted and reproduced here with the permission, from *Modest House Designs,* metric edition, published by Central Mortgage and Housing Corporation, Ottawa KIA OP7.

Canada adopted the metric system and contractors building this house in Canada would conform to metric

measurements. For the purpose of the following estimating exercise, use the scale of ⅛″ to 1′-0″.

Many jurisdictions may require masonry construction in all places in specifications where staggered walls are stated and shown.

17.4 GIRTHS

These are horizontal members secured between studs in the middle section of interior walls. The material is the same dimension as the framing. Their function is to stiffen the wall and obviate deflection. They also afford a fire stop.

17.5 GROUNDS

An example is shown under the baseboard of the typical wall section, Fig. 9.1, p. 170. This member is a plaster ground and is sometimes called a rough baseboard. It is set true to line so the plasterer can screed his finishing coat to a good straight line and, finally, the finishing carpenter can cover with a baseboard, also true to line.

Problem

Take off all the plates and studs for house design no. 133.

Method No. 1—Plate Take-Off

Step 1: Using an architect's scaled tape (the scale is ⅛″ to 1′-0″ on one side and ¼″ to 1′-0″ on the other side), measure off all the plates, both exterior and interior. Many estimators use an ordinary 12′-0″ tape.

Step 2: Start measuring with the tape at the top left-hand corner of the drawing and measure off the rear wall; do not retract the tape but turn the corner and add to the already payed-out tape the length of the right-hand wall. The tape should now read 36′-4″ plus 24′-4″, which is 60′-8″.

Again do not retract the tape, but pay out more tape and add to the measure already taken the length of the front wall, then the left wall. This concludes the measuring of the perimeter, which may be read on the tape as 121′-4″.

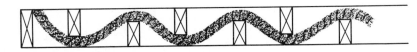

Fig. 17.1 Staggered stud partition with blanket roll insulation.

This three-bedroom bungalow, with basement and carport is planned to recognize the needs and constraints of a *homemaker in a wheelchair* by removing most of those barriers to movement and perception usually found in a conventional house.

The entrances are reached from the carport without the need to negotiate steps, and all doors in the house are slightly wider than usual to permit easy wheelchair passage. The kitchen includes the laundry centre and the bathroom allows ready access to the fixtures and shows a walk-in shower with seat instead of a bathtub. Other restrictions would be eased by lowering light switches, raising wall plugs; providing easily reached and operating windows, and lowering closet hanging rods.

Design

1.06 M

Area 96.0 m²

Left Side

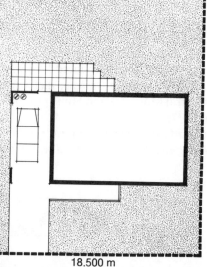

18.500 m

Lot Plan

All dimensions are given in millimetres unless otherwise
indicated.

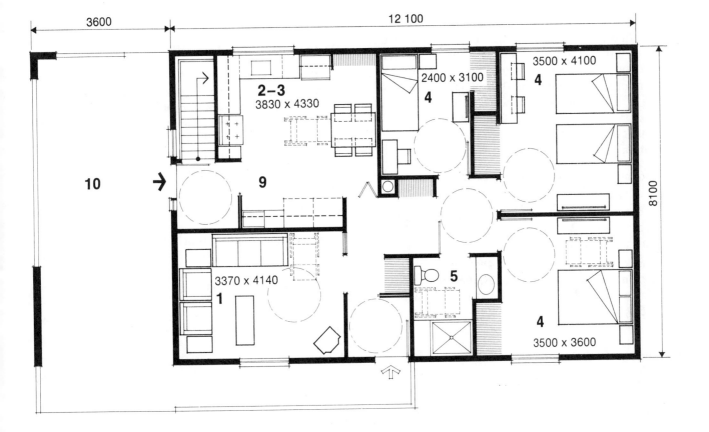

3600

12 100

8100

2–3
3830 x 4330

2400 x 3100

3500 x 4100

4

4

10

9

3370 x 4140

1

5

4

3500 x 3600

Wheelchair

1500 wheelchair turning circle

Again, do no retract the tape, but keep paying it out until all the interior walls in addition to the perimeter may all be read as one final figure on the tape.

Some estimators put a colored check mark (✔) on each wall after the length is measured.

Step 3: Remember that all the walls have one bottom plate and two top plates. The lin ft of walls, therefore, must be multiplied by 3 for the sum total of the lin ft of all the wall plates for the house. Remember to take off the plumbing wall for the bathroom.

Step 4: Order the material in random lengths, unless for economy of labor in construction it is advantageous to use specific lengths. Convert to fbm.

Step 5: In one lin ft of 2 × 4, there is ⅔ fbm of lumber. In one lin ft of 2 × 4 framed wall with one bottom plate and two top plates, there are two lin ft of lumber.

Therefore, the number of fbm of lumber required for the 2 × 4 plates (in all the walls) is the sum total of lin ft as read on the tape at Step 2 multiplied by 2.

Assume a small building has 156 lin ft of 2 × 4 framed walls.

(a) 156 lin ft of bottom plate plus two top plates in 468 lin ft of plates.

(b) Order in 12'-0" lengths for economy of construction: 468 ÷ 12 = 39. Order ³⁹⁄ 2 × 4—12'-0", which is 312 fbm.

(c) Check! In every lin ft of 2 × 4 frame wall with one bottom plate and two top plates, there are 2 bd ft of lumber. Therefore, in 156 lin ft of wall there are 156 × 2 = 312 fbm.

Method No. 1—Stud Take-Off

Once the actual lin ft of plates has been taken, divide this figure by the OCs of the studs. Assuming the total lin ft of plates is 156 and the studs are placed 16 OC, the number of studs required is 156 ÷ 1⅓, plus two extra studs for every corner, both internal and external, plus two extra studs for all openings, such as doors and windows. It is always surprising to most junior estimators to discover how many studs are needed around the bathroom area in a small house. Count the number of studs required to form the bathroom, the two linen closets, and the coat closet of the bathroom area of the house design in Fig. 17.2.

Points to Remember When Estimating Frame Walls

(a) All sides of openings in the walls will have to be doubled, so add two studs per opening.

(b) All header sizes will have to conform to either the specifications or local building code, whichever is the better standard.

(c) Make no deductions of studs or plates for standard-sized door and window openings.

(d) Remember to take off the different dimensioned lumber walls separately; for example, bathroom walls in residential construction are usually thicker to permit the installation of the plumbing services.

Method No. 2—Plate Take-Off

Use the same method as in Method No. 1.

Method No. 2—Stud Take-Off

Allow one stud for every lin ft of wall framing. *This method is only a rough approximation.* For a small building cut up into small rooms, this approximation would not allow for sufficient studs. On the other hand, for a building cut up into large rooms, this method would allow for too many studs. Although this method is sometimes used, Method No. 3 is much more accurate.

Method No. 3—Plate Take-Off

Use the same method as in Method No. 1.

Method No. 3—Stud Take-Off

For engineered prefabricated construction, the plan should be drawn to a scale of at least ¾" to 1'-0".

Step 1: The outside walls should be drawn as an exploded view. Every stud should be drawn in position, as should all the detail of the framing around openings. This may require the help of the superintendent or the foreman carpenter.

Step 2: Every member may then not only be counted but can also be measured for length. This is the engineered construction method. Finally, when all the members have actually been counted (estimated), they may be precut and then assembled in complete walls on jigs.

Remember also to draw and account for the interior walls and all headers.

Problem

Refer to the house design shown in Fig. 17.2. Using local prices for lumber, compare the difference in costs of wood framing lumber to erect twenty such units by using the three different methods of estimating. Satisfy yourself that Method No. 2 is only a rough approximation.

17.6 FIBERBOARDS

The standard size of plywood and other fiberboards is $4'-0'' \times 8'-0''$ and is marketed by the 1,000 sq ft, but other sizes may be obtained by special orders. In residential construction, the fiberboards are usually nailed to the studs and plates while the framing is being semi-prefabricated on the subfloor of the building and before the walls are turned up into place.

Building Paper

This may be purchased in rolls of 400 sq ft or more according to weight; check local dealers. Deduct for openings more than 7 sq ft, but add 8 percent for lap.

Labor for Semi-Prefabricating Framing Walls

Using hand electric saws for semi-prefabricating outside framing walls on the subfloor, the following operations can be performed:

(a) cutting and forming all openings and assembling the framing;

(b) cutting and nailing fiberboard sheathing to the framing;

(c) securing building paper to the outside of the fiberboards;

(d) setting the frames true to line, plumb, and to angle corners specified.

The labor required will be about 22 carpenter hr and about 8 carpenter's helper hr per 1,000 sq ft of completed framing. This is equal to a time rate per sq ft of:

Carpenter time	0.22 sq ft per hr
Carpenter's helper time	0.08 sq ft per hr

With this data, assuming that the rate of pay for a carpenter is $6.50 per hr and that of the helper is $4.80 per hr, the cost to frame 180 sq ft of wall would be:

$$(180 \times 0.22 \times 6.50) + (180 \times 0.08 \times 4.80) = \$326.52$$

Labor for Semi-Prefabricating Interior 2×4 Framing

Using hand electric saws for semi-prefabricating interior walls on the subfloor and erecting plumb and true to line, it will require about 6 carpenter hr and

about 2 carpenter's helper hr per 100 sq ft of completed framing. This is equal to a time rate of:

Carpenter time	0.06 sq ft per hr
Carpenter's helper time	0.02 sq ft per hr

With this data, assuming that the rate of pay for a carpenter is $6.50 per hr and that of the helper $4.80 per hr, the cost to frame 120 sq ft of wall would be:

$$(120 \times 0.06 \times 6.50) + (120 \times 0.02 \times 4.80) = \$58.32$$

Make out suitable code-index cards for all the items you estimate.

Problem

Using the following specifications and local prices for labor and materials, make a complete proposal bid to frame the exterior and interior walls of the house design shown in Fig. 17.7.

Specifications

1. Exterior and interior 2×4 $8'-0''$ high frame walls with one bottom and two top plates. The quality of the lumber is to be that used locally.

2. The left-side bathroom wall is to be 2×6 framing.

3. All headers to conform to local dimensional requirements for spans.

4. Exterior walls to be sheathed with $\frac{5}{16}$ sheathing grade plywood.

5. Interior walls to have 2×4 girths in their middle sections.

6. Use a heavy-grade building paper.

7. Nailing is according to the schedule in Chap. 7.

Allow 6 percent overhead; allow 5 percent fringe benefits; allow 10 percent profit.

17.7 BUILDING PAPERS

Sheets of polyethylene vapor barrier are stapled to the underside of ceiling joists and to the inside of exterior frame walls. (See the typical wall section, Fig. 9.1, p. 170.) When estimating, find the total number of sq ft in the ceilings plus the outside walls. Deduct only the very large openings and add 8 percent for waste. Find the number of sq ft contained in one roll and order to the next higher roll required. Many building-paper rolls contain 400 sq ft per roll, but check! A workman should staple about 250 sq ft of vapor barrier per hr. *His speed will depend upon having a stapling machine that works and a suitable scaffold to work from.*

Outside felt and building papers (breather-type) are estimated by the number of sq ft required. Deduct for very large openings and allow for a 2″ lap and for waste, which will also be reckoned at about 8 percent. The outside papers are usually applied at the same time as the siding, stucco, or other finish and should be estimated as one such inclusive unit. *Remember to use your guide list of main headings when estimating.*

17.8 SIDING

The source for the information on Table 17.1 on red cedar siding is the Western Red Cedar Lumber Association.

A nominal 10″-wide board is machined down to 9½″ and a rabbet of ½″ is provided for lap. This leaves 9″ of coverage; in 1′-0″ of length, the area covered is $9 \times 12 = 108$ sq in. Applying the factor of 1.33, 108 $\times 1.33 = 143.64$ sq in—say, 1 sq ft.

The source of the following information is the Canadian Lumbermen's Association.

Table 17.1

Bevel, Colonial, and Rabbeted Siding Allowances

Type	Nominal Width	Finished Size	F.H.A. Recommended Lap	Feet Required (equals area in sq ft × the undernoted factor)
Bevel	5″	$\frac{15}{32} \times \frac{3}{16} \times 3\frac{1}{2}$	1″	1.76
Bevel	6″	$\frac{15}{32} \times \frac{3}{16} \times 4\frac{1}{2}$	1″	1.57
Bevel	6″	$\frac{15}{32} \times \frac{3}{16} \times 5\frac{1}{2}$	1″	1.46
Bevel	8″	$\frac{15}{32} \times \frac{3}{16} \times 7\frac{1}{2}$	$1\frac{1}{4}$″	1.41
Colonial	8″	$\frac{3}{4} \times \frac{3}{16} \times 7\frac{1}{2}$	$1\frac{1}{4}$″	1.41
Colonial	10″	$\frac{3}{4} \times \frac{3}{16} \times 9\frac{1}{2}$	$1\frac{1}{4}$″	1.33
Colonial	12″	$\frac{3}{4} \times \frac{3}{16} \times 11\frac{1}{2}$	$1\frac{1}{4}$″	1.28
Rabbeted*	6″	$\frac{15}{32} \times \frac{3}{16} \times 5\frac{1}{2}$	$\frac{1}{2}$″ rabbet	1.46
Rabbeted*	8″	$\frac{15}{32} \times \frac{3}{16} \times 7\frac{1}{2}$	$\frac{1}{2}$″ rabbet	1.41
Rabbeted*	10″	$\frac{15}{32} \times \frac{3}{16} \times 9\frac{1}{2}$	$\frac{1}{2}$″ rabbet	1.33

*Also known as bevel, colonial, or bungalow siding. The rabbet is $\frac{3}{16}$″ thick and $\frac{1}{2}$″ deep. See Fig. 17.2. Deduct for openings.

THE WHITE PINE BUREAU

Standard Patterns & Sizes

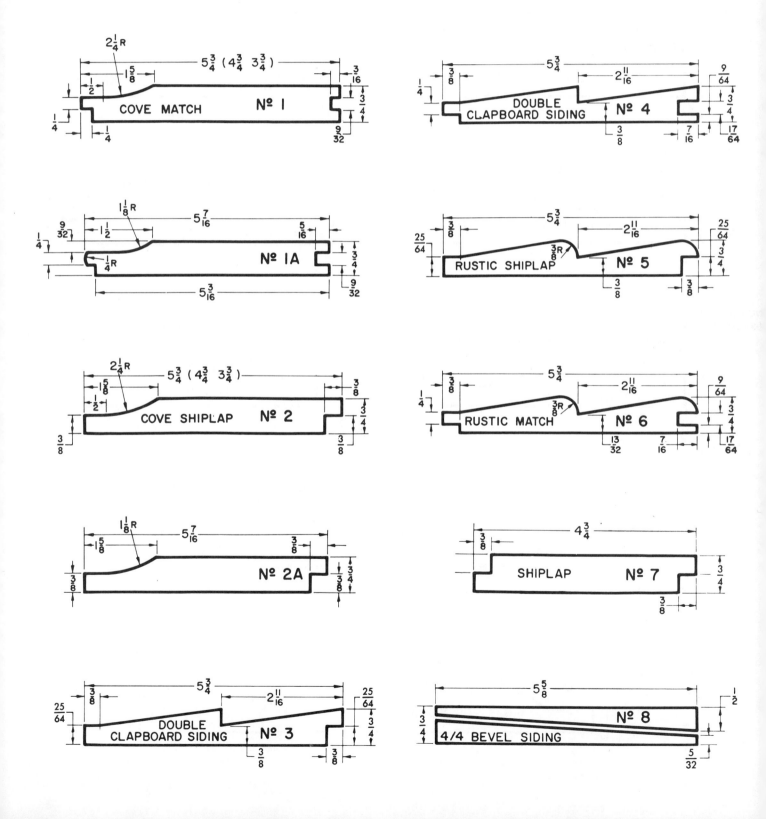

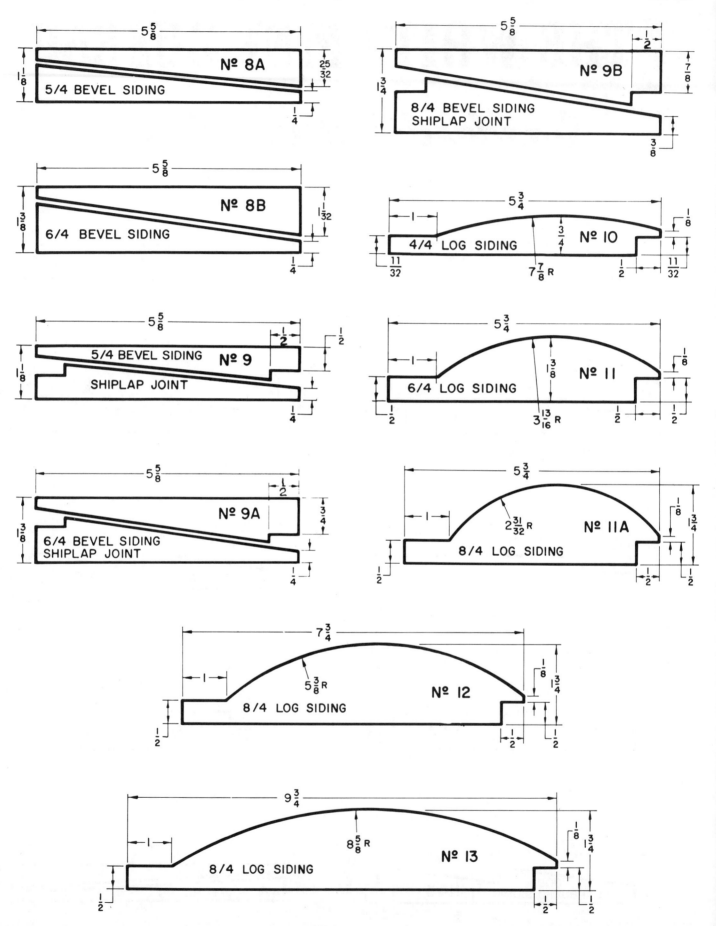

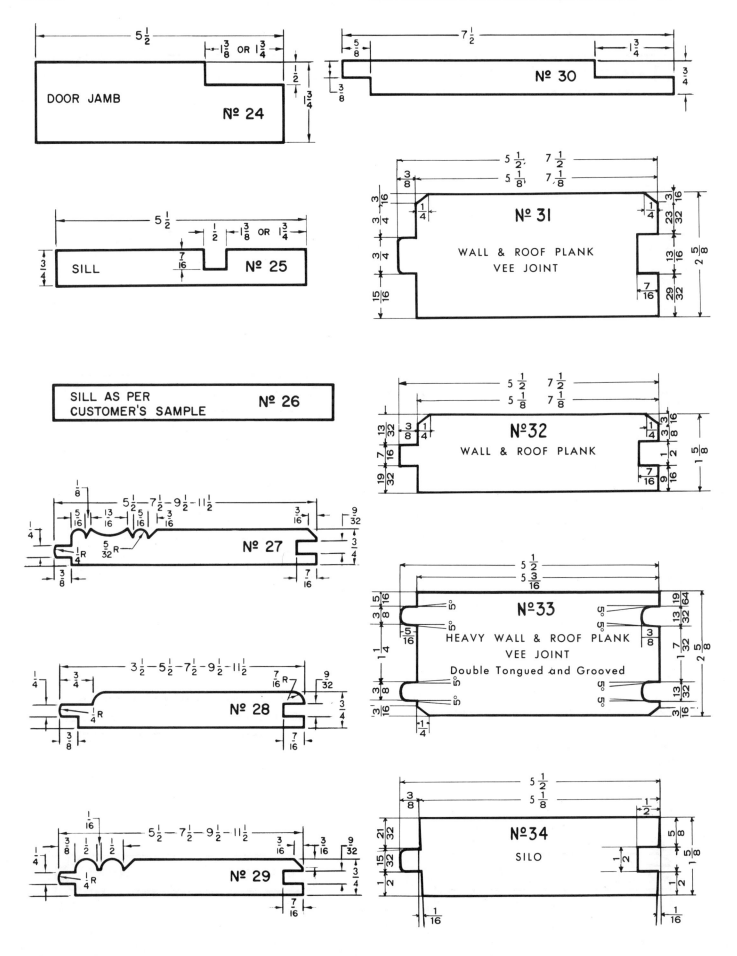

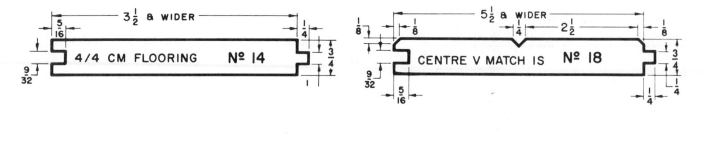

4/4 CM FLOORING   Nº 14

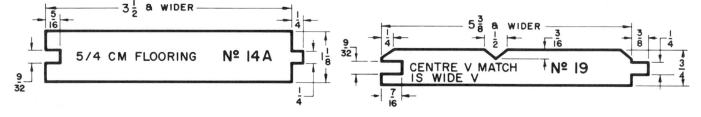

CENTRE V MATCH IS   Nº 18

5/4 CM FLOORING   Nº 14A

CENTRE V MATCH IS WIDE V   Nº 19

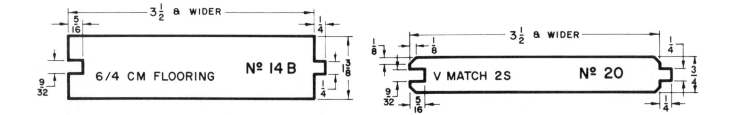

6/4 CM FLOORING   Nº 14B

V MATCH 2S   Nº 20

8/4 CM FLOORING   Nº 15

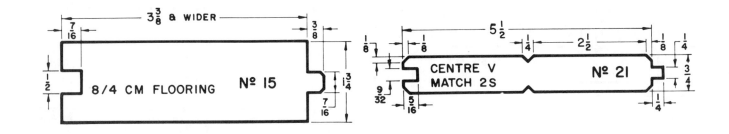

CENTRE V MATCH 2S   Nº 21

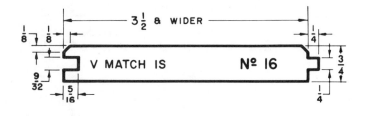

V MATCH IS   Nº 16

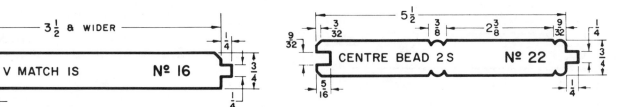

CENTRE BEAD 2S   Nº 22

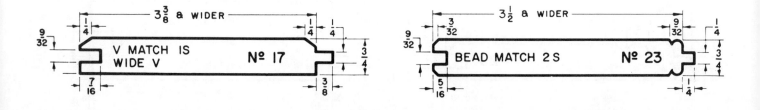

V MATCH IS WIDE V   Nº 17

BEAD MATCH 2S   Nº 23

17.9 LABOR FOR APPLYING BEVEL SIDING: FIRST-CLASS WORK

The approximate time required by a carpenter to make a first-class job of placing bevel siding with mitered corners is shown in Table 17.2.

Table 17.2

Carpenter Time for Placing Bevel Siding

Nominal Size	Actual Size	Actual Exposure	No. of sq ft per 8-hr day	No. of sq ft per hr
4	3¼″	2¾″	260	33
5	4¼″	3¾″	312	39
6	5¼″	4¾″	343	43
8	7¼″	6¾″	390	48
10	9¼″	8¾″	455	56
12	11¼″	10¾″	520	65

Estimating Labor for Bevel Siding

Step 1: Find the number of sq ft to be covered.

Step 2: Divide the number of sq ft to be covered by the number of sq ft (by size of siding) that a carpenter will place in 1 hr.

Step 3: Multiply the result of Step 2 by the carpenter's rate per hr.

Problem

1. Make a material and carpenter-time estimate to place 10″ bevel siding on the house shown in Fig. 17.2 the height of the walls to be covered is 8′-0″; deduct for all openings. Use your own judgment in allowing for waste in cutting and extra time required to fit the siding to the rake of the roof barge boards.

2. Make out a code-index card for number of sq ft of bevel siding a carpenter will place in 1 hr for each size.

17.10 SHINGLES

Both wood and asphalt shingles are estimated by the square. One square is equal to an area 10′-0″ × 10′-0″, or 100 sq ft. When you are estimating a roof, you should think in terms of squares. Without going outside think of your own home; approximately how many squares of shingles did it take to cover the roof? Check! Wooden shingles are made up in bundles of four per square; asphalt shingles are made up in bundles of three per square, and the latter are designated by the number of lb weight per square. The heavier the shingle, the greater the cost.

Wood Cedar Shingles

A very excellent handbook, comprising about 100 pages, is available for a nominal cost upon application to either of the following:

Red Cedar Shingle Bureau   Red Cedar Shingle Bureau
5510 White Building   550 Burrard Street
Seattle, Wash.   Vancouver 1, B.C.

No estimator's files can be completed without this information, and you are earnestly recommended to get it and study it.

Shingles Other Than Wood

Secure copies of manufacturers' brochures which give size, weight, coverage, colors, and application specifications. Architects usually state that they apply their specifications to meet manufacturers' specifications or the Bureau of Standards.

It is most important to remember that not only must you build up your A.I.A. file, but you must read the brochures before they are filed. Carefully discriminate between what you require as an estimator and what may be discarded as advertising material.

17.11 INSULATION

The following material has been excerpted from material supplied by Owens-Corning FIBERGLAS*, and is here reproduced with the permission of that corporation.

Insulation

Its performance is measured in terms of R-values. (The "R" stands for thermal resistance to heat flowing out in winter and flowing in during summer.) The higher the R-value, the better the resistance and performance of the insulation.

Pink Fiberglas* blankets contain millions of trapped air cells and offer superior resistance to heat flow as illustrated here. 6″ of Fiberglas insulation has an R-value of 19. And so do 15″ of wood and 7′ of brick.

Thickness Needed For R-Value Fiberglas
Blanket Type Insulations

R-Value	Thickness	R-Value	Thickness
R-38	12″ (two layers 6″)	R-22	6½″
R-33	10″ (3½″ plus 6½″)	R-19	6″
R-30*	9½″ (6″ plus 3½″)	R-13	3⅝″
R-26*	7¼″ (two layers 3⅝″)	R-11	3½″

*Also available as a single-layer R-26 or R-30 batt.

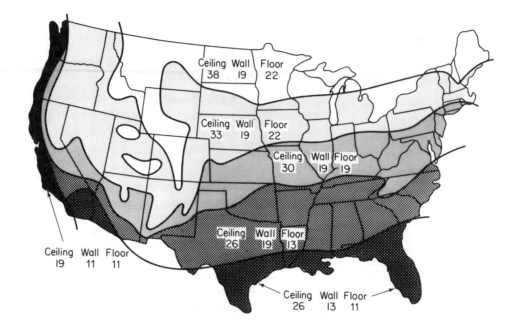

Ceiling Wall Floor
38   19   22

Ceiling Wall Floor
33   19   22

Ceiling Wall Floor
30   19   19

Ceiling Wall Floor
26   19   13

Ceiling Wall Floor
19   11   11

Ceiling Wall Floor
26   13   11

Fig. 17.3

The Recommended Insulation Standards for ceilings, walls and floors shown on this map were developed from a comprehensive computer analysis of 71 U.S. cities. Owens-Corning based the recommendations on insulation costs, weather data, current and projected heating and cooling costs, and a return on investment for 20 years with savings discounted 10% each year.

These recommendations represent the minimum insulation levels for homes in your area, using materials and techniques available today. Find your location on the map and you'll find out how much pink insulation you should use.

Insulation and Moisture

Moisture in your home from cooking, washing dishes, laundering, etc., can cause problems if it reaches a cold surface and condenses within wall or ceiling cavities. Properly applied insulation and good ventilation can help stop this problem.

Vapor barriers such as insulation facing (kraft or foil) or separate vapor barriers (polyethylene or foil-backed gypsum board) should be installed on the warm-in-winter side of walls or ceilings. This will help keep water vapor from reaching a cold surface where it can condense.

To prevent condensation, a positive movement of air out of the attic is essential. Eave vents, openings under the eaves, combined with gable vents or roof vents are effective. As a general rule, one square foot of free vent area should be provided for each 150 square feet of attic floor area when no vapor barrier is used. With a vapor barrier, one square foot of vent area per 300 square feet of floor area is recommended.

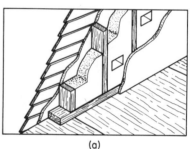

(a)

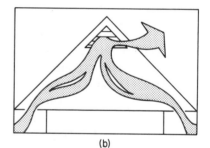

(b)

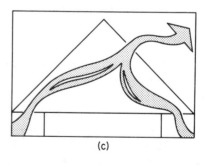

(c)

Fig. 17.4

Air-driven roof vents give an intermediate solution. Power roof or gable vents which are activated by a temperature and humidity control are also used.

Foundation walls of heated crawl spaces should be insulated with an R-11 pink blanket insulation with a polyethylene ground cover.

Fiberglas Perimeter Insulation, a rigid board product should be installed around slab floor edges to reduce heat loss and keep slab warm.

In choosing between these types of insulation, the following factors should be considered:

Batt/Blanket Types:

. in most circumstances these types are more easily handled and applied than loose fill.
. they are premanufactured, with the quality assured.
. they are the most suitable insulation materials for vertical surfaces in the attic (though rigid insulation could also be used.
. they can be installed with an attached vapor barrier if desired.
. the cost per unit of R-value is generally higher than for loose fill.
. the choice *between* batt and blanket types will depend upon the particular job to be done. Blanket insulation is often more awkward to install.

Fig. 17.5

Loose Fill Types:

. best suited for nonstandard or irregular joist spacing or when space between joists has many obstructions. Gets into small areas.
. loose fill generally costs less per unit of R-value than batts or blankets.
. if a vapor barrier is desired, it must be applied separately from the insulation.

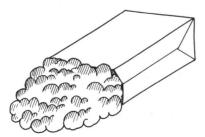

Fig. 17.6

If Using Batt/Blanket Type:

the differences between glass and mineral fiber are not large. Glass fiber is easier to handle and may fill the space more effectively than some mineral fiber batts. On the other hand, mineral fiber tends to have a higher R-value per inch. Make your choice accordingly.

If Using Loose Fill Type:

glass and mineral fiber are fire and moisture resistant. cellulose fiber has a higher insulation value for a given thickness. It is made from recycled newsprint, and as such reduces waste in other areas. It is less prone to undesirable settling than other loose fill insulations. It is treated with a fire retardant, though some formulations may not last the lifetime of the insulation. The insulation does absorb water, and therefore should not be used where water can come in direct contact with the insulation.

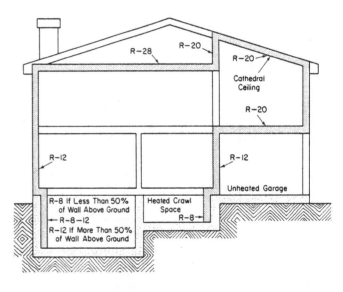

Fig. 17.7

Recommended Minimum Insulation Levels For Existing Buildings, After Upgrading

	Recommended Minimum R-Value of Insulation
Ceiling	28
"Cathedral" ceiling	20
Basement walls (less than 50% of the wall above ground)	8

The following material has been excerpted from a booklet written and published by the Department of Energy Mines and Resources, Canada.

Water Vapour

Cold, outdoor winter air is able to contain very little water vapor. The warmer the air gets, however, the more it can hold. Consequently, as winter air filters into a house and is heated, its "relative humidity" (that is, the amount of moisture it contains relative to what it *could* contain) drops. The air *feels* dry to us.

However, human activity changes that. Our day-to-day living adds water to the air. The following figures are approximations, but they give an idea of how extensive this impact is:

Quantity of Moisture Added to the Air by Normal Human Activity

Activity	*lb of moisture*
Washing clothes, per week	4.0
Drying clothes by hanging on a line indoors, per week	26.0
Cooking and dishwashing, per week	35.0
Each shower	.5
Each tub bath	.2
Normal respiration and skin evaporation, per person per 24 hour day	2.9

Heating Ducts

Heating ducts running through unheated or cool basements should be insulated.

Insulation wool batts of glass or mineral-rock base are manufactured in paper-enveloped batts; one side is vapor-resistant and is to be placed on the inside, or warm side of the room; the other side is perforated vapor-permeated paper. The batts are 2" or 3" in thickness; 2'-0" to 8'-0" or blanket roll in lengths; and in widths of 15", 19", and 23", designed to fit snugly in frame walls with studs 16", 20", and 24" OC, respectively. They are packaged in cartons or rolls. (See Fig. 17.8 and Fig. 17.9.)

A 15"-wide batt together with the stud will cover a total width of 16".

Batts 2'-0" long will cover with the stud

$$2 \times 1\tfrac{1}{3} = 2.66 \text{ sq ft}$$

Batts 4'-0" long will cover with the stud

$$4 \times 1\tfrac{1}{3} = 5.33 \text{ sq ft}$$

Batts 8'-0" long will cover with the stud

$$8 \times 1\tfrac{1}{3} = 10.66 \text{ sq ft}$$

The number of 2'-0"-long batts required to cover 100 sq ft of frame wall with studs 16" OC is: $100 \div 2.66 = 38$ batts.

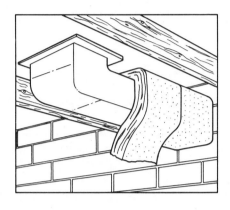

Fig. 17.8

Problem

Complete the following table:

Number of Glass, Mineral, or Rock-Wool Insulation Batts Required for 100 sq ft of Frame Wall		
Wall Studs OC's	Size of Batts Required	Batts Required for 100 sq ft of Frame Wall
16"	15" × 24"	38
16"	15" × 48"	19
16"	15" × 96"	9.5
20"	19" × 24"	
20"	19" × 48"	
24"	23" × 24"	
24"	23" × 48"	

Note the coverage is the same for both 2" and 3" batts.

Ceilings

Ceilings may be insulated with wool batts, wool blanket roll, or with loose fill applied between the ceiling joists and supported by the ceiling. Loose fill is

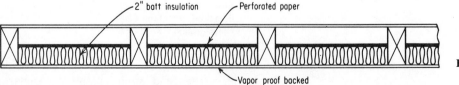

Fig. 17.9 Wool insulation

marketed by the 4-cu-ft-capacity sack and the sq-ft coverage of ceilings is dependent upon the depth of fill required. *Remember to estimate for a vapor barrier secured to the ceiling joists before the ceiling and the fill are applied.* All vapor barriers should be fixed to the warm side of the room.

Labor for Placing Wool Batts

A workman should place about 150 to 180 sq ft of 2″ wool batts per hr. You must inquire locally as to what trade does this work. In some states, it is carpenters' work.

Reflective Insulation

There are a number of different types of reflective insulation. Many types are made up in rolls of 250 to 500 sq ft. A workman should place about 250 sq ft of reflective insulation per hr.

Rigid Insulation

This type of insulation, as the name implies, is rigid enough to be marketed in panels of varying thickness and sizes. All insulating materials are light and none of them is very demanding on labor time. When estimating roofing you will have to be specific with the subtrade roofing company as to who will supply and place rigid insulation. Rigid insulation for roofs, walls, and floors is perhaps the easiest of all estimating problems. Take the area required in sq ft and convert to the sizes of panels in which it is supplied. There is very little waste in any type of insulating material.

Square-Foot Areas of Walls Plus Ceilings for Various Sized Rooms

Table 17.3 shows the number of sq ft of surface area of the walls plus the ceilings for various sized rooms. Make deductions of 18 sq ft for door openings and the actual area of other openings. It is as important to discard dated material as to add new.

17.12 STUCCO WIRE MESH

In sheathed frame wall construction (breather type) the building paper prevents most of the absorption of moisture from fresh mortar. Use large mesh stucco wire of a minimum net weight of 1.8 lb per sq yd or 20 lb per 100 sq ft. Openings should not be less than ¾″ in the small dimension or larger than 3″ in the large dimension, and should not exceed 4 sq in. in area. The stucco wire should be nailed or stapled securely with special furring nails to hold the reinforcing at least ¼″ from the wall. The knurls should be next to the sheathing paper.

Estimate in sq ft or sq yd. Check locally for method of marketing and convert to merchantable units.

TABLE 17.3

Room Wall and Ceiling Areas

Dimension of Room	8 ft Ceiling Requires Square Feet	9 ft Ceiling Requires Square Feet	10 ft Ceiling Requires Square Feet	Dimension of Room	8 ft Ceiling Requires Square Feet	9 ft Ceiling Requires Square Feet	10 ft Ceiling Requires Square Feet
8′ × 8′	320	352	384	11′ × 16′	608	662	716
8′ × 10′	368	404	440	11′ × 18′	662	720	778
8′ × 12′	416	456	496	12′ × 12′	528	576	624
8′ × 14′	464	508	552	12′ × 14′	584	636	688
8′ × 16′	512	560	608	12′ × 16′	640	696	752
8′ × 18′	560	612	664	12′ × 18′	696	756	816
9′ × 10′	394	432	470	13′ × 14′	614	668	722
9′ × 12′	444	486	528	13′ × 16′	672	730	788
9′ × 14′	494	540	586	13′ × 18′	730	792	854
9′ × 16′	544	594	644	14′ × 14′	644	700	756
9′ × 18′	594	648	702	14′ × 16′	704	764	824
10′ × 10′	420	460	500	14′ × 18′	764	828	892
10′ × 12′	472	516	560	15′ × 16′	736	798	860
10′ × 14′	524	572	620	15′ × 18′	798	864	930
10′ × 16′	576	628	680	16′ × 16′	768	832	896
10′ × 18′	628	684	710	16′ × 18′	832	900	968
11′ × 12′	500	546	592	17′ × 17′	833	901	969
1′ × 14′	554	604	654	18′ × 18′	900	972	1044

Labor for Placing Stucco Reinforcing Mesh

Two men working together should place about 100 sq yd of stucco reinforcing mesh in 5 hr at a rate for one man of 10 sq yd per hr.

17.13 BRICK VENEER FOR FRAME CONSTRUCTION

The source of the following is Central Mortgage and Housing Corporation, Ottawa, Canada.

Many jurisdictions state the following for brick veneer:

(a) Height of brick-veneer finish for frame walls shall not exceed 35'-0" above foundation walls.

(b) Purpose of brick veneer shall be considered as weathering surface only. It shall not be considered as contributing to the strength of the frame. Brick veneer shall not be less than 3¾" thick.

(c) Brick veneer shall be secured to framing members by corrosion-resistant metal ties every 16" in height and width ties. Ties shall be ³⁄₁₆ corrosion-resistant steel rods, not less than 1" × 7" no. 28 gage galvanized sheet steel.

(d) Wood frame walls shall first be covered with acceptable building paper before the brick veneer is erected and the veneer shall be at least 1" from the sheathing of the frame wall.

(e) Weep holes, spaced approximately 2'-0" apart, are required and may be formed by leaving open vertical joints in the bottom course or by other means acceptable to the authority having jurisdiction.

Note carefully: All building standards should be constantly checked for amendments.

Estimating Brick Veneer

Estimate as for running bond and allow for a mason to lay about 600 to 700 face bricks in brick-veneer in 8 hr of construction, depending upon the season of the year and the height above ground. See Chapter 16, page 283.

Problem

On correct stationery, using local prices for labor and materials, make a commercially acceptable proposal bid for erecting brick veneer to all the walls (8'-0" high) for our house design, Fig. 17.2. The specifications are as follows:

1. Use a good quality (standard size) local face brick;

2. brick to be laid up with ½" horizontal cement mortar concave joints and ¼" vertical joints;

3. metal ties to comply with your local building code.

Your work should cover the following items:

(a) number of bricks (remember to add for waste);

(b) price of brick delivered to the job;

(c) amount of mortar mixed on the job;

(d) price of components of mortar;

(e) number and price of noncorrosive metal ties;

(f) scaffold: temporary carpenter-built trestles, metal scaffold rentals, or whatever lends itself to the most expeditious, inexpensive, but safest method;

(g) mason and mason's helper labor costs;

(h) workmen's fringe benefits;

(i) workmen's compensation;

(j) overhead expenses;

(k) profit.

Stone and Other Masonry Unit Veneer Construction

Secure manufacturers' brochures for types, quality, coverage specifications, and prices. Read your local building code. The method of estimating will conform to the foregoing information.

17.14 GLASS BLOCKS

Standard hollow units of glass are 4" thick and 6", 8", and 12" square. They are not load-bearing but are laid up by regular masons using cement-lime mortar. They should be erected with an asphalt emulsion to break bond between the mortar and sill. An expansion joint is made at the sides and head with a resilient fiber pad 4⅛" × ⅜" × 25"; or a no. 20 gage galvanized perforated steel strip 24" × 1¼" may be secured and imbedded. Galvanized reinforced wire ties are imbedded in mortar joints every 24" in the height of the wall regardless of the size of the blocks. See Table 17.4 on p. 317.

Labor for Glass-Block Construction

A mason with his helper should lay up 100 sq ft of glass block at the following time rates:

100 sq ft of 6" glass block in 7 to 8 hr
100 sq ft of 8" glass block in 6 to 7 hr
100 sq ft of 12" glass block in 5 to 6 hr

Table 17.4

Quantities Required for 100 Square Feet of Glass block

	6″ Blocks	8″ Blocks	12″ Blocks
Number of blocks	400	225	100
Asphalt emulsion	$\frac{1}{6}$ gal	$\frac{1}{3}$ gal	$\frac{2}{3}$ gal
Expansion strips	38 pieces	67 pieces	150 pieces
Panel anchors	20 pieces	36 pieces	80 pieces
Wall ties	13 pieces	23 pieces	100 pieces
Mortar	11 cu ft	14 cu ft	22 cu ft

17.15 ORNAMENTAL MASONRY UNITS

When specified, they may be complete manufactured units or they may be constructed units of a combination of standard-sized bricks with ornamental units. Secure prices for ornamental units from the manufacturer, make special allowance for the mason's erecting time, and deduct from the brickwork the amount displaced.

17.16 CARPENTRY, JOINERY, AND MILLWORK

Below is an extract from a typical specification of millwork.

The work under this heading shall include the furnishing of all labor, materials, and services necessary and reasonably incidental thereto for all carpentry, joinery, and millwork as shown on the drawings and herein specified.

All clauses set out in the General Conditions and the Conditions of the Contract shall apply to and govern this trade.

Carpentry

Carpentry work should be true and framed together in the best manner of the trade. All boards shall be nailed at each bearing and all joints shall be made over bearings. All T and G material shall be blind-nailed. (Note that *T and G* means "tongue and groove.")

Finishing Material

(a) Work shall be assembled at the mill insofar as practical and delivered ready for erection. When it is necessary to cut and fit on the job, the material shall be made with ample allowance for cutting.

Work shall be made in accordance with the measurements taken on the job.

(b) Moldings shall be true to detail, cleanly cut, and sharp.

(c) Exposed surfaces shall be machine sanded to an even, smooth surface ready for finish.

(d) Mill assemblies shall be joined with concealed nails and screws where practical or with mortise and tenon joints with glue blocks where practical. Exposed nails shall be countersunk. Glue shall be waterproof.

(e) Scribing, mitering, and joining shall be done neatly and shall conform to details.

(f) Flat members of trim and base and so on shall be backed out to prevent warping.

(g) All finish shall be hand cleaned and hand sanded in addition to machine sanding in the mill.

Note how often the word *all* appears. This pattern of wording should be used in all written matter between you and all subtrades; it is emphatic and all-inclusive.

The millwork manufacturers may supply all the dimension lumber as well as the finishing material and made-up units such as cupboards, stairs, and so on.

Some estimators for small, inexpensive, and easily assessed jobs will add a percentage of the cost of finishing materials for the carpenters to install. This would include such items as fitting baseboard, hanging doors, and so on. This method is not recommended except for well-established costs such as inexpensive and similar housing units that conform to set patterns of construction. Where this method is used, the cost of installation would run between 60 percent and 90 percent of material cost.

All labor costs will vary according to the type of finish and the class of materials used. For average job conditions, the labor times discussed below are offered as a guide. You are invited to disagree with these times only after you have made your own on-the-job time study. This you should always be doing. It is most important to remember that these suggested times can only be maintained when the rooms are kept clean and the materials, trim, doors, hardware, and so on are placed in the room ready for the carpenters to work. Whenever a tradesman can be seen using a broom on a job, the contractor is losing money. Keeping a job clean is almost a trade in itself and should be so recognized.

Baseboard Estimating

Take the perimeter of the room and make no allowance for door openings. The smaller the room, the higher the labor cost; the longer the walls of rooms, the lower the labor costs.

A carpenter should fit about 200 lin ft of 4″ baseboard in an 8-hr day at a rate of 25 lin ft for 1 hr of carpenter time.

When estimating be sure to give careful consideration to the size of rooms. I know of a medical center where a plastering contractor lost money because he overlooked the fact that the rooms were of small area with high walls. This required more changes of scaffold for height and constant dismantling and reassembling of scaffold from room to room. In addition, it was just as big a problem to clean out the plaster droppings from a small room as a large room. The carpentry trade was affected in much the same way.

Carpet Strip

Estimate the quantities in the same manner as for baseboard and use the same figures. A carpenter should fit about 500 lin ft of carpet strip in an 8-hr day at a rate of about 60 lin ft for 1 hr of carpenter time.

Door Casings

Estimate by the lin ft by type. A carpenter should fix a good-quality residential inside-door casing true to size, square, and plumb in about ¾ hr. He can only do this if the rough carpentry has been accurately built.

Door Trim

Estimate by the lin ft. A carpenter should trim around one side of a good-quality residential door casing (say about 18 lin ft) in about ½ hr. He can only do this if the casing has been accurately placed.

Door Hanging

Using machine tools, a carpenter should hang a good-class residential door and fit the lock to it in about 1½ hr. He can only do this providing the door casing is erected true to size, square, and plumb.

Note how every operation in building construction is dependent upon the preceding one.

To hang heavy hardwood doors in public buildings with two carpenters working as a team will take 3 hr of carpenter time (or more) for the average-size door.

Window Trim

A carpenter should fit the window stool, apron, and trim around a window up to 16 sq ft in about 1½ hr of carpenter time. There would be very little difference in cost for larger windows of the same type. For thermopane windows—say, 48 sq ft—a carpenter should trim around in not more than 1½ hr, as there is no apron or stool on many of these assemblies. Talk with your foremen. Give out time sheets and get all the data that you can from your own company. In the final analysis, it is your company that has to build up its own time-study system.

Glazing

This is a subtrade and in many places may be handled only by members of the glaziers' union. There is always some breakage of glass during construction, and a minimum loss of about 5 percent should be allowed. The window cleaning on a large job is also a subtrade.

QUESTIONS

1. Make a neat drawing (to the scale of 1½″ = 1′-0″) of a horizontal section of frame wall with 2″ × 6″ plates and staggered 2″ × 4″ studs, with blanket roll insulation woven between the staggered studs. Note that the dimension lumber is full size.

2. State briefly how 2 × 4 wall plates for housing units should be estimated and ordered.

3. State briefly how 2 × 4 framing studs for housing units should be estimated and ordered.

4. How is building paper estimated and ordered in your area?

5. Define and make a neat sketch where necessary of the following:

    (a) girth in wall framing
    (b) plaster ground (rough baseboard)
    (c) siding (wood)
    (d) brick veneer to a frame wall

6. Define the following:

    (a) fiberboard
    (b) wood and asphalt shingles
    (c) four different types of insulation
    (d) stucco wire
    (e) glass block
    (f) ornamental masonry
    (g) millwork
    (h) wire mesh

7. How would you take off for an estimate of $8'-0'' \times 15''$ wool insulating batts?

8. In what main unit is fiberboard marketed for industrial construction?

9. In what units are shingles estimated?

10. What is the coverage of one bundle of wood shingles?

11. What is the coverage of one bundle of #210 lb asphalt shingles?

12. In what bond is brick veneer (to wood frame walls) laid up?

13. How is window cleaning at the completion of a large project allowed for on the estimate?

18

ROOFS: DEFINITIONS OF ROOF MEMBERS, BASIC ROOF DATA, AND ROOFING MATERIALS

This chapter is designed to help you toward a complete understanding of the different types of roofs and the basic data required for you to estimate quickly and accurately the quantities required to frame and cover any roof.

With accurate basic data the following estimates may very easily be made:

(a) the roof rafters;

(b) the roof sheathing;

(c) the roof building paper;

(d) the area to be covered by shingles or other roofing materials;

(e) the area to be covered by paint;

(f) quantities required for flat roofs.

It is most earnestly recommended that you make the drawings and do the problems presented in this chapter. Make them to any convenient scale. *A few hours of careful thought and study will enable you to find the area of any roof in matter of minutes.*

18.1 TYPES OF ROOFS

The Flat Roof

This is a roof in which heavy ceiling joists are used as rafters. It has a minimum slope for drainage. It must be well supported at the walls. It must be heavily waterproofed. This type of roof is used extensively in commercial construction. (See Fig. 18.1.)

The Shed or Lean-to Roof

This is a roof which has one sloped surface only. The slope is across the width of the building. The horizontal distance over which the slope passes is the "run." The lean-to roof is so named because it leans against another building or wall. It saves the expense of building one wall. It is the simplest roof that a roof framer may have to build, and is used for simple extensions to existing buildings. (See Fig. 18.2.)

The Gable Roof

This is a roof which has two sloped surfaces meeting at the *ridge.* The horizontal distance from the foot of one rafter to the foot of the opposite rafter is called the *span.* Half the span over which each rafter passes is the *run.* This roof is like two lean-to roofs placed together, and is used extensively in residential and light commercial construction. (See Fig. 18.3.)

The Hip Roof

This is a roof which has a sloped surface from each wall toward the ridge. The rafters that fit against the ridge are called "common rafters" (CR's). The main part of the roof, whose members are CR's, is like two lean-to roofs placed together. The rafters that lie at 45° (on plan) are called "hip rafters" and the rafters that fit against the hip rafters are called "jack rafters," (See 18.4.)

The Gambrel or Barn Roof

This is a gable-type roof which has more than one slope on one face. It is used extensively by farmers. The upper portion is like two lean-to roofs placed to-

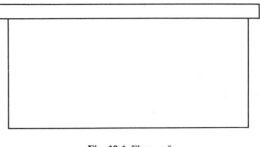

Fig. 18.1 Flat roof.

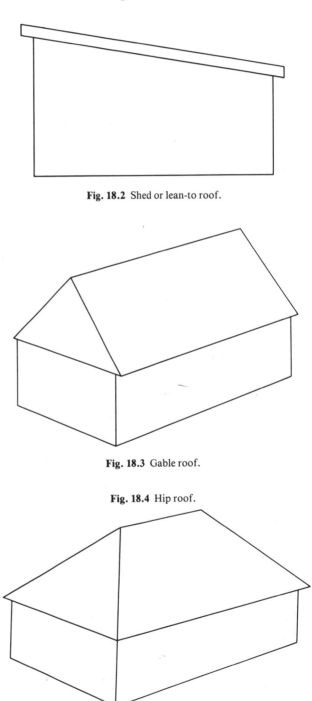

Fig. 18.2 Shed or lean-to roof.

Fig. 18.3 Gable roof.

Fig. 18.4 Hip roof.

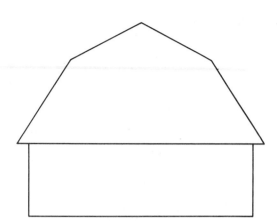

Fig. 18.5 Gambrel or barn roof.

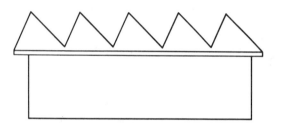

Fig. 18.6 Factory or sawtooth roof.

gether. The lower slopes are like lean-to-roofs. (See Fig. 18.5.)

The Sawtooth Roof

This is a series of roofs which, when viewed from the ends, resemble the angles of sawteeth. It is a factory-type roof. It allows for the placing of glass on one slope to light the floor below. Each pair of slopes resembles two lean-to roofs of different runs and slopes but of the same height. (See Fig. 18.6.)

The basic principle of sloped-roof framing is right-angle triangulation, where the base of the right triangle is called the *run* the altitude is called the *rise* and the hypotenuse is called the *line length.*

18.2 DEFINITIONS AND ROOF TERMS AS APPLIED TO REGULAR ROOFS

Span: The distance between two opposite common rafter (CR) birdsmouths.
Run: One-half the span.
Wall Plate: The member on the top of the wall to which the rafters are secured.
Rise: The theoretical height of the roof over the tops of the wall plates.

Pitch: The slope of the roof. It is the relationship of the rise of the roof to the span. A quarter-pitch roof has a rise equal to one-quarter of its span.
Ridge: The uppermost horizontal member of a roof against which the CRs fit.
Line Length: The theoretical length of any rafter, excluding the tail. This length is used by carpenters. The estimator uses the total length of the rafter, which includes the overhang at the eaves.
Plumb and Level Lines: When a rafter is in position, any vertical line is a plumb line and any horizontal line is a level line.
Birdsmouth: The part of a rafter with a plumb and level cut that fits and is secured to the wall plates.
Rafter Overhang: That part of a rafter that extends beyond the wall or building line.
Tail Cuts: The shape of the cut or cuts at the overhang end of a rafter.

18.3 THE SPAN

The span is the distance between two opposite rafter seats. It is the total rise divided by the pitch.

Example 1

Rise 6'-0" divided by a pitch of ¼:

$$6'\text{-}0'' \div \frac{1}{4} = \frac{6'\text{-}0'' \times 4}{1} = 24'\text{-}0'' \text{ (span)}$$

Example 2

Rise 8'-0" divided by a pitch of ⅓:

$$8'\text{-}0'' \div \frac{1}{3} = \frac{8'\text{-}0'' \times 3}{1} = 24'\text{-}0'' \text{ (span)}$$

18.4 THE RUN

The run is one-half the span. It is the horizontal distance over which any rafter passes.

18.5 THE RISE

The rise is the theoretical height of the roof over the plates. The rise is span multiplied by pitch.

Example 1

Span 24'-0" multiplied by a pitch of ¼:

$$\frac{24'\text{-}0'' \times 1}{4} = 6'\text{-}0'' \text{ (rise)}$$

Example 2

Span 24'-0" multiplied by a pitch of ⅓:

$$\frac{24'\text{-}0'' \times 1}{3} = 8'\text{-}0'' \text{ (rise)}$$

18.6 THE PITCH

The pitch is the slope of a roof. It is rise divided by span.

Thus:

$$\frac{\text{Rise over}}{\text{span}} \text{ (remember if you can)}$$

Example 1

$$\frac{6'\text{-}0'' \text{ (rise)}}{24'\text{-}0'' \text{ (span)}} = ¼ \text{ pitch}$$

Example 2

$$\frac{8'\text{-}0'' \text{ (rise)}}{24'\text{-}0'' \text{ (span)}} = ⅓ \text{ pitch}$$

18.7 TO FIND THE AREA OF A GABLE ROOF

Note that estimating roof areas includes the overhang at the *verges* (sometimes called *cornices* or *eaves*).

Example 1

Step 1: Draw, to the scale of ¼" to 1'-0", the plan of a rectangular roof 24'-0" × 36'-0" with a ⅓ pitch as at *A–B–C–D*, Fig. 18.7.

Step 2: On the line of the ridge *e–f*, project the rise of the roof *g* and *h*. The rise is ⅓ of the span, and ⅓ of 24'-0" is 8'-0" (scaled 8'-0").

Draw the slopes of the roof *g–A*, *g–D*, *h–B*, and *h–C*.

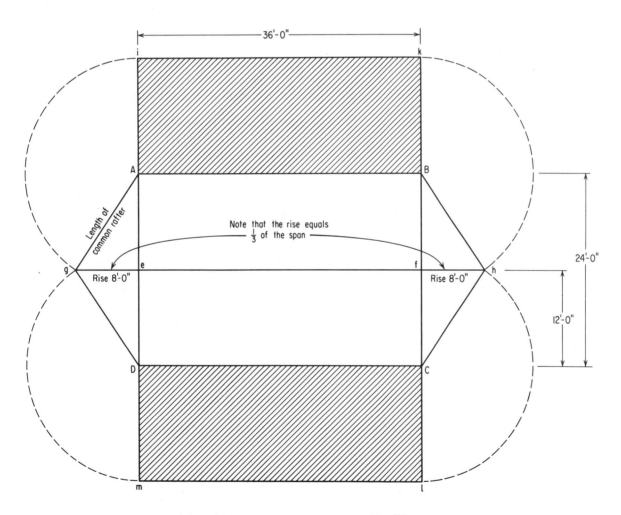

Fig. 18.7 An exploded view of a gable roof.

Scale the slope. *On the sectional drawings of plans, the length of the slope may be measured directly.*

Step 3: With *A* as a center and *A–g* (the slope of the roof) as a radius, describe an arc to intersect with *i.* Complete the projection *A–i–k–B* and repeat at the opposite side for *C–l–m–D.*

Step 4: Make a tracing of the drawing and with a sharp knife cut on all the outside solid lines. Fold on lines *A–B, B–C, C–D,* and *D–A.* Examine the model.

Problem

1. Using mathematics, find the percentage increase of the ⅓-pitched area over that of the ground-plan area of the roof in Fig. 18.7.
2. Draw the same plan area as for Fig. 18.7 and develop the drawing with a ⅜-pitch roof.
3. Find the percentage increase of the ⅜-pitched area over that of the ground-plan area for the ⅜-pitch roof.

18.8   TO FIND THE AREA OF A REGULAR HIP ROOF

A regular hip roof has all the slopes of equal pitch. Where there are hips and valleys, they have slopes similar to each other.

Example

Step 1: Draw, to the scale of ¼″ to 1′-0″, the plan of a regular hip roof 24′-0″ × 36′-0″ as at *A–B–C–D,*

Fig. 18.8. Note very carefully that the roof includes the overhang. This roof also is to have a ⅓ pitch.

Step 2: Draw the runs (plans) of the skeleton framing showing the common rafters, hips, and the ridgeboard. The hip-rafter runs are at 45° from the wall plates. *This is very important to remember.* Letter all the intersections of rafters with plates and the ridge as in Fig. 18.9.

Step 3: With a scaled rise of the roof (8′-0″) as a radius, and *g* as a center, describe an arc *g₃* (see the shaded triangle in Fig. 18.10). Draw the line of the hypotenuse *g₃* and *g₂.*

This is the length of the common rafters. To find the length, measure it! *On sectional drawings of plans, the length of the common rafter may be scaled directly.*

Step 4: At each end of the drawing continue line *e–f* of indefinite length (see Fig. 18.11). With *g₂–g₃* as a radius, describe an arc *e–g₄.* Join *g₄* with *A* and *D.* Repeat at the opposite side for *f–g₅* and join *g₅* with *B* and *C. The projections show the actual roof areas* of the triangles *A–g–D* and *B–h–C* on the plan.

Step 5: With *g₂–g₃* as a radius (hypotenuse of the shaded triangle), draw the front and rear exploded sections as shown on the complete drawing in Fig. 18.12.

Step 6: Take a sharp knife and cut on *g–g₃–g₂* and on all the outside lines. Fold the exploded portions away from you on lines *A–B, B–C, C–D,* and *D–A,* and on *g–g₂;* examine the model.

Step 7: By measuring, find the area of the exploded hip roof. Compare the area with the area of the gable

Fig. 18.8 Plan area.

Fig. 18.9 Plan area showing the runs of the rafters.

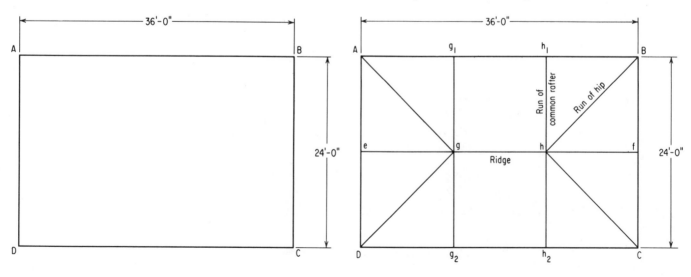

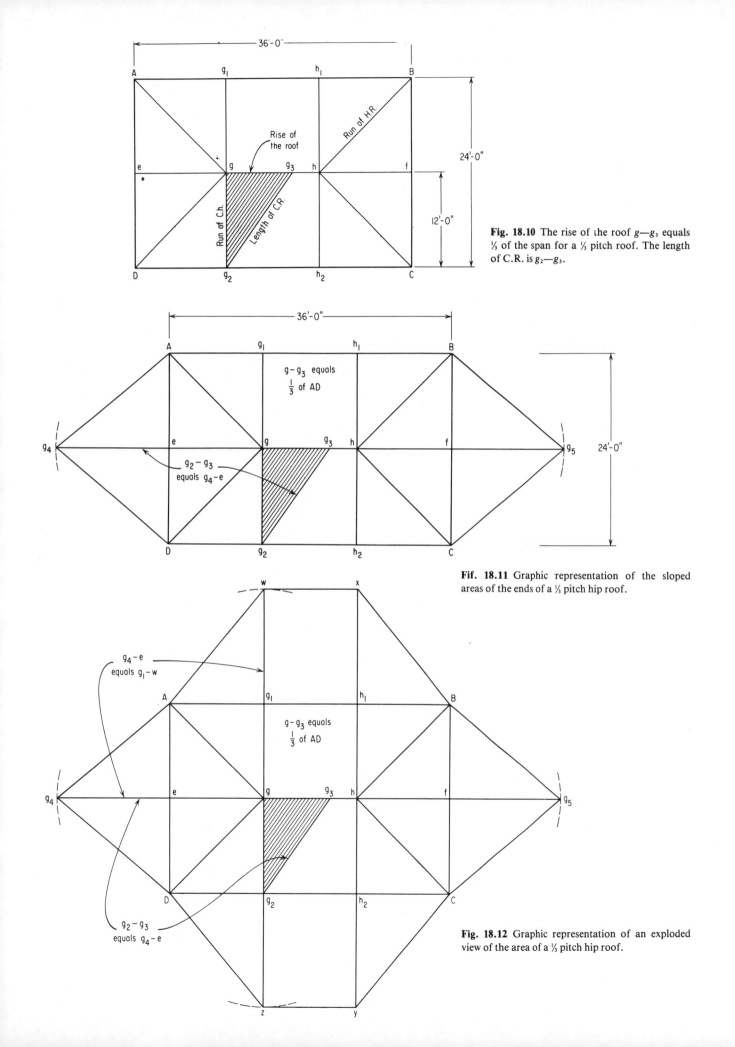

Fig. 18.10 The rise of the roof g—g_3 equals ⅓ of the span for a ⅓ pitch roof. The length of C.R. is g_2—g_3.

Fif. 18.11 Graphic representation of the sloped areas of the ends of a ⅓ pitch hip roof.

Fig. 18.12 Graphic representation of an exploded view of the area of a ⅓ pitch hip roof.

roof drawn to the same plan, dimensions, and pitch (Fig. 18.7).

Step 8: Satisfy yourself that the sloped area of a regular hip roof has the same area as that of a gable roof, providing both have the same plan, shape, and dimensions, and that they both have the same pitch (slope).

Problem

Draw the same plan area as for Fig. 18.8 and develop the drawing with a ⅜-pitch hip roof. Find the percentage difference in area of the ⅜-pitch roof with a flat roof of the same plan.

18.9 TO FIND THE AREA OF A ROOF: FRAMING-SQUARE METHOD

Once the foregoing methods are understood, the framing-square method is much faster; *it only takes a few minutes.*

Step 1: Examine a carpenter's framing square. It has a body 2″ wide and 24″ long; the tongue is 1½″ wide and 16″ long. On one side, at the 24″ end, is the rafter table information (see Fig. 18.13, where a framing square is shown with the rafter tables).

The framing square is devised on a unit span of 24″ and a unit run of 12″. This is very important and must be remembered. (See Fig. 18.14.)

Step 2: To understand how to read the rafter tables, one must know:

(a) how to find the pitch of a roof;

(b) how to find the rise in in. per ft run of the slope of the roof.

Step 3: Find the pitch of a roof with a span of 24′-0″ and a rise of 6′-0″.

(a)
$$\frac{\text{Rise}}{\text{span}} : \frac{6'\text{-}0'' \ (\text{rise})}{24'\text{-}0'' \ (\text{span})}$$

Cancel to ¼ pitch.

(b) Now try a roof with 36′-0″ span and 12′-0″ rise.

Step 4: A roof with a span of 24′-0″ and a run of 12′-0″ may be considered to have a run of twelve units, with each unit being 1 ft. *This is the framing-square conception.*

The rise of the roof is 6′-0″ (72″). For every ft of run the "rise per ft run" is 72″ ÷ 12 = 6″.

Example

(a) A roof with a ⁵⁄₁₂ pitch and a span of 36′-0″ (and a run of 18′-0″) may be considered to have a run of eighteen units, with each unit being 1 ft.

(b) The rise of the roof is ⁵⁄₁₂ of the span of 36′-0″; ⁵⁄₁₂ of 36 is 15′-0″ rise.

(c) For every ft of run, the rise in in. per ft run is 15 × 12 = 180″ ÷ 18 (run) = 10″ rise per ft run.

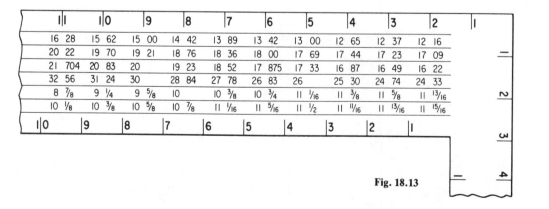

2|3	2|2	2|1	2|0	1|9	1|8	1|7	1|6	1|5	1|4	1|3	1|2
Length common rafters per foot run					21 63	20 81	20 00	19 21	18 44	17 69	16 97
Length hip or valley per foot run					24 74	24 02	23 32	22 65	22 00	21 38	20 78
Diff. in length of jacks 16 inches centers					28 84	27 74	26 66	25 61	24 585	23 588	22 625
Diff. in length of jacks 2 feet centers					43 27	41 62	40	38 42	36 08	35 38	33 94
Side cut of jacks use					6 ¹¹⁄₁₆	6 ¹⁵⁄₁₆	7 ³⁄₁₆	7 ½	7 ¹³⁄₁₆	8 ⅛	8 ½
Side cut hip or valley use					8 ¼	8 ½	8 ¾	9 ¹⁄₁₆	9 ⅜	9 ⅝	9 ⅞
2|2	2|1	2|0	1|9	1|8	1|7	1|6	1|5	1|4	1|3	1|2	1|1	1|0

1|1	1|0	9	8	7	6	5	4	3	2	1
16 28	15 62	15 00	14 42	13 89	13 42	13 00	12 65	12 37	12 16	
20 22	19 70	19 21	18 76	18 36	18 00	17 69	17 44	17 23	17 09	
21 704	20 83	20	19 23	18 52	17 875	17 33	16 87	16 49	16 22	
32 56	31 24	30	28 84	27 78	26 83	26	25 30	24 74	24 33	
8 ⅞	9 ¼	9 ⅝	10	10 ⅜	10 ¾	11 ¹⁄₁₆	11 ⅜	11 ⅝	11 ¹³⁄₁₆	
10 ⅛	10 ⅜	10 ⅝	10 ⅞	11 ¹⁄₁₆	11 ⁵⁄₁₆	11 ½	11 ¹¹⁄₁₆	11 ¹³⁄₁₆	11 ¹⁵⁄₁₆	
1|0	9	8	7	6	5	4	3	2	1	

Fig. 18.13

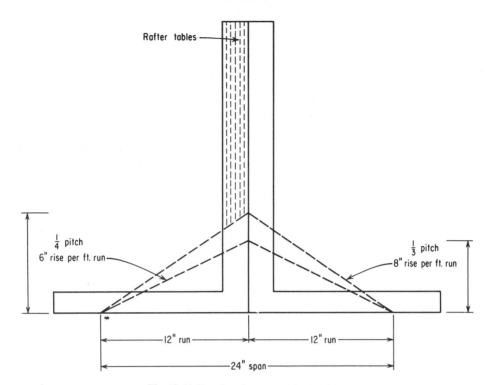

Fig. 18.14 Two framing squares back to back.

(d) The rise per ft run with the framing-square conception of a unit span of 24″ is as follows:

Pitch Span	Rise in inches per ft run
$\frac{1}{4}$ of 24″ =	6″ rise per ft run
$\frac{1}{3}$ of 24″ =	8″ rise per ft run
$\frac{5}{12}$ of 24″ =	10″ rise per ft run
and so on	

Step 5: Reexamine Fig. 18.14, where two framing squares are shown back-to-back with two roof pitches drawn in. *This depicts the unit run of 12″ and the unit span of 24″, upon which the framing-square rafter tables are based.*

Step 6: Measure across the left-hand framing square from 12″ on the tongue to 6″ on the blade and read 13⁷⁄₁₆″, or 13.42 inches.

On the first line of the rafter tables (see Fig. 18.13) read under the 6″ (rise per ft run) on the blade the figure 13.42. This is the ready-reckoned hypotenuse of a unit of 1 ft of run for 6″ of rise. It is ¼ pitch.

Now measure across from the 12″ on the tongue to the 8″ on the blade and read 14⁷⁄₁₆″ or 14.42 inches.

Read the rafter table (on the framing square). This is the ready-reckoned hypotenuse of a unit of 1 ft of run for 8″ of rise. It is a ⅓ pitch.

Step 7: Examine and complete the table of relative pitches on p. 328. The first line has been completed as a guide and you are to fill in the blanks.

Problem

Study the first completed line of the incomplete table on p. 328. Pay particular attention to cancelling the figure in the first column to produce the roof pitch in the second column. In the third column the span referred to is the framing-square unit span of 24″. Thus ¹⁄₁₂ of 24″ = 2″ of rise per ft of run. Read under 2″ of the framing-square tables and find 12.16. Complete the following table:

18.10 INCREASED PERCENTAGE OF ROOF AREA OVER GROUND AREA

Problem

Using the framing square, find the percentage increased of roof area over that of the ground area for a roof with a ⁵⁄₂₄ pitch.

Framing-Square Unit of Span 24"	Cancel Down for	Rise in in. per ft run	Rafter Tables Length of Rafter
Rise over span	Pitch	Pitch times Span	per ft run
$\frac{2}{24}$	$\frac{1}{12}$	2"	12.16
$\frac{3}{24}$	$\frac{1}{8}$		
$\frac{4}{24}$	$\frac{1}{6}$	4"	
$\frac{5}{24}$		5"	
$\frac{6}{24}$			
$\frac{7}{24}$	$\frac{7}{24}$	7"	
$\frac{8}{24}$	$\frac{1}{3}$		
$\frac{9}{24}$		9"	
	$\frac{5}{12}$		
$\frac{11}{24}$	$\frac{11}{24}$		
$\frac{12}{24}$	$\frac{1}{2}$		

Example

Step 1: Find the pitch of the roof (unless it is already given).

Step 2: Find the number of in. of rise per ft run of the roof.

Step 3: Read on the first line of the rafter table (on the framing square) under the rise per ft run (as in Step 2, above) and find the ready-reckoned length of the slope of the roof for 1 ft of run. The first line of the framing-square rafter table reads "Length common rafters per ft run." *The length of rafter per ft of run may be read directly from the foregoing completed problem.*

Step 4: A roof with a ⁵⁄₂₄ pitch has a rise of ⁵⁄₂₄ of 24" per ft of run (24" is the unit of span upon which the framing square is based). In this case it is 5" rise per ft of run.

Step 5: Read under 5" on the rafter table and find 13.00, which is 13" slope for every 12" (or 1 ft) of run. See the foregoing completed problem.

Step 6: Find the ratio of the pitched area of the roof to the plan area of the roof. The unit framing-square run is 12" and the framing-square length of the slope per 12" of run for a ⁵⁄₂₄ pitch roof is 13. Thus the ratio is 12:13.

Step 7: Run is to slope as 100 percent is to x percent; 12 is to 13 as 100 percent is to x percent;

$$\frac{13 \times 100}{12} = 108\% \text{ plus}$$

The percentage increase of the roof area over that of the plan area for all regular ⁵⁄₂₄-pitch roofs, whether or not they have gables or intersecting roofs, is approximately 8 percent.

Roof Areas—Problem

Complete the second, fourth, and fifth columns of the following exercise; but before doing so, check the accuracy of the table you completed at the end of Sec. 18.9. You will be using your completed table to solve the following exercise. As an estimator, your time will be rewardingly spent in making yourself efficient in the use of the framing square.

Size of Roof Plan	Total Plan Area, sq ft	Pitch of Roof	% increased Roof Area over Plan	Total Pitched Area
14'-0" × 22'-0"		$\frac{5}{24}$	%	
24'-0" × 28'-0"		$\frac{1}{4}$	%	
26'-0" × 30'-0"		$\frac{7}{24}$	%	
28'-0" × 42'-0"		$\frac{1}{3}$	%	
30'-0" × 44'-0"		$\frac{3}{8}$	%	
32'-0" × 46'-0"		$\frac{1}{2}$	%	

18.11 INCREASED PERCENTAGE OF ROOF AREA OVER GROUND AREA FOR A ROOF OF UNEQUAL PITCH

Problem 1

Find the area of a gable roof which is to be erected with an unequal pitch. The minor roof run is 8'-0". The major roof is 16'-0". Both roofs have a rise of 8'-0". The total length of the roof is 40'-0". (See Fig. 18.15.)

Remember to calculate the pitches of both sides of the roof from hypothetical spans. Only the runs are given and

$$\frac{\text{rise}}{\text{span}} = \text{pitch}$$

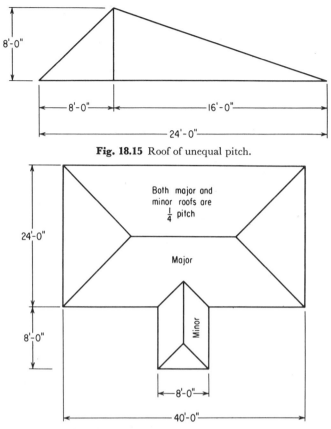

Fig. 18.15 Roof of unequal pitch.

Both major and minor roofs are $\frac{1}{4}$ pitch

Major

Minor

Fig. 18.16 Intersecting roof of equal pitch.

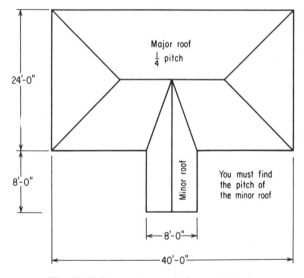

Major roof $\frac{1}{4}$ pitch

Minor roof

You must find the pitch of the minor roof

Fig. 18.17 Intersecting roof of unequal pitch.

Find:

(a) the plan area of the minor roof;

(b) the percentage increase of the area of the minor roof over that of the ground plan that it covers;

(c) the sloped area of the minor roof;

(d) the plan area of the major roof;

(e) the percentage increased area of the major roof over that of the ground-plan area which it covers;

(f) the sloped area of the major roof;

(g) the total area of the unequal gable pitch roof.

Problem 2

What is the area of the ¼-pitch intersecting roof of equal pitch, as shown in Fig. 18.16? The dimensions are as shown and the length of the roof, including the verges (cornices), is 40′-0″.

Problem 3

What is the area of an intersecting roof of unequal pitch as dimensioned in Fig. 18.17? The major roof has a pitch of ¼ and you are to determine the pitch of the minor roof.

It is very important to note that both the major and the minor roofs have the same rise, and that the percentage increase of roof areas applies only to that portion of the total plan area which is covered by each respective pitched roof.

Problem 4

Using the rafter tables, complete the following table. The first line has been completed as a guide. Study Sec. 18.10.

Pitch of roof	Rise in inches per foot run	Percentage increase of sloped area over ground area
$\frac{1}{8}$	3″	3%
$\frac{1}{6}$		
$\frac{5}{24}$		
$\frac{1}{4}$		
$\frac{7}{24}$		
$\frac{1}{3}$		
$\frac{3}{8}$		
$\frac{1}{2}$		
$\frac{5}{8}$		
$\frac{2}{3}$		
$\frac{3}{4}$		

18.12 TO FIND THE MERCHANTABLE LENGTH OF A GABLE ROOF RAFTER

Dimension lumber, unless specially ordered, is marketed only in multiples of $2'-0''$, starting at $8'-0''$.

Example

Estimate the number (in merchantable lengths) of 2×6 common rafter stock required for a gable roof with a span (including the eaves) of $24'-0''$. The length of the roof (including the verges) is $36'-0''$ and the roof has a $\frac{1}{3}$ pitch. Rafters are to be spaced at $16''$ OC. *The plan area of a roof is equal to the area of ground that it covers.*

Step 1: Determine the length of the slope. The pitch of a roof is usually shown on one of the elevation drawings.

Step 2: Remember that it is a common practice for estimators to scale from one of the sectional drawings, and that dimension lumber is marketed in even-numbered foot lengths.

Step 3:

(a) The number of CRs required is equal to the overall length of the roof divided by the OCs, plus one for the end, and then multiplied by two (two sides).

(b) The length of the roof is $36'-0''$ and the rafters are $16''$ OCs. The number of CRs required is 36 divided by $1\frac{1}{3}$ feet ($16''$), plus one rafter for the end, times two (two sides).

$$(36 \div 1\tfrac{1}{3}) + 1 \quad \times 2 = \quad \frac{(36 \times 3)}{4} + 1$$

Problem 1

In the preceding example the actual pitch of the roof would be uneconomical for the span. Why? Try it!

What would be the merchantable length of lumber required using a $\frac{5}{24}$ pitch for the same roof?

Problem 2

Estimate the number of rafters required for the roof in the previous example if the rafters were placed at $2'-0''$ OCs.

Problem 3

Estimate the merchantable length and number of rafters required for the roof in Problem 1 if the roof was $\frac{3}{8}$ pitch and the rafters placed at $16''$ OCs.

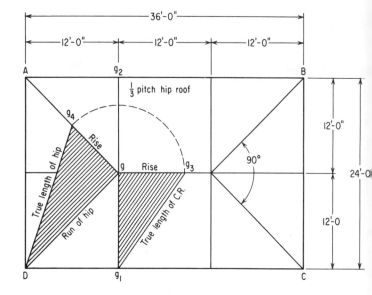

Fig. 18.18 Development of the lengths of CR's and HR's for a $\frac{1}{3}$ pitch roof.

18.13 TO FIND THE LENGTH OF HIP RAFTERS FOR REGULAR HIP ROOFS

Example

Step 1: Draw to the scale of $\frac{1}{4}''$ to $1'-0''$ a rectangular plan of a $\frac{1}{3}$-pitch hip roof $24'-0'' \times 36'-0''$ showing the common and hip rafter runs. (See Fig. 18.18.) Hip rafters are abbreviated HRs.

Step 2: From a g lay off the rise of the roof as at g_3. In this case, it is a $\frac{1}{3}$ pitch. One-third of the roof span is $8'-0''$. Join g_3 and g_1, which is the length of the CR stock required. Measure it and order in merchantable lengths.

Step 3: With g–g_3 as a radius, describe an arc g_3–g_4. Join g_4–D. Then g_4–D is the length of the hip rafter. Measure it and order in merchantable lengths. Remember that hip rafter runs are $45°$ from CRs on plan.

Step 4: Make a tracing and then take a sharp knife and cut from D to g_4 and then from g_4 to g. Turn the right triangle up on line D–g.

Step 5: Cut on lines g–g_3 and g_3–g_1. Turn the right triangle up on line g–g_1. Examine the model. the hip rafters and the common rafters both have the same rise but different runs. *You will never waste your time as an estimator by making drawings or models.*

Problem 1

Find the merchantable lengths of common rafters and hip rafters required for Fig. 18.18, assuming the pitch to be $\frac{5}{24}$ and the CRs placed at $24''$ OCs.

Problem 2

Find the merchantable lengths of common rafters and hip rafters required at 16″ OC for a regular hip roof 28′-0″ × 32′-0″ with a ¼-pitch roof. Draw and develop a scaled plan.

18.14 TO FIND THE LENGTH OF JACK RAFTERS REQUIRED

Step 1: Draw the runs of the jack rafters at 16″ OCs as in Fig. 18.19. Jack rafters are abbreviated to JRs.

Step 2: Examine the drawing very carefully. It will be seen that the longest jack rafter no.1 plus the shortest no. 8 equal in length the common rafter. Jack rafter no. 2 plus no. 7 together also equal in length the CR, and so on. Theoretically, it requires the same number and lengths of pieces of common-rafter material for either a gable or a hip roof where both have the same plan area, shape, and pitch.

Step 3: To estimate the common- and jack-rafter material for a regular hip roof, treat the roof as a gable roof but add one extra piece of CR stock for every hip or valley to allow for waste in cutting jack rafters.

Step 4: Remember to add for hip and/or valley stock.

Problem

Use any available plan and develop it (on the actual plan) with the few fine lines required to find the lengths of the rafters required. Even if the plan is for a gable roof, you may still develop it with light development lines as for a regular hip roof. Most plans are drawn to the scale of ¼″ to 1′-0″.

18.15 LABOR COSTS FOR ROOF FRAMING

Labor time will depend upon the job conditions, the height and pitch of the roof, the weather, the kind of cutting equipment used, the job organization, and the morale of the men. Where men have to carry materials for thirty to forty yards, their morale will be low and they will be spending too much time walking instead of working. All building materials should be placed on the job as close as possible to where it will be used. It is on this supposition that all estimates are made.

Gable Roof Framing

Labor: Allow 18 to 20 hr of carpenter time and 6 to 8 hr carpenter's helper time for cutting and erecting 1,000 bd ft of gable roof framing. *Remember that there are no published figures like your own for estimating labor time. Keep good accurate records and use your own figures.*

Where a roof is cut up with intersecting roofs or dormer windows, allow an extra 7 to 10 percent more time for each.

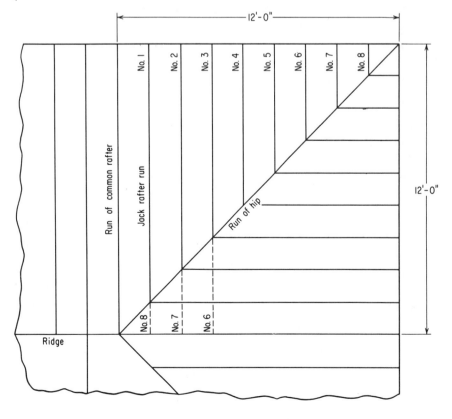

Fig. 18.19 Plan of jack rafter runs. Note carefully that the length of the run of No. 1 plus No. 8 and No. 2 plus No. 7 respectively equal in length the run of the common rafter.

Gable-Roof Sheathing

Labor: Estimate 10 to 12 hr of carpenter time and 5 to 7 of carpenter's helper time to sheath 1,000 sq ft of gable roof with 1 × 8 close-sheathed shiplap.

Where a gable roof is cut up with intersecting roofs or dormer windows, add 7 to 10 percent more time.

Problem

Estimate the material and labor cost to erect a simple $\frac{5}{24}$-pitch gable roof covering 30'-0" × 60'-0" ground area. Use 2 × 6 rafter stock spaced 16" OC's and 1 × 8 shiplap close-sheathed. Use local labor and material costs and do not forget the nails. Make no allowance for profit on this exercise.

Hip-Roof Framing

Labor: Allow 22 to 24 hr of carpenter time and 7 to 9 of carpenter's helper time for cutting and erecting 1,000 bd ft of simple hip roof framing.

Where a roof is cut up with intersecting roofs or dormer windows, allow an additional 8 to 10 percent more time.
Note: Taking into account that a simple gable roof must have gable end walls complete with cladding and a finishing barge board, there is little difference in cost either for materials or labor from that of a simple hip roof framing.

Where a hip roof is cut up with intersecting roofs or dormer windows, allow an additional 12 to 16 percent more time.

Problem

Estimate the material and labor costs to erect a simple $\frac{5}{24}$-hip roof covering 30'-0" × 60'-0" of ground area. Use 2 × 8 for the hips and 2 × 6 for the rafters spaced at 16" OC and 1 × 6 shiplap close-sheathed. Use local labor and material costs. Do not forget nails. For nail allowances, see Chap. 7.

18.16 FLAT ROOFS

In general, the quantities required for flat roofs are among the easiest parts of a building estimate. The following flat roof systems are usually erected by the contractor's own men.

Flat roof decks supported on:

(a) solid wood beams;

(b) glue-laminated wood beams;

(c) I-beams;

(d) one-way systems of tile and reinforced concrete deck;

(e) one-way metal-pan formed reinforced concrete deck;

(f) the grid system of metal-pan formed reinforced concrete deck.

The following flat roof systems are subcontracted:

(a) prestressed and postressed reinforced concrete roof decks;

(b) precast concrete roof slabs.

A typical roof assembly that may be erected by the contractor's own men is one which is supported on steel I-beams. (See Fig. 18.20.)

This type of assembly may consist of:

(a) an I-beam, which may be subcontracted for a price including hoisting into place;

(b) wood members bolted to the flanges of the I-beam;

(c) ceiling joists of a design-strength to carry the roof framing, the roofing, and the snow load;

(d) wood-framed dwarf walls (roof saddles), which are designed to carry the roof sheathing with a fall of 1" to 1'-0" away from the parapet walls;

(e) the roof sheathing (usually plywood);

(f) rigid insulation.

The roofing (roof covering) is a subtrade.

Estimating Rough Carpentry for Flat Roofs

The cost will depend upon the height of the roof above ground, the hoisting equipment, the weather conditions, and the job runner's organizing ability.

The estimate would include:

(a) the lin ft of dimension lumber required for bolting to the flanges of the I-beams;

(b) the bolts for the flanges of the I-beams;

(c) the ceiling joists and the bridging for the ceiling joists;

(d) dimension lumber for the dwarf walls (saddles)—take the average height of the studs and estimate as for wall framing;

(e) plywood sheathing by the thousand sq ft—add 5 to 8 percent for waste according to the amount of openings in the roof deck for skylight and so on;

(f) rigid insulation in panels—reckon by the thousand sq ft.

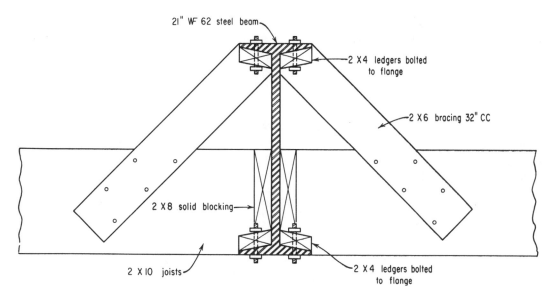

Fig. 18.20 Detail of steel beam and joist assembly.

Carpenter Labor for Flat Roofs: Dimension Lumber

A carpenter should erect about 450 to 500 bd ft of dimension lumber per 8-hr day, and on most roof jobs he would require the same amount of time for a helper.

Carpenter time	about 60 bd ft per hr
Carpenter's helper time	about 60 bd ft per hr

Carpenter Labor for Flat Roofs: Rough Plywood Sheathing

Carpenters working in pairs on this class of work should place and nail about 1,000 sq ft of 16″ OC's in about 5 hr at a rate of

Carptenter time	100 sq ft per hr
Carpenter's helper time	200 sq ft per hr

Labor Time Placing Rigid Insulation

Rigid insulation is very light and there are no problems for jointing, and the waste is negligible. Check locally as to who is responsible for placing this material. Two workmen working together should place about 2,000 sq ft in 5 hr at a rate of 200 sq ft per workman per hr.

18.17 ONE-WAY COMBINATION TILE AND CONCRETE ROOF ASSEMBLY

The one-way combination tile and concrete floor, also used for roof construction, is essentially a flat slab in which a large percentage of solid concrete has been replaced by structural clay tile. (See Fig. 18.21.)

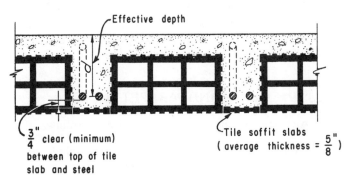

Fig. 18.21 One-way combination tile and concrete floor (Courtesy Structural Clay Products Institutue).

Estimating Concrete Floor and Roof Assemblies

(a) Estimate as for solid concrete and deduct for the displacement per lin ft of structural clay tile in the floor.

(b) Assume the tiles are 8 × 12 × 12; the displacement in cu ft for each lin ft is $\frac{2}{3} \times 1 \times 1$, which is $\frac{2}{3}$ cu ft or 0.67 cu ft per lin ft.

(c) Reckon the reinforcing by designated bar numbers in weight.

(d) Estimate the amount of concrete delivered per cu yd; allow for hoisting and placing.

(e) Estimate the quantity of clay tile delivered and allow for hoisting.

(f) The reinforcing will probably be a subtrade.

18.18 METAL-PAN FORMED REINFORCED CONCRETE FLOORS AND ROOFS

Metal pans are cheaper than clay tile. They may be rented in various sizes and in gages of 14, 16, and 18. They are very easy to place, but check locally to see who is responsible for this part of the work. There are two types: the heavy type, which is used over and over again, and the light type, which is used only once and left in place. There are a number of manufacturers of metal forms. Secure manufacturers' specifications and make out cards showing the displacement of each type by size. One-way metal pan forms are estimated the same way as clay tiles. Allow for either rental or purchase of the light-weight type left in place. (See Fig. 13.8.)

For the grid-placed metal forms, estimate the roof deck as for solid concrete and deduct the displacement of the grid metal pan forms. Make a very close study of the types of openings in the roof (skylights and pipework for ventilators and so on). There are more losses incurred in this class of work, through not accounting for sleeves and openings and allowances for other trades, than for any other reason. Allow for reinforcing both ways and for more wood formwork to support the floor assembly until such time as it is self-supporting. Make out a series of code-index cards showing the amount of displacement for each type grid metal form.

18.19 COPING AND PARAPET WALLS

An extract from a suggested specification follows:

1. Contractor shall provide and place copings and parapet walls as detailed on the plans.
2. After coping is in place, the flashing shall be installed as shown in details on plans. All joints in flashing shall be soldered and watertight. Coping units shall be attached to masonry wall with dowels as shown on plans. Dowels shall be placed at least 12 in. from joints in coping.
3. Construction at junction of parapet wall and roof shall be as detailed on plans, with counterflashing imbedded in the masonry walls at the time they are laid. All exposed masonry-wall surfaces on the inside (roof side) of the parapet wall shall be given the same surface application as the outside-wall surface.

18.20 PARAPET WALLS AND FLAT DECK ASSEMBLIES

Study the four sections of parapet walls shown in Fig. 18.22, which are reproduced by courtesy of the National Concrete Masonry Association.

1. At the top left is shown a section through a one-way system of tile and reinforced concrete roof. The roofing subcontractor's price would cover: (a) insulation; (b) roof flashing; (c) the built-up roofing laid to manufacturer's specifications.

   Note the precast concrete coping and the metal dowels. In all your estimating you will have to pay very close attention to miscellaneous metals. There is often a conflict as to who will supply the material, and where any item has not been considered it still has to be supplied, probably at the expense of the general contractor.
2. At the top right elevation notice the precast-concrete joists supporting a reinforced-concrete deck. Pay close attention to the flashing and counterflashing. Parapet walls are subject to very severe weather conditions and must be carefully built.
3. At the lower left observe the rigid insulation between the one-way system of tile and reinforced-concrete deck and the finished-concrete deck. Notice the insulated expansion joint between the floor bearing at the wall and the outside wall.
4. At the lower right is shown a further refinement of precast-concrete joists supporting a reinforced-concrete deck sandwiched with insulation.

All these assemblies are easy to estimate for quantities. The cost of placing the materials will depend upon the location of the job. (In large cities transportation to and from the site may cause very serious delays, which must be allowed for in the estimate.) Further cost estimates will depend upon the height of the building, hoisting arrangements, the season of the year, and above all the ability of the job supervisors.

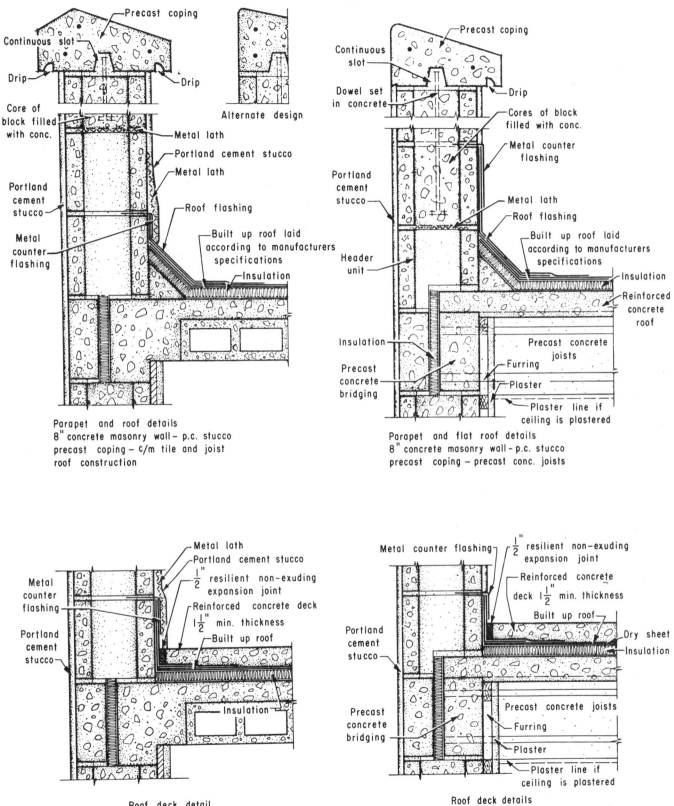

Precast coping

Continuous slot

Drip

Drip

Core of block filled with conc.

Alternate design

Portland cement stucco

Metal counter flashing

Metal lath

Portland cement stucco

Metal lath

Roof flashing

Built up roof laid according to manufacturers specifications

Insulation

Parapet and roof details
8" concrete masonry wall – p.c. stucco
precast coping – c/m tile and joist
roof construction

Precast coping

Continuous slot

Dowel set in concrete

Drip

Portland cement stucco

Cores of block filled with conc.

Metal counter flashing

Metal lath

Roof flashing

Header unit

Built up roof laid according to manufacturers specifications

Insulation

Reinforced concrete roof

Insulation

Precast concrete bridging

Precast concrete joists

Furring

Plaster

Plaster line if ceiling is plastered

Parapet and flat roof details
8" concrete masonry wall – p.c. stucco
precast coping – precast conc. joists

Metal lath
Portland cement stucco
$\frac{1}{2}$" resilient non-exuding expansion joint
Reinforced concrete deck $1\frac{1}{2}$" min. thickness
Built up roof

Metal counter flashing

Portland cement stucco

Insulation

Roof deck detail
c/m tile and joist roof construction

Metal counter flashing

$\frac{1}{2}$" resilient non-exuding expansion joint

Reinforced concrete deck $1\frac{1}{2}$" min. thickness

Built up roof

Dry sheet

Insulation

Portland cement stucco

Precast concrete bridging

Precast concrete joists

Furring

Plaster

Plaster line if ceiling is plastered

Roof deck details
precast concrete joists

Fig. 18.22 Four sections of parapet walls.

QUESTIONS

1 Define the following types of roofs and make a freehand sketch of each.
   (a) flat roof
   (b) shed or lean-to-roof
   (c) gable roof
   (d) hip roof
   (e) gambrel or barn roof
   (f) sawtooth roof

2. Define the following:
   (a) span
   (b) run
   (c) wall plates
   (d) rise
   (e) pitch
   (f) ridge
   (g) line length of a rafter
   (h) plumb cut of a rafter
   (i) birdsmouth
   (j) rafter overhang
   (k) tail cut of a rafter

3. A roof has a span of 24′-0″ and a rise of 6′-0″; what is the pitch?

4. A roof has a span of 36′-0″ and a rise of 12′-0″; what is the pitch?

5. A roof has a ¼-pitch and a span of 28′-0″; what is the rise?

6. A roof has a $\frac{5}{12}$-pitch and a span of 36′-0″; what is the rise?

7. Using the framing-square tables given, complete the following table. The first item has been completed as a guide.

Framing-square unit of span	Cancel down	Rise in inches per foot run	Read on first line of rafter tables
Rise over span	*Pitch*	*Pitch times span*	*Length of rafter per foot run*
$\frac{2}{24}$	$\frac{1}{12}$	2″	12.37
$\frac{3}{24}$	$\frac{1}{8}$		
$\frac{4}{24}$	$\frac{1}{6}$	4″	
$\frac{5}{24}$		5″	
$\frac{6}{24}$			
$\frac{7}{24}$	$\frac{7}{24}$		
$\frac{8}{24}$	$\frac{1}{3}$	8″	
$\frac{10}{24}$			
$\frac{12}{24}$	$\frac{1}{2}$		

8. Draw a section of a one-way combination tile and concrete floor.

19
STAIRS: WOOD AND CONCRETE

This chapter is designed to help you toward an understanding of the construction and estimating of stairs. You will learn, or refresh your memory about, some of the definitions of parts of stairs which, as an estimator, you should know. The real deciding factors in the cost of stairs are the architectural concept and the class of workmanship required.

While every stairway is different and has its own problems to be solved, an attempt has been made in this chapter to give you the feel of a medium-class wooden stair for a small apartment block and also for a concrete stair suitable for a warehouse. In both cases, the actual layout of the work is just as important for these buildings as it would be for a public building of great beauty.

19.1 WOODEN STAIRS

In effect, a stairway may be considered to be an inclined hall leading to other areas of a building. Remember that a good house is designed around the stairway. Stairs are provided for the purpose of passage of *persons of all ages in varying degrees of health,* and should be easy of ascent and constructed strong enough to carry the weight of any predetermined load, such as persons carrying heavy pieces of furniture or equipment.

19.2 STAIR DEFINITIONS

Stairwell opening The opening in a floor through which ascent is gained to an upper floor.

Fascia board The finished trim around the inside of a stairwell which covers the rough flooring assembly.

Total rise of a stair The perpendicular height from the *finished* lower floor to the top of the *finished* floor above.

Riser The perpendicular height from the top of one tread to the top of the next tread above. Note that there is one more riser than tread to every staircase because the top riser leads to a floor, not a tread.

Run of stairs The horizontal distance from the front of the first riser at the foot of a stairway to the front of the top riser.

Run of stair tread The horizontal distance from the front of one riser to the front of the next riser above. The nosing is additional to this. See Fig. 19.1.

Angle of flight The incline (slope) of a stair.

Headroom The minimum allowable perpendicular height from a point on the angle of flight to the finished underside of the floor, or stair soffit above.

Stair soffit The finished underside of a stair.

Story rod A straight wooden measuring rod, which is placed perpendicular from the finish of one floor to the finish of the floor above, on which is marked the total height of the stairs. It is used by stair builders. See Fig. 19.2.

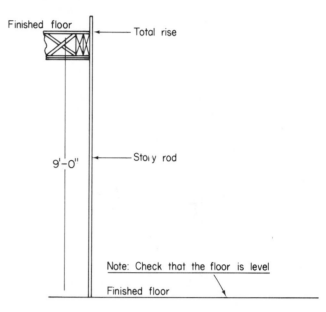

Fig. 19.2. The story rod in position for measuring the total rise from finished floor to finished floor.

Stringers

(a) The sides of stairs into which the treads and risers are secured.

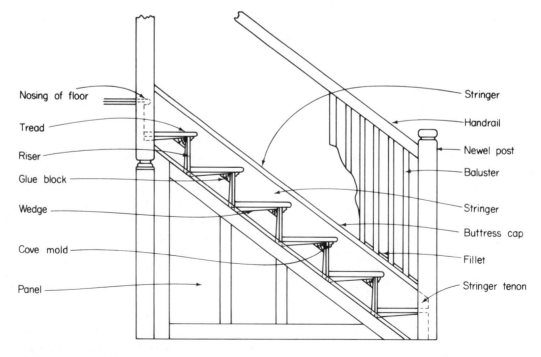

Fig. 19.1 Wooden stair with open stringer.

(b) A mitered stringer is one into which the risers are mitered to an open stringer, (i.e.) a stringer that is not covered by a wall.

(c) A notched stringer is one from which (on the upper side) the profile of the tread and riser is cut away. It is similar in respects to a stair carriage.

(d) A housed stringer has grooves (trenches) cut into the stringer to receive the risers and treads. They are called housings in wood stringers and are usually half an inch deep.

Glue blocks Triangular pieces of wood about three inches long and two inches wide on each side of a 90° angle. They are glued to the underside of wooden stairs. See Fig. 19.3.

Handrail This is what its name implies. It should be placed thirty-two inches perpendicular above the face of the risers. At landings it should be placed thirty-six inches above the level of the floor.

Landings Horizontal platforms between flights of stairs (usually to change the direction of the flight). They should be provided with a handrail.

19.3 STAIRWAY TYPES

There are three basic types of stairway:

(a) An *enclosed stairway* is separated from hallways and living units by means of walls or partitions and made accessible to such hallways or living units by means of a door or doors.

(b) An *interior stairway* is within the exterior walls of a building.

(c) An *open stairway* is one in which the floor landings are part of the public hallways.

19.4 LABOR FOR ERECTING A CLOSED-STRINGER STAIRCASE

All the members for a housed and wedged stringer, straight-flight staircase are usually prepared and supplied by the millwork subtrade. (See Fig. 19.3.)

On the job, a carpenter should erect a straight-flight stair covering a plan area of say 3'-0" × 12'-0" = 36 sq ft in 8 hr of carpenter time at a rate of 4½ plan sq ft per 1 hr of time. The general contractor would have to supply all fastening devices. Make out a code-index card for this item and check all future jobs for verification of time.

19.5 SMALL APARTMENT BLOCK STAIRS

Examine the drawings in Figs. 19.4 and 19.5 and the following typical specifications for the wooden stairs designed for a small apartment block.

Small Apartment Block Stairway Specifications

1. The millwork contractor shall check all dimensions for his work at the building and shall be responsible for the correctness of same.
2. (a) All stairs shall be supported on four rough stair

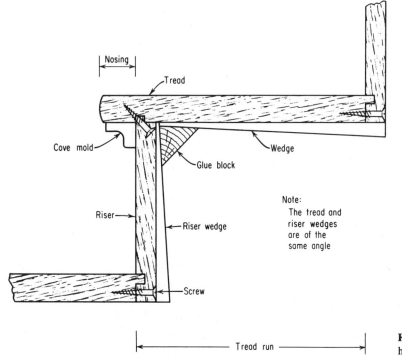

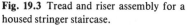

Fig. 19.3 Tread and riser assembly for a housed stringer staircase.

Fig. 19.4 Small apartment stair plan.

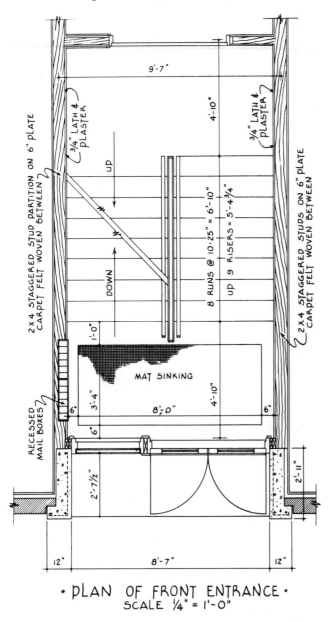

· PLAN OF FRONT ENTRANCE ·
SCALE ¼" = 1'-0"

horses (strings) of 2″ nominal stock material with an effective depth of not less than 4″ and solid bearings at the top and bottom ends. (See Fig. 19.6.)

(b) The finished strings shall be ⅜″ thick with molding at the top to detail.

(c) All the lumber for the stairs shall be kiln-dried, construction-grade West-Coast Fir, except the rough strings, which shall be first-grade Mountain Fir.

3. (a) To all treads of interior stairs, supply and install ³⁄₁₆″ sheet rubber.

(b) To all risers of interior stairs, supply and install ⅛″ sheet rubber.

(c) All landings to be covered with ³⁄₁₆″ sheet rubber; allow for a border.

(d) All colors to be chosen by the architect.

(e) All nosings to be finished with AF-517 Miralite Aluminium Nosings. (Note that there now is a tendency to use more rubber nosings than metal.)

4. A continuous handrail shall be provided to detail on both sides of the stairway.

Small Apartment Block Stairway Drawings—Front Entrance

Note the following important points about Figs. 19.4 and 19.6:

(a) The folding doors open outward to comply with public building fire regulations.

(b) Provision has to be made for a mat sinking at the foot of the stairs. (This would have to be remembered on the estimate too.)

(c) There is one more riser than tread to each flight, and the first flight shows eight runs of 10.25″ (which does not include the nosing). There are nine risers, which would be taken out of selected ⅜″ boards.

(d) The landing is called a *half-space landing*.

(e) The walls on either side of the stairway are constructed of 2 × 4 staggered studs on 2 × 6 plates with carpet felt woven between. This is to eliminate some of the transmission of sound and also satisfies a fire regulation in this type of construction.

Section Through Front Entrance Stairs

(a) The flight down to the lower suites has seven risers and six treads. (See Fig. 19.5.)

(b) The flight from the main entrance to the first floor has nine risers and eight treads.

(c) The flight from the first floor to the landing (shown dotted) has eight risers and seven treads. Read the plans again; the landing is not a perfect rectangle.

(d) The uppermost flight has eight risers and seven treads. Observe that (a) has seven risers and six treads, (b) has nine risers and eight treads, (c) and (d) each have eight risers and seven treads. *Note carefully that all stairs have one more riser than treads.*

(e) The critical measurement on these stairs is the height of the half-space landing immediately over the entrance floor. This determines the equal riser height for both flights which it serves.

(f) The stringers are cut from 2 × 12 material. (See the detail drawing, Fig. 19.7.)

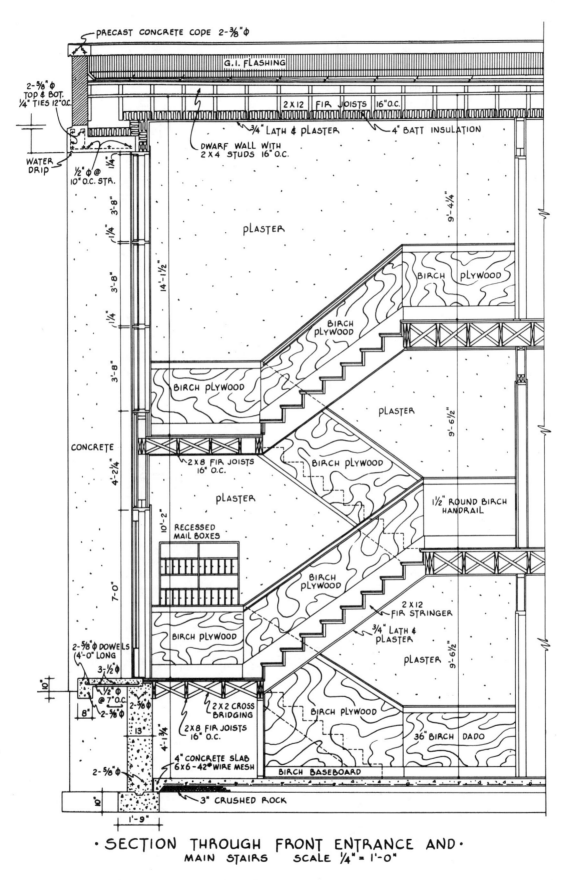

PRECAST CONCRETE COPE 2-⅜"∅

G.I. FLASHING

2-⅜"∅ TOP & BOT. ¼" TIES 12" O.C.

WATER DRIP

½"∅ @ 10" O.C. STR.

2 X 12 FIR JOISTS 16" O.C.

¾" LATH & PLASTER

4" BATT INSULATION

DWARF WALL WITH 2 X 4 STUDS 16" O.C.

3'-8"

¼"

14'-1½"

3'-8"

¼"

3'-8"

4'-2¼"

10'-2"

7'-0"

PLASTER

BIRCH PLYWOOD

BIRCH PLYWOOD

BIRCH PLYWOOD

2 X 8 FIR JOISTS 16" O.C.

PLASTER

RECESSED MAIL BOXES

CONCRETE

BIRCH PLYWOOD

BIRCH PLYWOOD

PLASTER

1½" ROUND BIRCH HANDRAIL

BIRCH PLYWOOD

BIRCH PLYWOOD

2 X 12 FIR STRINGER

¾" LATH & PLASTER

PLASTER

9'-4¼"

9'-6½"

9'-6½"

2-⅝"∅ DOWELS (4'-0" LONG)

3½"

½"∅ @ 7" O.C.

2-⅝"∅

2-⅝"∅

13"

4'-1¾"

10"

8"

2-⅝"∅

10"

2 X 2 CROSS BRIDGING

2 X 8 FIR JOISTS 16" O.C.

4" CONCRETE SLAB 6X6-42# WIRE MESH

BIRCH BASEBOARD

36" BIRCH DADO

3" CRUSHED ROCK

1'-9"

· SECTION THROUGH FRONT ENTRANCE AND ·
MAIN STAIRS   SCALE ¼" = 1'-0"

Fig. 19.5 Small apartment stair elevation.

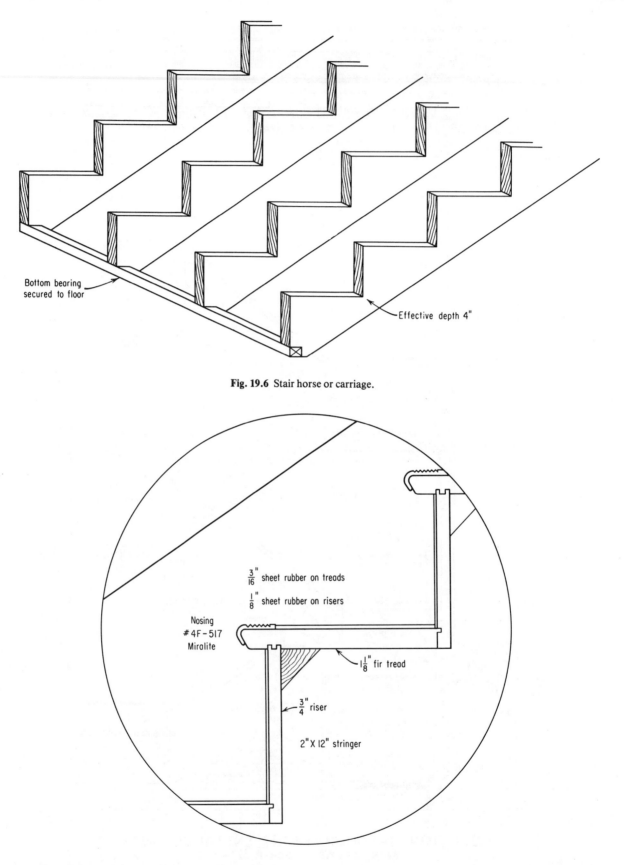

Fig. 19.6 Stair horse or carriage.

Bottom bearing
secured to floor

Effective depth 4"

$\frac{3}{16}$" sheet rubber on treads

$\frac{1}{8}$" sheet rubber on risers

Nosing
#4F-517
Miralite

$1\frac{1}{8}$" fir tread

$\frac{3}{4}$" riser

2" X 12" stringer

Fig. 19.7 Main tread and riser.

Detail of Tread and Riser Assembly

(a) The treads are 1⅛″ fir.

(b) The risers are kiln-dried select ¾″ boards.

(c) The sheet rubber on the treads is 3/16″ thick.

(d) The sheet rubber on the risers is ⅛″ thick.

(e) The nosings are #4F-517 Miralite.

(f) The stringers are 2 × 12 lumber.

Estimating Materials for Stairs

Depending upon the size of the contracting company, the stairs in the preceding pages may be made by the general contractor's own men working in the temporary carpenter's shop on the job. There would be very little machine work required for this class of stairs.

Stringers for stair horses are to be taken from 2 × 12 stock. Scale the length on the drawings and remember that the specifications state that there shall be four stair horses to each flight. As an example, the length of 2 × 12 required for the flight to the lower suites is 7′-0″, and since lumber is marketed by even foot lengths (except special orders), the stair-horse material for this one flight would be 2/2 × 12—14′-0″.

Tread length may be either scaled or read directly from the plan. The length required is about 4′-4″. In this case 1/14-0″ × 2 × 12 will cut into three treads. The total width of a tread is 10¼″ tread plus about 1¼″ nosing = 11½″, and the thickness is 1⅛″. Nominal 2 × 12 stock will be finished as 1⅛″ × 11½″. The tread material for this flight of stairs would be 2/2 × 12—14′-0″, which will cut into 6/2 × 12—4′-4″ treads.

Riser lengths are the same as the tread lengths. These will be taken from ¾″ finished select boards. Examine the detail drawing closely and you will see that the risers must be taken from the 1 × 10 (nominal size) select boards, which are marketed as such but whose finished size is ¾″ × 9½″. The riser material for this flight would be:

1/1 × 10—14′-0″
1/1 × 10—18′-0″   which will cut 7/1 × 10 = 4′-4″ risers

Finish stringer will be of the same length as the stair-horse material; there is one for each side. The finish stringers for this flight would be: 1/1 × 10—14′-0″, which will cut 2/1 × 10—7′-0″ finish stringers.

Metal nosings will be required for each tread, *plus one extra for the nosing to the landing* at the main entrance. Seven metal nosings are required for this flight.

Fastening devices will be nails, screws, and glue and also sandpaper will be required. Draw a full-size tread and riser assembly, measure the type of screw required,

and estimate all these items, which not only have to be purchased but also have to be applied.

Labor for Layout and Cutting

The stair estimate should be made by the sq-ft plan area covered by any one flight. For layout and cutting the material to size for this flight of stairs allow 4 hr of carpenter time. The plan area of this flight is about 4′-0″ × 5′-0″—say, 20 sq ft. For this class of work, a carpenter should lay out and cut all the members at a rate of about 5 sq ft per hr of carpenter time.

Labor for Erecting

For erecting each flight of these stairs allow one carpenter about 3 hr. The plan area is about 20 sq ft. For this class of work, a carpenter should erect about 6½ sq ft per hr of carpenter time.

Sheet rubber and metal nosing for this class of work would probably be done by an inclusive bid by the linoleum and rubber tile subtrade.

Millwork will include the layout and cutting of the materials but not the erection of the stairs. The mill in this case should be bound by the first paragraph of the specifications.

It is important to remember that a carpenter could make and erect a 4′-0″-wide wooden stair in about the same time as that of a 3′-0″ stair. A similar situation exists for the forms for a concrete stair.

Problem

1. Make out three code-index cards: one for the carpenter time per sq ft to lay out and cut the material size, one for the carpenter time per sq ft to erect the stair, and one inclusive card for the carpenter to lay out, cut the members to size, and erect the stair. Exclude the rubber sheathing and metal nosings.

2. Using local prices for labor and materials (on the correct stationery), make a complete estimate to supply and erect:

   (a) all the stair-horse material;
   (b) all the treads;
   (c) all the risers;
   (d) all the finish stringers;
   (e) all the fastening devices;
   (f) all the labor and fringe benefits;
   (g) all the overheads;
   (h) all the profit;
   for all the flights of stairs shown in Figs. 19.4 and 19.5 (excluding landings).

19.6 CONCRETE STEPS

This is an exercise on the estimating of concrete for a set of concrete steps. You may have to design and construct such steps and you should study your local building code on the subject.

Authorities state that all stairs shall be constructed allowing for a maximum of safety and a minimum of effort in use.

Several rules are given for the relations of the depth of the riser to the width of the tread. One such rule states that where public stairs are provided, the riser shall be not less than 6″ nor more than 7″, and the sum of the riser and the tread shall be approximately 17″.

The Rule of 17″ for Stairs

Three examples of this rule are:

Riser	Tread	Sum
6″	11″	17₀
6½″	10½″	17″
7″	10″	17″

A further refinement states that it is desirable to have an uneven number of risers to a stair. It may be noted that it is rhythmically easier to climb a stair with an uneven number of risers. The ancient Greeks knew it. Try it!

Problem

Estimate the cost for a set of concrete steps and landing which is to be provided in front of a public building. The specifications are as follows:

1. The perpendicular height of the steps from the finished sidewalk to the finished landing is 2′-8″.

2. The steps are to be built with an uneven number of risers, each riser as near to 6″ as possible. Use the rule of 17″ in determining the riser and tread sizes. (See Fig. 19.8.)

3. The width of the stair is 5′-0″, and the plan of the landing is 5′-0″ × 5′-0″. *Note: Where a landing is square it is known as a quarter-space.*

4. The stairs are to be reinforced with four no. 5 reinforcing rods placed at right angles to the treads 12″ OC; no. 2 temperature rods are to be placed and tied at 12″ OC at right angles to the reinforcing rods.

5. The side walls and the wall abutting the building are to be 6″ thick.

6. The landing is to be 4″ thick.

7. The soffit is to be 3″ thick.
   Note: The soffit is the underneath portion of the stair.

8. The center of the stair is to be made up of inert material.

9. When placing the concrete, provision is to be made for the feet of the posts of an ornamental metal handrail assembly.

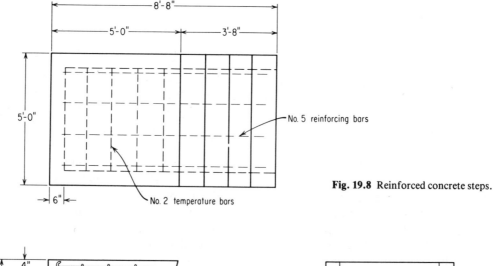

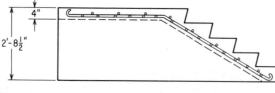

Fig. 19.8 Reinforced concrete steps.

Example

Use the work-up sheets and keep the work exceptionally neat.

Note:

(a) Use semipermanent forms for the sides and allow for riser material and sufficient lumber to secure the forms. Very little, if any, of the lumber except the semipermanent forms can be salvaged.

(b) Use ready-mixed 3,000 psi concrete delivered.

(c) Only the concrete requirements are dealt with in this exercise.

Step 1: The height of the stair is 2'-8½". By inspection it may be seen that there are five risers: 32½" ÷ 5 = 6½" for each riser.

Step 2: There is one more riser than treads to every stair. This is very important and must be remembered. Check Fig. 19.8 and count them. The width of the tread is 17" − 6½" = 10½".

Step 3: Estimate the seps and landings as for solid concrete and then deduct for the inert core.

Landing: 5'-0" × 5'-0" × 1'-9" high
$$5 \times 5 \, _{2}\tfrac{3}{4} = 68\tfrac{3}{4} \text{ cu ft}$$
Steps: 2'-9" rise × 3'-6" run divided by two, times 5'-0" wide
$$\frac{2\tfrac{3}{4} \times 3\tfrac{1}{2} \times 5}{2} = 24\tfrac{1}{16} \text{ cu ft} + 68\tfrac{3}{4} \text{ cu ft}$$
$$= 93 \text{ cu ft (approx.)}$$

Step 4: Deduct the inert core.

Landing: 4'-0" × 3'-0" × 2'-4" height (estimate)
$$4 \times 4 \times 2\tfrac{1}{2} = 37\tfrac{1}{3} \text{ cu ft}$$
Steps: End section core of steps area, say, 3'-6" long × 1'-10" high to underside of soffit divided by two, times 4'-0" width
$$\frac{3\tfrac{1}{2} \times 1\tfrac{5}{6}}{2} \times 4 = 12\tfrac{5}{12} \text{ cut ft} + 37\tfrac{1}{3} \text{ cu ft}$$
$$= 50 \text{ cu ft (approx.)}$$

Step 5: From the total solid volume in Step 2 deduct the volume of the inert core: 93 − 50 = an estimated 43 cu ft or 1½ cu yd of 3,000 psi ready-mixed concrete.

As an estimator you must get the feel of quantities; concrete steps are usually a surprise to most junior estimators in that they require less concrete than is at first thought.

19.7 REINFORCED-CONCRETE STAIR PROBLEM

Specifications

1. Provide an open reinforced-concrete single-flight stairway for a warehouse in the country using on-the-job mixed concrete.

2. The perpendicular height from the finished floor to the finished floor is 8'-2½".

3. The stairs are to have thirteen risers. Use the rule of 17".

4. The width of the stair is 4'-0".

5. The thickness of the soffit at right angles to the rake of the stair is 3".

6. The soffit is to be reinforced with four no. 6 reinforcing rods, and no. 2 temperature rods are to be placed laterally at 12" OC.

Problems

1. Draw the stair plan, front and side elevations, to the scale of ¾" to 1'-0". *As an estimator, if you cannot draw an item, it is very doubtful that you can estimate it.*

2. Using the work-up sheets, estimate:
   (a) The amount of wet concrete required for the job.
   (b) the weight of the reinforcing rods.
   (c) the weight of the finished stair (reckon solid concrete and exclude the weight of the reinforcing).
   (d) Using a 5-gal paste and a 1:2½:3 mix, how much water, cement, sand, and gravel will be required?
   (e) How much water, cement, sand, and gravel should be fed into a 3-cu-ft capacity mixer for each batch?
   (f) If each batch takes 4 min to mix and place, how long will it take to place the concrete for the stairs?
   (g) Using local prices for materials and labor, one man at a carpenter's rate of pay, and two men at carpenter's helper rates, what would be the estimated cost excluding the formwork for the stairs? Allow for fringe benefits, overhead expenses, the use of equipment (either your own or rented), and profit.

19.8 STEEL STAIRS

Stair details have been shown in Fig. 19.8 as a part of the structural drawings. Some architects prefer to have them on the architectural and some on the structural drawings. The reasons for showing them on the structural

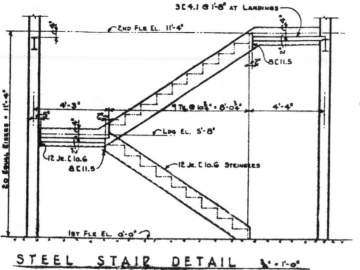

STEEL STAIR DETAIL ¾"=1'-0"

Fig. 19.9

TYPICAL TREAD
AND RISER 1½"=1'-0"

drawings is that in complicated stair layouts, there may be many unusual structural problems for which the structural engineer should be responsible. Stairs are ordinarily furnished by a steel fabricator as part of the structural steel contract and are allied closely with the other structural steel.

The stair treads and risers in Fig. 19.9 are fabricated of bent plate and the treads are covered with a two-inch concrete fill. Stair stringers are junior channels. Landings are plate stiffened with channels and covered with concrete.

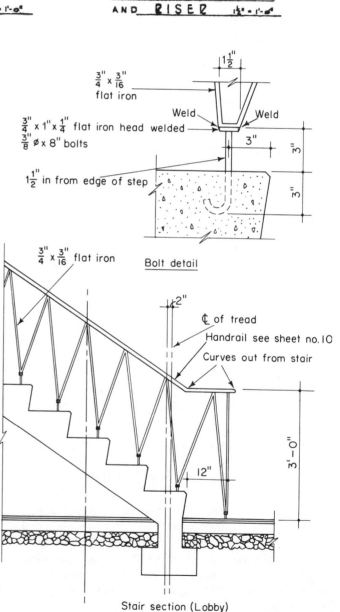

Fig. 19.10 Stair section and detail

QUESTIONS

1. Define the following wooden stair terms:
   (a) tread
   (b) nosing
   (c) riser
   (d) nose molding
   (e) stringer
   (f) newel post
   (g) soffit
   (h) stairwell
   (i) headroom clearance
   (j) stair horse or carriage

2. Make a sectional drawing (scale 1½" to 1'-0") of *one* tread and rise assembly for a housed stringer (wooden) stair and show the following:
   (a) riser
   (b) tread
   (c) tread nosing
   (d) wood wedges
   (e) screws
   (f) glued wood blocks

3. How many risers has a stair that has nine treads?

4. State any rule for determining the relationship of the height of a riser to the width of a tread.

5. List ten main headings with which an estimator would be concerned in determining the cost of a *volume mixed* reinforced open concrete stair between two floors in a warehouse.

20
STRUCTURAL STEEL AND METAL UNITS

20.1 STRUCTURAL STEEL

This section is designed to make you aware of some of the most frequently used kinds of steel structural members in modern buildings. In all estimates there is some miscellaneous metal, and great care must be taken in apportioning the responsibility for procurement and placing. As an example, assuming that flat-brick arches are to be supported on flat-steel strips, who would be responsible for the procurement of this item if the brickwork is subcontracted? Be very careful of miscellaneous metal on all your estimates. Heavy structural steel construction is a specialized field, but the estimator should know the names of most of the common members, which are as follows (see Fig. 20.1).

(a) wide-flange beams, abbreviated;

(b) American standard I-beam;

(c) American standard channel;

(d) angles: equal-leg angles (bar sizes up to 2½″ × 2½″ × ⅜″ with larger structural sizes);

(e) angles: unequal-leg (bar sizes up to 2½″ × 2″ × ⅜″ with larger structural sizes);

(f) structural Tee's;

(g) Structural Zee's;

(h) miscellaneous light columns;

(i) steel lally columns;

(j) base plates (flat-steel sections over 6″ in width and ¼″ and over in thickness);

(k) flat-steel strips 6″ or less in width and ¹³⁄₁₆″ or over in thickness

(l) round steel bars;

(m) square steel bars;

(n) half-round steel bars;

(o) half-oval steel bars.

For a full treatment of structural steel, get catalogues from the following sources:

Concrete Reinforcing Steel Institute
Chicago, Illinois

Steel Company of Canada Limited
525 Dominion Street
Montreal, Que., Canada

20.2 MISCELLANEOUS METAL UNITS

In all buildings there are some miscellaneous metal units to be installed. The following list is offered as a guide; it should be added to and kept up to date. You cannot have too many guide lists; and remember, it is very expensive to forget any item.

Metal units

1. adjustable legs for concrete-column supports;
2. adjustable shelf supports;
3. adjustable (telescopic) steel columns for light construction;
4. canopies;
5. chimneys;
6. fences and gates;
7. firegrates;
8. flashing;
9. flower boxes;
10. galvanized sheets;
11. galvanized-steel straps for bridging;
12. handrails;
13. joist hangers;
14. mailboxes;
15. metal doors and windows;
16. metal trim around doors and windows;
17. milk and package receptacles;
18. nuts and bolts;
19. stairs;

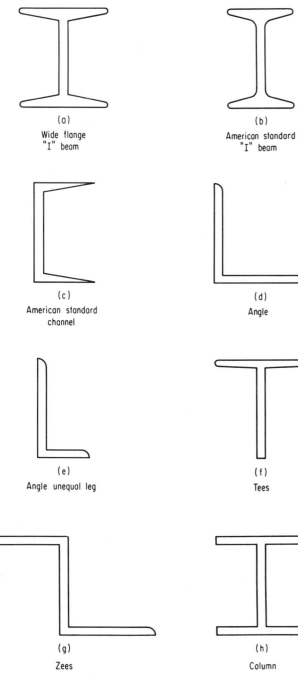

(a) Wide flange "I" beam

(b) American standard "I" beam

(c) American standard channel

(d) Angle

(e) Angle unequal leg

(f) Tees

(g) Zees

(h) Column

Fig. 20.1 Examples of structural steel.

20. stair nosings over wood or concrete stairs;

21. screws—special types;

22. suspension straps perforated for convenience in bending and hanging;

23. screens and grills;

24. steel door saddles (thresholds);

25. weatherstripping;

26. wire products;

27. ventilators and louvers.

Remember that there is often a conflict of opinion as to whether or not subcontractors will supply and fix miscellaneous metal. Unless this is made quite clear and definite to all subtrades, it will be the responsibility of the general contractor. This can be a very expensive item, and may result in a general contractor's loss on his estimate unless explicit instructions are given to the subcontractors before they submit their bids.

20.3 COLUMN BASE PLATES

Base plates for heavy columns are usually shipped loose and assembled in the field, especially for fully

riveted construction. In a heavy building, the friction of the column on the base will often provide all of the resistance to lateral force that is necessary, so that rivets are not required between the base and the column. In this case, the only connection for the riveted design is between the base angles and the anchor bolts. (See Fig. 20.2.)

The base plate for the welded design is integral with the steel column and is welded to the column. Note that the weld is shown only on one side of the flanges so that it will not be necessary to turn the column over while the base plate is being welded. Again, it is assumed that friction will take care of lateral forces.

26.4 WALL ANCHORS FOR BEAMS

Connections of steel beams to walls are not always clearly shown on structural drawings. Since this is one of the principal structural elements, a sketch such as that shown in Fig. 20.3 should be made.

A standard detail should be included on each set of plans involving wall-bearing masonry construction. Steel beams should always be firmly anchored at all masonry walls, not merely held by the brick. If the

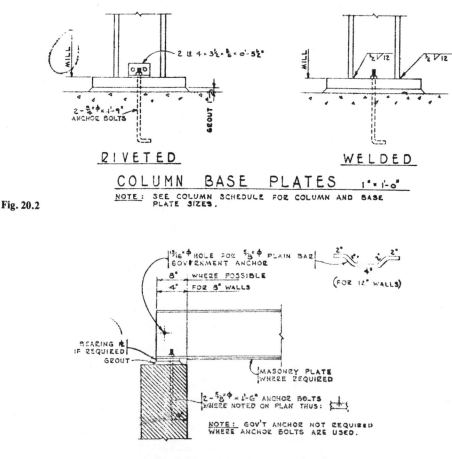

Fig. 20.2

Fig. 20.3 Standard wall anchorage for beams

masonry is placed around the beam, then anchor bolts are not necessary and "government" rod anchors only are required. The rod is bent in the shape shown so that it can be adjusted to fit into the brick coursing. Anchor bolts should be provided if the steel beam sits on top of a wall and does not have masonry to hold it down. Bearing plates may be called for on the framing plan, but are necessary only if the required bearing area can-not be developed by the beam. Bearing plates are difficult to work into the brick courses and should be avoided if possible. The masonry plates under the steel beams should not be extended into the bearing area, but should be cut off at the wall line. They are seldom satisfactory as bearing plates, being too thin. Anchor bolts should be placed to avoid the out course of brick. Otherwise, the brick must be cut to fit around the bolt.

QUESTIONS

1. Make a sectional drawing (scale $1\frac{1}{2}'' \times 1'\text{-}0''$) of each of the following structural steel members.
   (a) wide flange I-beam
   (b) American standard I-beam
   (c) American standard channel
   (d) angle equal leg
   (e) angle unequal leg
   (f) structural Tees
   (g) structural Zees
   (h) column

2. List ten miscellaneous metal units.

21

FINAL INSPECTION AND HANDOVER OF THE PROJECT

The final inspection, the maintenance of the completed building for a stated time, and the method of final payments to the contractor are all part of the general conditions of the contract set forth in the original specifications and contract documents. The following extract of specifications (as used by a school board) are typical of the clauses covering the final handover of the building to the owner, the final payments to the contractor, and the maintenance period of liability of the contractor.

21.1 EXTRACT OF SPECIFICATIONS

Payments

A typical contract statements is as follows:

On the twentieth of each month, the contractor shall submit to the architect an estimate on all work done and materials supplied onto the job. Contractor shall attach to his progress estimate receipted accounts and wage sheets covering the items entering into his estimate. Upon this total the board will allow ten percent (10%) to cover the contractor's overhead and operational expenses. When approved by the architect, the board shall pay to the contractor eighty-five percent (85%) of the amount of his estimate. The final payment of balance shall be paid to the contractor forty days after completion and acceptance of the building, provided satisfactory evidence is shown that all accounts for material and labor have been satisfied.

Correction of Work After Final Payment

(a) Neither the final certificate of payment nor any provisions of the contract documents shall relieve the contractor from responsibility for any faulty materials or workmanship that may appear within a period of one year from the date of the completion of the work, and he shall remedy any defects and pay for any damage to other work resulting therefrom which may appear within such period of two years, but beyond that the contractor shall not be liable.

(b) No certificate, payment, partial or entire use of the building or its equipment by the owner shall be construed as an acceptance of defective work or material.

Completion Date

The completion date shall be established in writing by the architect and contractor, after notice is presented to the architect that work has been completed.

Progress Schedule

During the construction of the building the general contractor must keep very close watch on the actual work accomplished against the amount estimated on the progress schedule, so that the building may be handed over at the contracted time. If for any reason the building has been falling behind schedule, the contractor must consider the wisdom of paying premium rates of pay for overtime work to his men. In many contracts there is a penalty clause for each day of delay in handing over the completed project.

21.2 SIDEWALK AND LANDSCAPING

When concrete sidewalks are not placed in the late fall because of extremely cold weather, the architect may retain a holdback of cash sufficient to do that work in the spring. Landscaping is usually an entirely separate contract outside the scope of the general contractor's work.

21.3 GENERAL CONTRACTOR'S INSPECTION

A short time before the final inspection it is usual for the general contractor, in company with the superintendent and the foremen, to make a careful inspection of the whole work, at which time notations are made of all details to be completed and a time set for these to be done. It is most likely that some broken glass will have to be replaced, door locks adjusted, paint work retouched, and so on.

21.4 WINDOW-CLEANING

On a large job this is a subtrade; for smaller jobs the contractor's own men may do the work, but in either case it must be allowed for in the estimates.

21.5 JANITOR SERVICES

Some jobs will require the services of janitors to thoroughly clean the whole building. Many specifications state that all fingerprints shall be removed from all paint and polished work and that the building shall be left ready for immediate occupation. This may also be a subtrade.

21.6 ARCHITECT'S FINAL INSPECTION

When the general contractor is quite sure that the building is completed, he will notify the architect in writing and request a date and time for the final inspection.

The final inspection is made by the architect with the clerk of the works, the owner, and the general contractor and his superintendent.

Any adjustments are noted and corrected and the building is then ready for handing over, provided there are no liens or other legal encumbrances against the property.

21.7 MAINTENANCE

Many contracts state that after a period of six months the contractor will examine the building and ease all doors and windows and make good any settlement cracks in any of the walls. It is usual for building to settle; and the better the class of workmanship in the fitting of doors, the more likely it is that such doors will require a little easing after a few months. The actual maintenance period is stated in the contract. It is your duty as an estimator to make a close study of the general conditions of the contract so that you will be in a position to advise your company on what provision should be made for the expense of maintenance.

Finally, use your schedules and remember that if any item is forgotten on the estimate, such item must still be purchased and installed.

QUESTIONS

1. When the owner has taken into use the whole or part of a newly erected building, does this exonerate the contractor from remedying any defects that appear within the time limit of maintenance?

2. By what type of communication is the actual completion date of a newly erected building established?

3. Under what conditions would a contractor consider paying premium rates to the workmen?

4. If, because of extremely cold weather, it was found impractical to complete the placing of concrete sidewalks until the following spring, what would the architect advise the owner concerning progress payments?

5. How are window-cleaning and janitoring services estimated on:
   (a) house construction
   (b) commercial buildings

6. Name six persons who usually make up the final inspection party of a newly completed building.

7. List four main documents (originating in the contractor's office) that are used in the preparation of a tender.

GLOSSARY

Alligatoring Coarse checking pattern characterized by a slipping of the new paint coating over the old coating to the extent that the old coating can be seen through the fissures.

Anchor bolts Bolts to secure a wooden sill plate to a concrete or masonry floor or wall.

Apron The flat member of the inside trim of a window, placed against the wall immediately beneath the stool.

Areaway An open subsurface space adjacent to a building, used to admit light or air or as a means of access to a basement.

Air-dired lumber Lumber that has been piled in yards or sheds for any length of time. For the United States as a whole, the minimum moisture content of thoroughly air-dried lumber is 12 to 15 percent and the average is somewhat higher. In the south, air-dried lumber may be no lower than 19 percent.

Airway A space between roof insulation and roof boards for movement of air.

Asphalt Most native asphalt is a residue from evaporated petroleum. It is insoluble in water but soluble in gasoline and melts when heated. Used widely in building for waterproofing roof coverings of many types, exterior wall coverings, flooring tile, and the like.

Astragal A closure between the two leaves of a double-swing or double-slide door to close the joint. This can also be a piece of molding.

Attic ventilators In houses, screened openings provided to ventilate an attic space. They are located in the soffit area as inlet ventilators and in the gable end or along the ridge as outlet ventilators. They can also consist of power-driven fans used as an exhaust system. (See also Louver.)

Axial Anything situated around, in the direction of, or along an axis.

Backband A simple molding sometimes used around the outer edge of a plain rectangular casing as a decorative feature.

Backfill The replacement of excavated earth into a trench around and against a basement foundation.

Balusters Usually small vertical members in a railing used between a top rail and the stair treads or a bottom rail.

Balustrade A railing made up of balusters, top rail, and sometimes bottom rail, used on the edge of stairs, balconies, and porches.

Bank or pit run gravel Natural aggregates as dug from a pit in the ground.

Barge board A decorative board covering the projecting rafter (fly rafter) of the gable end. At the cornice, this member is a fascia board.

Base or baseboard A board placed against the wall around a room next to the floor to finish properly between floor and plaster.

Base molding Molding used to trim the upper edge of interior baseboard.

Baseplate A plate attached to the base of a column that rests on a concrete or masonry footing.

Base shoe Molding used next to the floor on interior baseboard. Sometimes called a carpet strip.

Battern Narrow strips of wood used to cover joints or as decorative vertical members over plywood or wide boards.

Batter board One of a pair of horizontal boards nailed to posts set at the corners of an excavation, used to indicate the desired level; also a fastening for stretched strings to indicate outlines of foundation walls.

Bay The space between column center lines or primary supporting members, lengthwise in a building. Usually the crosswise dimension is considered the *span* or *width module,* and the lengthwise dimension is considered the *bay spacing.*

Bay window Any window space projecting outward from the walls of a building, either square or polygonal in plan.

Beam A structural member that is normally subjected to bending loads and is usually a horizontal member carrying vertical loads. (An exception to this is a *purlin.*)
There are three types of beams:

(a) *continuous beam:* A beam that has more than two points of support.
(b) *cantilevered beam:* A beam that is supported at only one end and is restrained against excessive rotation.
(c) *simple beam:* A beam that is freely supported at both ends, theoretically with no restraint.

Beam and column A primary structural system consisting of a series of beams and columns; usually arranged as a continuous beam, supported on several columns with or without continuity, that is subjected to both bending and axial forces.

Beam-bear plate Steel plate with attached anchors that is set in top of a masonry wall so that a purlin or a beam can rest on it.

Bearing partition A partition that supports any vertical load in addition to its own weight.

Bearing wall A wall that supports any vertical load in addition to its own weight.

Bed molding A molding in an angle, as between the overhanging cornice, or eaves, of a building and the sidewalls.

Bench mark A fixed point used for construction purposes as a reference point in determining the various levels of floor, grade, etc.

Bill of materials A list of items or components used for fabrication, shipping, receiving, and accounting purposes.

Bid Proposal prepared by prospective contractor specifying the charges to be made for doing the work in accordance to the contract documents.

Bid Bond A surety bond guaranteeing that a bidder will sign a contract, if offered, in accordance with his proposal.

Bid security A bid bond, certified check, or other forfeitable security guaranteeing that a bidder will sign a contract, if offered, in accordance with his proposal.

Bird screen Wire mesh used to prevent birds from entering the building through ventilators or louvers.

Blind-nailing Nailing in such a way that the nailheads are not visible on the face of the work—usually at the tongue of matched boards.

Blind stop A rectangular molding, usually ¾ by 1-⅜ inches or more in width, used in the assembly of a window frame. Serves as a stop for storm and screen or combination windows and to resist air infiltration.

Blue stain A bluish or grayish discoloration of the sapwood caused by the growth of certain moldlike fungi on the surface and in the interior of a piece of wood, made possible by the same conditions that favor the growth of other fungi.

Bodied linseed oil Linsee oil that has been thickened by suitable processing with heat or chemicals. Bodied oils are obtainable in a great range of viscosity from a little thicker than raw oil to just short of a jellied condition.

Bolster A short horizontal timber or steel beam on top of a column to support and decrease the span of beams or girders.

Bond Masonry units interlocked in the face of a wall by overlapping

Bonded roof A roof that carries a printed or written warranty, usually with respect to weather-tightness, including repair and/or replacement on a prorated cost basis for a stipulated number of years.

Bonus and penalty clause A provision in the proposal form for payment of a bonus for each day the project is completed prior to the time stated, and for a charge against the contractor for each day the project remains uncompleted after the time stipulated.

Boston ridge A method of applying asphalt or wood shingles at the ridge or at the hips of a roof as a finish.

Brace An inclined piece of framing lumber applied to wall or floor to stiffen the structure. Often used on walls as temporary bracing until framing has been completed.

Brace rods Rods used in roofs and walls to transfer wind loads and/or seismic forces to the foundation (often used to plumb building but not designed to replace erection cables when required).

Brick veneer A facing of brick laid against and fastened to sheathing of a frame wall or tile wall construction.

Bridging The structural member used to give lateral support to the weak plane of a truss, joist, or purlin; proves sufficient stability to support the design loads, sag channels, or sag rods.

Buck Often used in reference to rough frame opening members. Door bucks used in reference to metal door frame.

Building paper Used in many areas of building construction, it may form a dust trap between a hardwood floor and its subfloor.

Built-up roofing Roofing consisting of layers of rag felt or jute saturated with coal tar pitch, with each layer set in a mopping of hot tar or asphalt; ply designates the number of layers.

Butt joint The junction where the ends of two timbers or other members meet in a square-cut joint.

Camber A permanent curvature designed into a structural member in a direction opposite to the deflection anticipated when loads are applied.

Cant strip A triangular-shaped piece of lumber used at the junction of a flat deck and a wall to prevent cracking of the roofing which is applied over it.

Canopy Any overhanging or projecting structure with the extreme end unsupported. It may also be supported at the outer end.

Cantilever A projecting beam supported and restrained only at one end.

Cap The upper member of a column, pilaster, door cornice, molding, and the like.

Cap plate A horizontal plate located at the top of a column.

Casement frames and sash Frames of wood or metal enclosing part or all of the sash, which may be opened by means of hinges affixed to the vertical edges.

Casing Molding of various widths and thicknesses used to trim door and window openings at the jambs.

Cash allowances Sums that the contractor is required to include in his bid and contract sum for specific purposes.

Caulk To seal and make weathertight the joints, seams, or voids by filling with a waterproofing compound or material.

Cavity wall Two separate leaves of masonry units forming a wall with a space of 2″ minimum between them and secured together with metal ties or bonding units. The leaves are called wythes or withes and may be of different sized units.

Cement, Keene's A white-finish plaster that produces an extremely durable wall. Because of its density, it excels for use in bathrooms and kitchens and is also used extensively for the finish coat in auditoriums, public buildings, and other places where walls may be subjected to unusually hard wear or abuse.

Certificate of occupancy Statement issued by the governing authority granting permission to occupy a project for a specific use.

Certificate of payment Statement by an architect informing the owner of the amount due a contractor for work accomplished and/or materials suitably stored.

Change order A work order, usually prepared by the architect and signed by the owner or his agent, authorizing a change in the scope of the work and a change in the cost of the project.

Channel A steel member whose formation is similar to that of a "C" section without return lips; may be used singly or back-to-back.

Checking Fissures that appear with age in many exterior paint coatings, at first superficial, but which in time may penetrate entirely through the coating.

Checkrails Meeting rail sufficiently thicker than a window to fill the opening between the top and bottom sash made by the parting stop in the frame of double-hung windows. They are usually beveled.

Clip A plate or angle used to fasten two or more members together.

Clip angle An angle used for fastening various members together.

Collar beam Nominal 1- or 2″-thick members connecting opposite roof rafters. They serve to stiffen the roof structure.

Collateral loads A load, in addition to normal live, wind, or dead loads, intended to cover loads that are either unknown or uncertain (sprinklers, lighting, etc.).

Column In architecture: A perpendicular supporting member, circular or rectangular in section, usually consisting of a base, shaft, and capital. In engineering: A vertical structural compression member which supports loads acting in the direction of its longitudinal axis.

Combination doors or windows Doors or windows used over regular openings that provide winter insulation and summer protection and often have self-storing or removable glass and screen inserts. This eliminates the need for handling a different unit each season.

Concrete plain Concrete either without reinforcement, or reinforced only for shrinkage or temperature changes.

Condensation In a building: beads or drops of water (and frequently frost in extremely cold weather) that accumulate on the inside of the exterior covering of a building when warm, moisture-laden air from the interior reaches a point where the temperature no longer permits the air to sustain the moisture it holds. Use of louvers or attic ventilators will reduce moisture condensation in attics. A vapor barrier under the gypsum lath or dry wall on exposed walls will reduce condensation in them.

Conduit, electrical A pipe, usually metal, in which wire is installed.

Construction, dry wall A type of construction in which the interior wall finish is applied in a dry condition, generally in the form of sheet materials or wood paneling, as contrasted to plaster.

Construction, frame A type of construction in which the structural parts are wood or depend upon a wood frame for support. In codes, if masonry veneer is applied to the exterior walls, the classification of this type of construction is usually unchanged.

Contract documents Working drawings, specifications, general conditions, supplementary general conditions, the owner-contractor agreement and all addenda (if issued).

Coped joint See Scribing.

Corbel out To build out one or more courses of brick or stone from the face of a wall, to form a support for timbers.

Corner bead A strip of formed sheet metal, sometimes combined with a strip of metal lath, placed on corners before plastering to reinforce them. Also, a strip of wood finish three-quarters round or angular placed over a plastered corner for protection.

Corner boards Used as trim for the external corners of a house or other frame structure against which the ends of the sidings are finished.

Corner braces Diagonal braces at the corners of a frame structure to stiffen and strengthen the wall.

Crawl space Space between the lowest member of a floor and the ground beneath, affording two or three feet of clearance.

Curb A raised edge on a concrete floor slab.

Curtain wall Perimeter walls that carry only their own weight and wind load.

Dampproofing Material used to render a surface impervious to the passage of dampness by an emulsion or with special mortar for masonry units near ground level.

Datum Any level surface to which elevations are referred (see Bench Mark).

Dead load The weight of the structure itself, such as floor, roof, framing and covering members, plus any permanent loads.

Deflection The displacement of a loaded structural member or system in any direction, measured from its no-load position, after loads have been applied.

Demurrage Fee charged for detention of vessel, vehicle or carrier over allowable time.

Design loads Those loads specified by building codes, state or city agencies, or owner's or architect's specifications to be used in the design of the structural frame of a building. They are suited to local conditions and building use.

Door guide An angle or channel guide used to stabilize and keep plumb a sliding or rolling door during its operation.

Downspout A hollow section such as a pipe used to carry water from the roof or gutter of a building to the ground or sewer connection.

Drain Any pipe, channel, or trench through which waste or other liquids are carried off; i.e., to a sewer pipe.

Drip (a) A member of a cornice or other horizontal exterior-finish course that has a projection beyond the other parts for throwing off water. (b) A groove in the underside of a sill or drip cap to cause water to drop off on the outer edge instead of drawing back and running down the face of the building.

Drip cap A molding placed on the exterior top side of a door or window frame to cause water to drip beyond the outside of the frame.

Dry-wall Interior covering material, such as gypsum board or plywood, which is applied in large sheets or panels.

Ducts In a house, usually round or rectangular metal pipes for distributing warm air from the heating plant to rooms, or air from a conditioning device or as cold air returns. Ducts are also made of asbestos and composition materials.

Eave The line along the sidewall formed by the intersection of the inside faces of the roof and wall panels; the projecting lower edges of a roof, overhanging the walls of a building.

Erection The assembly of components to form the completed portion of a job.

Expansion joint A connection used to allow for temperature-induced expansion and contraction of material.

Fabrication The manufacturing process performed in the plant to convert raw material into finished metal building components. The main operations are cold forming, cutting, punching, welding, cleaning, and painting.

Fascia A flat, broad trim projecting from the face of a wall, which may be part of the rake or the eave of the building.

Field The jobsite or building site.

Field fabrication Fabrication performed by the erection crew or others in the field.

Field welding Welding performed at the jobsite, usually with gasoline-powered machines.

Filler (wood) A heavily pigmented preparation used for filling and leveling off the pores in open-pored woods.

Fire-resistive In the absence of a specific ruling by the authority having jurisdiction, applies to construction materials that are not combustible in the temperatures of ordinary fires and that will withstand such fires without serious impairment of their usefulness for at least one hour.

Fire-retardant chemical A chemical or preparation of chemicals used to reduce flammability or to retard spread of flame.

Fire stop A solid, tight closure of a concealed space, placed to prevent the spread of fire and smoke through such a space. In a frame wall, this will usually consist of 2 × 4 cross blocking between studs.

Fishplate A wood or plywood piece used to fasten the ends of two members together at a butt joint with nails or bolts. Sometimes used at the junction of opposite rafters near the ridge line.

Flagstone (flagging or flats) Flat stones, from 1″ to 4″ thick, used for rustic walks, steps, floors, and the like.

Flashing Sheet metal or other material used in roof and wall construction to protect a building from water seepage.

Flat paint An interior paint that contains a high proportion of pigment and dries to a flat or lusterless finish.

Flue The space or passage in a chimney through which smoke, gas, or fumes ascend. Each passage is called a flue, which together with any others and the surrounding masonry make up the chimney.

Flue lining Fire clay or terra-cotta pipe, round or square, usually made in all ordinary flue sizes and in 2-ft lengths, used for the inner lining of chimneys with the brick or masonry work around the outside. Flue lining in chimneys runs from about a foot below the flue connection to the top of the chimney.

Fly rafters End rafters of the gable overhang supported by sheathing and lookouts.

Footing A masonry section, usually concrete, in a rectangular form wider than the bottom of the foundation wall or pier it supports.

Foundation The supporting portion of a structure below the first-floor construction, or below grade, including the footings.

Framing, balloon A system of framing a building in which all vertical structural elements of the bearing walls and partitions consist of single pieces extending from the top of the foundation sill plate to the roofplate and to which all floor joists are fastened.

Framing, platform A system of framing a building in which floor joists of each story rest on the top plates of the story below or on the foundation sill for the first story, and the bearing walls and partitions rest on the subfloor of each story.

Frieze In house construction, a horizontal member connecting the top of the siding with the soffit of the cornice.

Frostline The depth of frost penetration in soil. This depth varies in different parts of the country. Footings should be placed below this depth to prevent movement.

Fungi, wood Microscopic plants that live in damp wood and cause mold, stain, and decay.

Fungicide A chemical that is poisonous to fungi.

Furring Strips of wood or metal applied to a wall or other surface to even it; furring normally serves as a fastening base for finish material.

Gable In house construction, the portion of the roof above the eave line of a double-sloped roof.

Gable end An end wall having a gable.

Gloss enamel A finishing material made of varnish and sufficient pigments to provide opacity and color, but little or no pigment of low opacity. Such an enamel forms a hard coating with maximum smoothness of surface and a high degree of gloss.

Gloss (paint or enamel) A paint or enamel that contains a relatively low proportion of pigment and dries to a sheen or luster.

Girder A large or principal beam of wood or steel used to support concentrated loads at isolated points along its length.

Girths or Girts Wood members tightly fitted between wood walls as stiffeners and as fire traps.

Grain The direction, size, arrangement, appearance, or quality of the fibers in wood.

Grain, edge (vertical) Edge-grain lumber has been sawed parallel to the pith of the log and approximately at right angles to the growth rings; i.e., the rings form an angle of 45° or more with the surface of the piece.

Grain, flat Flat-grain lumber has been sawed parallel to the pith of the log and approximately tangent to the growth rings, i.e., the rings form an angle of less than 45° with the surface of the piece.

Grain, quartersawn Another term for edge grain.

Grounds Guides used around openings and at the floorline to strike off plaster. They can consist of narrow strips of wood or of wide subjambs at interior doorways. They provide a level plaster line for installation of casing and other trim.

Grout Mortar made of such consistency (by adding water) that it will just flow into the joints and cavities of the masonry work and fill them solid.

Gusset A flat wood, plywood, or similar type of member used to provide a connection at intersection of wood members. Most commonly used at joints of wood trusses. They are fastened by nails, screws, bolts, or adhesives.

Gutter or eave trough A shallow channel or conduit of metal or wood set below and along the eaves of a house to catch and carry off rainwater from the roof.

Gypsum plaster Gypsum formulated to be used with the addition of sand and water for base-coat plaster.

Head The top of a door, window, or frame.

Header (a) A beam placed perpendicular to joists and to which joists are nailed in framing for chimney, stairway, or other opening. (b) A wood lintel.

Hearth The inner or outer floor of a fireplace, usually made of brick, tile, or stone.

Heartwood The wood extending from the pith to the sap-wood, the cells of which no longer participate in the life processes of the tree.

Hip The external angle formed by the meeting of two sloping sides of a roof.

Hip roof A roof that rises by inclined planes from all four sides of a building.

Humidifier A device designed to increase the humidity within a room or a house by means of the discharge of water vapor. They may consist of individual room-size units or larger units attached to the heating plant to condition the entire house.

I-beam A steel beam with a cross section resembling the letter I. It is used for long spans as basement beams or over wide wall openings, such as a double garage door, when wall and roof loads are imposed on the opening.

IIC A new system utilized in the Federal Housing Administration recommended criteria for impact sound insulation.

Impact load The assumed load resulting from the motion of machinery, elevators, cranes, vehicles, and other similar moving equipment.

INR (Impact Noise Rating) A single figure rating which provides an estimate of the impact sound-insulating performance of a floor-ceiling assembly.

Instructions to bidders A document stating the procedures to be followed by bidders.

Insulation Any material used in building construction for the protection from heat or cold.

Insulation board, rigid A structural building board made of coarse wood or cane fiber in ½- and ²⁵⁄₃₂-inch thicknesses. It can be obtained in various size sheets, in various densities, and with several treatments.

Insulation, thermal Any material, high in resistance to heat transmission, that, when placed in the walls, ceiling, or floors of a structure, will reduce the rate of heat flow.

Interior finish Material used to cover the interior framed areas, or materials of walls and ceilings.

Invitation to bid An invitation to a selected list of contractors furnishing information on the submission of bids on a subject.

Jack rafter A rafter that spans the distance from the wall-plate to a hip, or from a valley to a ridge.

Jamb The side of a door, window, or frame.

Joint The space between the adjacent surfaces of two members or components joined and held together by nails, glue, cement, mortar, or other means.

Joint cement A powder that is usually mixed with water and used for joint treatment in gypsum-wallboard finish. Often called *spackle.*

Joist Closely-spaced beams supporting a floor or ceiling. They may be wood, steel, or concrete.

Kiln-dried lumber Lumber that has been kiln dried often to a moisture content of 6 to 12 percent. Common varieties of softwood lumber, such as framing lumber, are dried to a somewhat higher moisture content.

Kip A unit of weight, force, or load equal to 1,000 pounds.

Knot In lumber, the portion of a branch or limb of a tree that appears on the edge or face of the piece.

Landing A platform between flights of stairs or at the termination of a flight of stairs.

Lath A building material of wood, metal, gypsum, or insulating board that is fastened to the frame of a building to act as a plaster base.

Lattice A framework of crossed wood or metal strips.

Lavatory A bathroom-type sink.

Ledger strip A strip of lumber nailed along the bottom of the side of a girder on which joists rest.

Liens Legal claims against an owner for amounts due those engaged in or supplying materials for the construction of the building.

Light Space in a window sash for a single pane of glass. Also, a pane of glass.

Lintel The horizontal member placed over an opening to support the loads (weight) above it.

Liquidated damages An agreed-to sum chargeable against the contractor as reimbursement for damages suffered by the owner because of contractor's failure to fulfill his contractural obligations.

Live load The load exerted on a member or a structure due to all imposed loads except dead, wind, and seismic loads. Examples include snow, people, movable equipment, etc. This type of load is movable and does not necessarily exist on a given member or structure.

Loads Anything that causes an external force to be exerted on a structural member. Examples of different types are:

(a) *dead load:* in a building, the weight of all permanent constructions, such as floor, roof, framing, and covering members.

(b) *impact load:* the assumed load resulting from the motion of machinery, elevators, craneways, vehicles, and other similar kinetic forces.

(c) *roof live load:* all loads exerted on a roof (except dead, wind, and lateral loads) and applied to the horizontal projection of the building.

(d) *seismic load:* the assumed lateral load due to the action of earthquakes, acting in any horizontal direction on the structural frame.

(e) *wind load:* the load caused by wind blowing from any horizontal direction.

Lookout A short wood bracket or cantilever to support an overhang portion of a roof or the like, usually concealed from view.

Louver An opening provided with one or more slated, fixed, or movable fins that allow flow of air, but exclude rain and sun or provide privacy.

Lumber Lumber is the product of the sawmill and planing mill, processed only by sawing, resawing, and passing lengthwise through a standard planing machine, crosscutting to length, and matching.

Lumber, board Yard lumber less than 2 inches thick and 2 or more inches wide.

Lumber, dimension Yard lumber from 2 inches to, but not including, 5 inches thick and 2 or more inches wide. Includes joists, rafters, studs, plank, and small timbers.

Lumber, dressed size The dimension of lumber after shrinking from green dimension and after machining to size or pattern.

Lumber, matched Lumber that is dressed and shaped on one edge in a grooved pattern and on the other in a tongued pattern.

Lumber, shiplap Lumber that is edge-dressed to make a close rabbeted or lapped joint.

Lumber, timber Yard lumber 5 or more inches in least dimension. Includes beams, stringers, posts, caps, sills, griders, and purlins.

Lumber, yard Lumber of those grades, sizes, and patterns which are generally intended for ordinary construction, such as framework and rough coverage of houses.

Mantel The shelf above a fireplace. Also used in referring to the decorative trim around a fireplace opening.

Masonry Stone, brick, concrete, hollow-tile, concrete-block, gypsum-block, or other similar building units or materials or a combination of the same, bonded together with mortar to form a wall, pier, buttress, or similar mass.

Mastic A pasty material used as a cement (as for setting tile) or a protective coating (as for thermal insulation or waterproofing).

Metal lath Sheets of metal that are slit and drawn out to form openings. Used as a plaster base for walls and ceilings and as reinforcing over other forms of plaster base.

Millwork Generally all building materials made of finished wood and manufactured in millwork plants and planing mills are included under the term *millwork*. It includes such items as inside and outside doors, window-and doorframes, blinds, porchwork, mantels, panelwork, stairways, moldings, and interior trim. It normally does not include flooring, ceiling, or siding.

Mullion The large vertical piece between windows. (It holds the window in place along the edge with which it makes contact.)

Nonbearing partition A partition which supports no weight except its own.

Parapet That portion of the vertical wall of a building that extends above the roof line at the intersection of the wall and roof.

Partition A material or combination of materials used to divide a space into smaller spaces.

Performance bond A bond that guarantees to the owner, within specified limits, that the contractor will perform the work in accordance with the contract documents.

Pier A structure of masonry (concrete) used to support the bases of columns and bents. It carries the vertical load to a footing at the desired load-bearing soil.

Pilaster A flat rectangular column attached to or built into a wall masonry or pier; structurally, a pier, but treated architecturally as a column with a capital, shaft, and base. It is used to provide strength for roof loads or support for the wall against lateral forces.

Precast concrete Concrete that is poured and cast in some position other than the one it will finally occupay; cast either on the jobsite and then put into place or away from the site to be transported to the site and erected.

Prestressed concrete Concrete in which the reinforcing cables, wires, or rods are tensioned before there is load on the member.

Progress payments Payments made during progress of the work, on account, for work completed and/or suitably stored.

Progress schedule A diagram showing proposed and actual times of starting and completion of the various operations in the project.

Punch list A list prepared by the architect or engineer of the contractor's uncompleted or work to be corrected.

Purlin Secondary horizontal structural members, located on the roof, extending between rafters, and used as (light) beams for supporting the roof covering.

Rafter A primary roof support beam usually in an inclined position, running from the tops of the structural columns at the eave to the ridge or highest portion of the roof. It is used to support the purlins.

Recess A notch or cutout, usually referring to the blockout formed at the outside edge of a foundation, which provides support and serves as a closure at the bottom edge of wall panels.

Reinforcing steel The steel placed in concrete to carry the tension, compression, and shear stresses.

Retainage A sum withheld from each payment to the contractor in accordance with the terms of the owner-contractor agreement.

Rolling doors Doors that are supported on wheels that run on a track.

Roof overhang A roof extension beyond the end or side walls of a building.

Roof pitch The angle or degree of slope of a roof from the eave to the ridge. The pitch can be found by dividing the height, or rise, by the span; for example, if the height is eight feet and the span is sixteen feet, the pitch is 8/16 or ½ and the angle of pitch is 45°. (See *Roof Slope.*)

Roof Slope The angle that a roof surface makes with the horizontal. Usually expressed as a certain rise in 12 inches of run.

Sandwich panel An integrated structural covering and insulating component consisting of a core material with inner and outer metal or wood skins.

Schedule of values A statement furnished to the architect by the contractor reflecting the amounts to be allotted for the principal division of the book. It is to serve as a guide for reviewing the contractor's periodic application for payment.

Sealant Any material that is used to close up cracks or joints.

Separate contract A contract between the owner and a contractor other than the general contractor for the construction of a portion of a project.

Sheathing Rough boarding (usually plywood) on outside of a wall or roof which is placed siding or shingles.

Shim A piece of steel used to level or square beams or column baseplates.

Shipping list A list that enumerates by part, number, or description each piece of material to be shipped.

Shop drawings Drawings that illustrate how specific portions of the work shall be fabricated and/or installed.

Sill The lowest member beneath an opening such as a window or door; also, the horizontal framing members at floor level, such as sill girts or sill angles; the member at the bottom of a door or window opening.

Sill, slip A sill that is the same width as the opening—it will slip into place.

Sill, lug A sill that projects into the masonry at each end of the sill. It must be installed as the building is being erected.

Skylight An opening in a roof or ceiling for admitting daylight; also, the reinforced plastic panel or window fitted into such an opening.

Sleeper Usually, a wood member embedded in concrete, as in a floor, that serves to support and to fasten subfloor or flooring.

Soil pressure The allowable soil pressure is the load per unit area a structure can safely exert on the substructure (soil) without exceeding reasonable values of footing settlements.

Soil stack A general term for the vertical main of a system of soil, waste, or vent piping.

Solid bridging A solid member placed between adjacent floor joists near the center of the span to prevent joists from twisting.

Spall A chip or fragment of concrete that has chipped, weathered, or otherwise broken from the main mass of concrete. concrete.

Span The clear distance between supports of beams, girders, or trusses.

Spandrel beam A beam from column to column carrying an exterior wall and/or the outermost edge of an upper floor.

Specifications A statement of particulars of a given job as to size of building, quality and performance of men and materials to be used, and the terms of the contract. A set of specifications generally indicates the desing loads and design criteria.

Splash block A small masonry block laid with the top close to the ground surface to receive roof drainage from downspouts and carry it away from the building.

Square A unit of measure—100 square feet—usually applied to roofing material. Sidewall coverings are sometimes packed to cover 100 square feet and are sold on that basis.

Stain, shingle. A form of oil paint, very thin in consistency, intended for coloring wood with rough surfaces, such as shingles, without forming a coating of significant thickness or gloss.

Stair carriage Supporting member for stair treads. Usually a 2″ plank notched to receive the treads; sometimes called a *rough horse.*

Stair landing See Landing.

STC (Sound Transmission Class) A measure of the ability of a material to stop ordinary noise.

Stile An upright framing member in a panel door.

Stock A unit that is standard to its manufacturer. It is not custom-made.

Stool A shelf across the inside bottom of a window.

Storm sash or storm window An extra window usually placed on the outside of an existing one as additional protection against cold weather.

Story That part of a building between any floor and the next floor or roof above.

String, stringer A timber or other support for cross members in floors or ceilings. In stairs, the support on which the stairs treads rest; also stringboard.

Strip flooring Wood flooring consisting of narrow, matched strips.

Stucco Most commonly refers to an outside plaster made with portland cement as its base.

Stud A vertical wall member to which exterior or interior covering or collateral material may be attached. Load-bearing studs are those which carry a portion of the loads from the floor, roof, or ceiling above as well as the collateral material on one or both sides. Non-load-bearing studs are used to support only the attached collateral materials and carry no load from the floor, roof, or ceiling above.

Subfloor Boards or plywood laid on joists, over which a finish floor is to be laid.

Subcontractor A separate contractor for a portion of the work (hired by the general contractor).

Substantial completion For a project or specified area of a project, the date when the construction is sufficiently completed in accordance with the contract documents, as modified by any change orders agreed to by the parties, so that the owner can occupy the project or specified area of the project for the use for which it was intended.

Supplementary general conditions One of the contract documents, prepared by the architect, that may modify provisions of the general conditions of the contract.

Suspended ceiling A ceiling system supported by hanging it from the overhead structural framing.

Tail beam A relatively short beam or joist supported in a wall on one end and by a header at the other.

Temperature reinforcing Lightweight deformed steel rods or wire mesh placed in concrete to resist possible cracks from expansion or contraction due to temperature changes.

Termites Insects that superficially resemble ants in size, general appearance, and habit of living in colonies; hence, they are frequently called "white ants." Subterranean termites establish themselves in buildings not by being carried in with lumber, but by entering from ground nests after the building has been constructed. If unmolested, they eat out the woodwork, leaving a shell of sound wood to conceal their activities, and damage may proceed so far as to cause collapse of parts of a structure before discovery. There are about fifty six species of termites known in the United States; but the two major ones, classified by the manner in which they attack wood, are ground-inhabiting or subterranean termites (the most common) and dry-wood termites, which are found almost exclusively along the extreme southern border and the Gulf of Mexico in the United States.

Termite shield A shield, usually of noncorrodible metal, placed in or on a foundation wall or other mass of masonry or around pipes to prevent passage of termites.

Terneplate Sheet iron or steel coated with an alloy of lead and tin.

Threshold A strip of wood or metal with beveled edges used over the finish floor and the sill of exterior doors.

Time of completion The number of days (calendar or working) or the actual date by which completion of the work is or the actual date by which complete of the work is required.

Toenailing to drive a nail at a slant to the initial surface in order to permit it to penetrate into a second member.

Tread The horizontal board in a stairway on which the foot is placed.

Trim The finish materials in a building, such as moldings, applied around openings (window trim, door trim) or at the floor and ceiling of rooms (baseboard, cornice, and other moldings).

Trimmer A beam or joist to which a header is nailed in framing for a chimney, stairway, or other opening.

Truss A structure made up of three or more members, with each member designed to carry basically a tension or a compression force. The entire structure in turn acts as a beam.

Turpentine A volatile oil used as a thinner in paints and as a solvent in varnishes. Chemically, it is a mixture of terpenes.

Undercoat A coating applied prior to the finishing or top coats of a paint job. It may be the first of two or the second of three coats. In some usages of the word it may become synonymous with priming coat.

Underlayment A material placed under finish coverings, such as flooring or shingles, to provide a smooth, even surface for applying the finish.

Valley The internal angle formed by the junction of two sloping sides of a roof.

Vapor barrier Material used to retard the movement of water vapor into walls and prevent condensation in them. Usually considered as having a perm value of less than 1.0. Applied separately over the warm side of exposed walls or as a part of batt or blanket insulation.

Varnish A thickened preparation of drying oil or drying oil and resin suitable for spreading on surfaces to form continuous, transparent coatings, or for mixing with pigments to make enamels.

Veneer A thin covering of valuable material over a less expensive body; for example, brick on a wood frame building.

Vehicle The liquid portion of a finishing material; it consists of the binder (nonvolatile) and volatile thinners.

Vent A pipe or duct which allows flow of air as an inlet or outlet.

Vermiculite A mineral closely related to mica, with the faculty of expanding on heating to form lightweight material with insulation quality. Used as bulk insulation and also as aggregate in insulating and acoustical plaster and in insulating concrete floors.

Volatile thinner A liquid that evaporates readily and is used to thin or reduce the consistency of finishes without altering the relative volumes of pigments and nonvolatile vehicles.

Wainscot Protective or decorative covering applied or built into the lower portion of a wall.

Wall bearing In cases where a floor, roof, or ceiling rests on a wall, the wall is designed to carry the load exerted. These types of walls are also referred to as load-bearing walls.

Wall covering The exterior wall skin consisting of panels or sheets and including their attachment, trim, fascia and weather sealants.

Wall non-bearing Wall not relied upon to support a structural system.

Wane Bark, or lack of wood from any cause, on edge or corner of a piece of wood.

Water closet More commonly known as a toilet.

Water-repellent preservative A liquid designed to penetrate into wood and impart water repellency and a moderate preservative protection. It is used for millwork, such as sash and frames, and is usually applied by dipping.

Weatherstrip Narrow or jamb-width sections of thin metal or other material to prevent infiltration of air and moisture around windows and doors. Compression weather stripping prevents air infiltration, provides tension, and acts as a counter balance.

Wood rays Strips of cells extending radially within a tree and varying in height from a few cells in some species to four inches or more in oak. The rays serve primarily to store food and to transport it horizontally in the tree.

Working drawing The actual plans (drawings and illustrations) from which the building will be built. They show how the building is to be built and are included in the contract documents.

Index

A

Accuracy tests, 108-131
  circle, area, 105
  dimension figures, 107
  ellipse, 106
  multiplication check, 107
  percentages, 104
  plane figures, 105
  quick calculations, 103
  rapid mental calculations, 104
  sphere, area, 106
  sphere, volume, 106
  subtraction check, 108
  twenty-seven times table, 107
Advertising, 173
Alterations and additions, 86
American Institute of Architects (A.I.A.), 9, 51-78
Apprenticeship, 7
Architect, 42
  progress certificates, 42
Attorneys and lawyers, 43, 49, 86
  liens, 43, 49, 86

B

Bankruptcy, 11
Base board estimating, 317
Bathroom fixtures, 176
Beams, 174
  Glu-lam, 174
  I beam, 174
  reinforced concrete, 174

Beams *(cont.):*
  wood, laminated, 174
  wood, solid, 174
Bench marks, 84
Board measure, 132, 133, 134
  dimension lumber, 132, 133, 134
Bonds, 49, 50, 80, 81, 82, 85
  check list for contractors, 79
  bonds, contracts, and insurance, 48
  labor and material, 49
  liens, 49
  performance, 49
  tender or bid, 49
Brick, 317
  estimating veneer, 317
  veneer, 316
  common bond, 282
Brickwork, 287
  coursing tables, 287
  English bond, 283
  English or Dutch bond, 283
  estimating, 285, 294
  estimating: cubic foot method, 294
  estimating: square foot method, 294
  estimating tables, 286, 288
  Flemish bond, 283
  general estimate sheet, 295
  running bond, 282
  stack bond, 282
  table of brick and mortar for 100 square feet, 297
  thermo ratings, 292
Building inspectors, 7

Builder's license, 3
Builder's loan charges, 173
Building paper, 305
Building, 7
    technicians, 7

C

Cabinets, 176
Carpentry, joinery, and millwork, 317
Carport, 174
Chimney, 175
City offices, 5
Cladding, 175
    interior, 175
Concrete, 249
    American and the Old Canadian Gallon capacities, 249
    bank or pit-run aggregates, 243
    basement floors, 276
    blocks, 278
    block constructed warehouse, 279, 281
    blocks, sizes and shapes, 279
    measuring volume of dry materials and water, 244
    mesh, steel-welding reinforcing, 277
    metal pans, 235
    metal pan formed reinforced floors and roofs, 334
    mixing, 238
    one-way combination tile and concrete roof assembly, 333
    proportioning, mixing, and placing, 237
    ready mixed, 238
    reinforcing steel, 255
    semipermanent forms, 253, 254
    sizes and weights of reinforcing steel rods, 258
    sizes of concrete mixes, 249
    stair reinforcing, 345
    steel bars, 257
    steps, 344
    striking off, 240
    stripping and reconditioning of forms, 253
    surface area of aggregate, 243
    testing, 100
    trowelling, 241
    water, 242
    water/cement ratio, 242, 248
    water tank estimate, 274
    weight, 249
    work-up sheets, 245, 246, 247
Construction Associations, 3
Construction company, 40
    building superintendent, 44
    daily progress report, 45
    engineers, 43
    foreman, 46

Construction company *(cont.)*:
    general manager, 42
    manager, 42
    owner, 42
    technician, 43
    timekeeper, 44
Contract, 93-95
    agreement, 93-95
    completion date, 100
    construction by day labor, 79
    cost plus a fixed fee, 79
    cost plus a percentage, 50
    cost plus variable sum, 79
    documents, 9, 17
    example of tender, 84
    fees and progress payments for small jobs, 11
    notice to tenderers, 83
    stipulated sum, 50
Collateral, 2
Corporation, 16
Credit sources of government loans, 2
Credit bureaus, 3
Critical Path Method (CPM), 217

D

Daily progress report, 45, 213
Damp-proofing, 174, 178
Demurrage, 8, 250
Depreciation, 12
    declining-balance method, 13
    straight-line method, 12
    sum of years—digits, 14
Doors, 175
    casings, 318
    hanging, 318
Drain or weeping tile, 174, 177
Drawings and specifications, 18-37
    alterations and additions, 86
    arbitration, 85
    bench marks, 84
    bids and estimates, 41
    clean-up, 86
    contingency sum, 85
    delays, 88
    disputes, 38, 84
    documents, precedence of, 38, 84
    examining, 86
    excavations and backfill, 38, 84
    extras, 87
    general conditions, 38, 84
    instruction to bidders, 89-92
    notices to bidders, 38
    payments, 85

Drawings and specifications *(cont.)*:
    prime costs, 86
    progress reports and photographs, 85
    progress schedule and diary, 85
    site, examination of, 38
    subcontractors, 47, 86
    supervision, 85
    temporary office and store sheds, 85
    temporary services, 85
    tenders, examples of, 38
Driveways, 174

E

Estimates, 208
    comparative appraisal method, 208
    cubic foot method, 208
    daily progress report, 213, 214, 215
    detail, 207, 209
    preliminary, 207, 208
    progress schedule, 211
    quantity, 79
    square foot method, 208
    unit, 207
Estimate sheets, 135
Estimating, 317
    base board, 317
    bid depository, 4
    wood wall framing, 301
    work-up sheets, 135
    work-up sheet example, 140
Estimator, 43
    junior, 44
    professional, 1
Equipment, 173
Excavations, 163-167, 173
Extras, 177

F

Fiber boards, 132, 305
Field trips, 9
Finance, 3
    accounts receivable, 3
    assets, 3
    balance sheet, 2
    builder's loan, 2
    current assets, 2
    definitions, 2
    equity, 2
    fixed assets, 3
    income tax, 3
    inventory, 3
    liabilities, 3
    long-term liabilities, 3

Finance *(cont.)*:
    machinery, 3
    net fixed assets, 3
    net worth, 3
    notes payable, 3
    profit and loss, 3
    promissory note, 2
    reconciliation of net worth, 3
    statement, 3
    surplus paid in, 3
Floor assembly, 174
    bridging, 232, 233
    bridging, wood or metal, 174
    building paper, 174
    floor deadening, 174
    headers (rangers), 174
    joists, 174
    joist hangers, 174
    reinforced concrete, 174
    subfloor, 174
    supporting first floor joists, 229
    tail joists, 174
Floor joists, 231
    first floor, 231
Frame construction, 301
    girths, 301
    grounds, 301
    labor for semiprefabricating, 305
    rough carpentry stud wall, 301
    labor, 315
    loose fill types, 313
    moisture, 312
    recommended standards, map of, 312
    reflective, 315
    rigid, 315
    R-value, 311
    stucco wire mesh, 315
    tables of room and ceiling areas, 315
    water vapor, 314
Insurance, 50, 86, 173

J

Joists, 176
    ceiling, 176
    first floor, 229

L

Land, 5
    abstract of title, 5
    appraisal, 180
    batter boards, 6
    caveat, 5
    datum and bench marks, 6

Land *(cont.):*
  easements, 5
  encumbrances, 4
  mechanic's lien, 5
  power of attorney, 5
  purchase, 4
  registry office, 4
  survey, 6
  title searching, 4
  writ or judgement, 5
Land grading, 185-187
  batter boards, 185-187
  bench mark, 187
  cut and fill, 184, 195
  datum lines, interpolation of, 191-193
  decimal equivalents of inches and feet, 188
  estimating cut, 188-191
  estimating sheet, cut and fill, 194, 196, 197, 198, 199
  sanitary fill, 184
  swell percentage, 187
  terms, 183, 184
Lumber, 140
  pricing, 140
Lumber standards, 135-139

M

Masonry units, 317
  ornamental, 317
Mesh: steel welding reinforcing, 277
Mortar data, 291
Mortar tables, 297
Mortgage companies, 3
Moldings, 307-310

N

Nails, 132, 158, 159

O

Overhead expenses, 97, 177
  attending on other trades, 100
  concrete testing, 100
  equipment, 100
  general, 98
  guided list of, 98
  individual job, 97
  major, guide list, 100, 101
  progress reports, 100
  separate job, 98
  temporary buildings, 100

P

Parapet walls, 334
  coping, 334

Parapet walls *(cont.):*
  flat deck assemblies, 334, 335
Partnerships, 15
  agreement, 16
  general, 15
  individual proprietorship, 15
  limited, 15
Perimeters, 162
Plastering, 301
  grounds, 301
Plywood and fiberboards, 132, 158
Professionals, 98
Profits, 97, 98, 101, 177
Progress reports, 174
Progress schedule, 211
  Gantt bar chart, 212
Project handover, 353
  architect's final inspection, 353
  general contractor's inspection, 353
  janitor services, 353
  maintenance, 353
  window cleaning, 353
Property, hand over of, 177

Q

Quantity surveying, 221

R

Railways and freightage, 8
Roofing materials, 174
Roofs, 320
  area framing, square method, 326
  area of gable roof, 323
  area of hip roof, 324
  carpenter labor, 333
  definitions, 322
  eaves, 174
  estimating for flat roofs, 332
  flat, 172, 332
  framing, 176
  framing squares, 327
  gable roof rafter, 330
  gable roof sheathing, 332
  hip roof rafter, 330
  labor costs, 331
  pitches, 172, 328
  types, 321, 322
Royal Architectural Institute of Canada, (R.A.I.), 9

S

Screws, 158, 160
Shingles, 311

Siding, 311
    labor for applying, 311
Special fixtures, 177
Stairs, 175, 337
    definitions, 338
    estimating materials, 343
    labor for erecting, 339
    small apartment block, 339, 340, 341
    stair horse or carriage, 342
    stairway types, 339
    sttel, 345, 346
    wooden, 338
Statistical Abstract of the U.S., 3
Steel, 350
    column base plates, 350
    metal units, 349
    structural, 348, 349
    wall anchors for beams, 350
Stock moldings, 132, 142-157
Stucco wire mesh, 315
Subcontractors, 47, 86
Subtrades, 8, 177
Supervision, 173

T

Tiles, 289
Tiling, 176
Trades, 176
    attending on others, 176
Trim, 176
    inside, 176
    outside, 175
Typical wall section, 169
    basic data, 170
    interior, 172

Typical wall section *(cont.)*:
    perimeter, 170
    plan areas, 172

U

Underpinning, 174
Unions, 8
    fringe benefits, 8
    labor legislation, 9
    local labor, 8
Utilities, 173
    permanent, 173

V

Vapor barrier, 305
    polyethylene, 305

W

Walls, 174
    concrete block and reinforcing, 174
    expanding metal, 174
    girths, 174
    metal strapping for plumbing, 174
    plates, 174
    studs, metal or wood, 174
Waterproofing, 174, 178, 179
Water tank, 274
Windows, 175
Wood, 141
    allowance for waste, 141
    hardwood flooring, 141
    lineal feet, 158
Workmen's Compensation Board, 8